Core Level Spectroscopies
for Magnetic Phenomena
Theory and Experiment

NATO ASI Series

Advanced Science Institutes Series

A series presenting the results of activities sponsored by the NATO Science Committee, which aims at the dissemination of advanced scientific and technological knowledge, with a view to strengthening links between scientific communities.

The series is published by an international board of publishers in conjunction with the NATO Scientific Affairs Division

A	Life Sciences	Plenum Publishing Corporation
B	Physics	New York and London
C	Mathematical and Physical Sciences	Kluwer Academic Publishers Dordrecht, Boston, and London
D	Behavioral and Social Sciences	
E	Applied Sciences	
F	Computer and Systems Sciences	Springer-Verlag
G	Ecological Sciences	Berlin, Heidelberg, New York, London,
H	Cell Biology	Paris, Tokyo, Hong Kong, and Barcelona
I	Global Environmental Change	

PARTNERSHIP SUB-SERIES

1. Disarmament Technologies	Kluwer Academic Publishers
2. Environment	Springer-Verlag
3. High Technology	Kluwer Academic Publishers
4. Science and Technology Policy	Kluwer Academic Publishers
5. Computer Networking	Kluwer Academic Publishers

The Partnership Sub-Series incorporates activities undertaken in collaboration with NATO's Cooperation Partners, the countries of the CIS and Central and Eastern Europe, in Priority Areas of concern to those countries.

Recent Volumes in this Series:

Volume 343—The Hubbard Model: Its Physics and Mathematical Physics
  edited by Dionys Baeriswyl, David K. Campbell, Jose M. P. Carmelo, Francisco Guinea, and Enrique Louis

Volume 344—Scale Invariance, Interfaces, and Non-Equilibrium Dynamics
  edited by Alan McKane, Michel Droz, Jean Vannimenus, and Dietrich Wolf

Volume 345—Core Level Spectroscopies for Magnetic Phenomena:
  Theory and Experiment
  edited by Paul S. Bagus, Gianfranco Pacchioni, and Fulvio Parmigiani

Series B: Physics

Core Level Spectroscopies for Magnetic Phenomena
Theory and Experiment

Edited by
Paul S. Bagus
University of Heidelberg
Heidelberg, Germany

Gianfranco Pacchioni
University of Milan
Milan, Italy

and

Fulvio Parmigiani
CISE SpA
Milan, Italy

Plenum Press
New York and London
Published in cooperation with NATO Scientific Affairs Division

Proceedings of a NATO Advanced Study Institute on
Core Level Spectroscopies for Magnetic Phenomena: Theory and Experiment,
held May 15–26, 1994,
in Erice, Italy

NATO-PCO-DATA BASE

The electronic index to the NATO ASI Series provides full bibliographical references (with keywords and/or abstracts) to about 50,000 contributions from international scientists published in all sections of the NATO ASI Series. Access to the NATO-PCO-DATA BASE is possible in two ways:

—via online FILE 128 (NATO-PCO-DATA BASE) hosted by ESRIN, Via Galileo Galilei, I-00044 Frascati, Italy

—via CD-ROM "NATO Science and Technology Disk" with user-friendly retrieval software in English, French, and German (©WTV GmbH and DATAWARE Technologies, Inc. 1989). The CD-ROM contains the AGARD Aerospace Database.

The CD-ROM can be ordered through any member of the Board of Publishers or through NATO-PCO, Overijse, Belgium.

Library of Congress Cataloging-in-Publication Data

On file

ISBN 0-306-45006-2

© 1995 Plenum Press, New York
A Division of Plenum Publishing Corporation
233 Spring Street, New York, N. Y. 10013

10 9 8 7 6 5 4 3 2 1

All rights reserved

No part of this book may be reproduced, stored in a retrieval system, or transmitted in any form or by any means, electronic, mechanical, photocopying, microfilming, recording, or otherwise, without written permission from the Publisher

Printed in the United States of America

PREFACE

For several years, core level spectroscopies and other, closely related, electron spectroscopies have provided very useful information about the atomic composition, the geometric structure, and the electronic structure of condensed matter. Recently, these spectroscopies have also been used for the study of magnetic properties; such studies have a great potential to extend our knowledge and understanding of magnetic systems.

This volume collects the lectures presented at the NATO Advanced Study Institute on "Core Level Spectroscopies for Magnetic Phenomena: Theory and Experiment" held at the Ettore Majorana Centre, Erice, Sicily, on 15 to 26 May 1994. The topics considered at the ASI covered a wide range of subjects involving the use of core-level and related spectroscopies to study magnetic phenomena. There are a large and growing number of applications of these spectroscopies to the study of magnetic materials; an important objective of the ASI was to stimulate further growth. The topics covered at the ASI can be placed into three general groups: 1) fundamental principles of core level spectroscopies; 2) basic aspects of magnetic phenomena; and, 3) the combination of the two previous topics embodied in applications of the spectroscopies to magnetism. In all three groups, theoretical interpretations as well as experimental measurements were presented, often both of these aspects were covered in a single lecture or series of lectures. The theoretical treatments of the spectroscopies as well as of the magnetic phenomena help to establish a framework for understanding many of the experimental measurements on magnetic materials.

Several of these spectroscopies are able to give valuable information about electronic and geometric structure of matter. High energy electron energy loss spectroscopy, EELS, is capable of giving sub-nanometer resolution for atomic positions and is also able to give information relevant to the nature of chemical bonds. Synchrotron radiation can be used for several core-level spectroscopies including: X-ray photoemission, XPS; X-ray emission, XES; and X-ray absorption, XAS, spectroscopies. Energy resolution which is as refined as 10 meV is presently possible, especially at the new generation of synchrotron facilities. The tunability of synchrotron radiation makes it possible to distinguish surface and bulk properties. It also makes it possible to separately identify and examine the specific atomic components of a system; this capability can be used to study the properties of sandwich materials consisting of magnetic layers separated by non-magnetic spacer layers. Inverse photoemission gives information on empty states which is complementary to the information on occupied states obtained with photoelectron spectroscopy. Information on magnetic materials can also be obtained from the analysis of the spin polarization of the Auger electrons. Dichroism, circular and linear, for XAS, XES, and XPS is used to identify the spin dependence of occupied and empty levels in magnetic materials.

Three topics of recurring interest at the ASI should be noted. First, the theoretical analyses of electron scattering were shown to provide valuable information about the relative importance of spin flip scattering as compared to scattering in the different potentials seen by the spin-up and spin-down electrons. Second, the temperature dependence of the magnetization was shown to be a probe of the local exchange interactions at the surfaces and interfaces of magnetic materials. Third, the unique aspects of circular and linear dichroism and spin polarized photoemission to relate microscopic electronic structure to macroscopic magnetic materials properties were covered by several lecturers. Overall, the outlook for the application of core-level spectroscopies to magnetism is very bright.

The School was financially supported by the NATO Scientific Affairs Division, and we gratefully acknowledge this generous support. The help of the Director of the International School of Solid State Physics, Professor Giorgio Benedek, and of the Director of the Ettore Majorana Centre, Professor Antonino Zichichi, were indispensable for the planning, organization, and operation of the School. We wish to express our sincere appreciation to the center staff, Dr. Alberto Gabriele, Dr. Jerry Pilarsky, Dr. Pinola Savalli, and Dr. Claudia Zaini for their expert assistance in all organizational matters and for their warm hospitality. The success of a school is ultimately determined by the interest and commitment of the lecturers and participants. We are particularly grateful to all the lecturers and to all the participants for their enthusiasms and collaboration.

Paul S. Bagus
Gianfranco Pacchioni
Fulvio Parmigiani

CONTENTS

Spin Polarized Electrons of Low Energy and Magnetism 1994
    H.C. Siegmann   1

Photoemission and Ferromagnetism
    P.D. Johnson   21

The Study of Empty Electron States of Solids with Core X-Ray Absorption
    and Inverse Photoemission
    L. Braicovich   41

The Spin Dependence of Elastic and Inelastic Scattering of Low Energy
    Electrons from Magnetic Substrates
    D.L. Mills   61

Spin-Resolved Core Level Photoemission Spectroscopy
    F.U. Hillebrecht, Ch. Roth, H.B. Rose, and E. Kisker   85

Spin Dependent Electron Mean Free Path in Ferromagnets
    H. Hopster   103

Exchange Interactions in Magnetic Surfaces and Layer Structures
    J. Mathon   113

Magnetic Circular Dichroism in Photoemission from Rare-Earth Materials:
    Basic Concepts and Applications
    G. Kaindl, E. Navas, E. Arenholz, L. Baumgarten, and K. Starke   131

Dichroic Photoemission for Pedestrians
    G. van der Laan   153

Spin Resolved Soft X-Ray Appearance Potential Spectroscopy
    V. Dose   173

Linear Magnetic Dichroism in Directional Photoemission from Core
    Levels and Valence Bands
    G. Rossi, F. Sirotti, and G. Panaccione   181

X-Ray Resonant Inelastic Scattering
    P. Carra and M. Fabrizio   203

Investigation of Local Electronic Properties in Solids by Transmission
    Electron Energy Loss Spectroscopy
      C. Colliex 213

d-like Quantum-Well States and Interface States of Paramagnetic
    Overlayers on Co(0001)
      D. Hartmann, A. Rampe, W. Weber, M. Reese, and G. Güntherodt 235

Electronic Correlations in the 3s Photoelectron Spectra of the Late
    Transition Metal Oxides
      F. Parmigiani and L. Sangaletti 249

Contributors 265

Index 267

Core Level Spectroscopies
for Magnetic Phenomena
Theory and Experiment

SPIN POLARIZED ELECTRONS OF LOW ENERGY AND MAGNETISM 1994

H.C. Siegmann[*]

Stanford Linear Accelerator Center
Stanford, California 94309

INTRODUCTION

The spin of the electron and its leading rôle in establishing the phenomenon of magnetism became a clear concept in the course of the development of quantum mechanics in the first half of our century. In the second half, one was concerned to what extent quantum theory could explain solid state phenomena, and the main focus was the development of electron spectroscopy on solids and their surfaces. The analysis of the spin polarization of the electrons makes it possible to distinguish spin up from spin down states in the solid even when they are degenerate in energy. Therefore, spin polarized electron spectroscopy became a major topic in magnetism and generated the new field of surface and two-dimensional magnetism[1,2]. Recently, the following discoveries as well as realisations of older ideas have produced a wealth of new opportunities for basic research and applications of magnetism.

New Developments

Sources of spin polarized electrons. GaAs with a negative electron affinity surface is an ideal source of electrons. It delivers a monchromatic, pulsable and intense electron beam when irradiated with light close to the gap energy of 1.5 eV. If the light is circularly polarized, the emerging electron beam is spin polarized. The degree of spin polarization may approach 1 when the degeneracy between heavy and light hole bands in the cubic crystal structure is removed by stress. Recently, it has been shown that this is possible by growing a strained GaAs-type layer on an appropriate substrate[3]. 80% polarization have been routinely reproduced and used in high energy experiments at the Stanford Linear Accelerator[4]. Sources of spin polarized electrons are essential to understand and investigate magnetism at surfaces[1] and to perform low energy spin polarized electron microscopy in which growth and appearance of spontaneous magnetization in atomic layers of magnetic materials can directly be imaged at video frequencies[5]. With the new strained layer photocathodes, the spin induced contrast will be obtained 10 times faster than with the old cubic GaAs-type sources. Therefore, the experiments employing elastic and inelastic scattering of polarized electrons from

Core Level Spectroscopies for Magnetic Phenomena: Theory and Experiment
Edited by Paul S. Bagus et al., Plenum Press, New York, 1995

surfaces, spin polarized inverse photoemission and low energy electron microscopy will be easier by one order of magnitude in the future.

Ultrathin magnetic structures. With monoatomic layers of atoms that are perfect over extended areas, one can build ultrathin magnetic structures in which some striking new magnetic phenomena are observed. Preparation and measurement of such structures initially posed great challenges to the experimentalist which however are increasingly overcome. The layer by layer growth of metals makes it possible to establish well defined magnetization profiles at a surface which have greatly advanced our understanding of the spectroscopy of low energy electrons. This will be one of the main topics of this paper. As a very striking simple example of an ultrathin magnetic structure consider one monolayer of Fe or Co deposited on W(110). Fe and Co grow epitaxially on W(110) surfaces with the easy direction of magnetization in plane. The absence of noticeable island formation is warranted by the fact that the surface energy of the W(110) substrate is higher than the one of Fe and Co, hence an energy gain arises from spreading Fe and Co as much as possible on the W surface. As soon as one monolayer is completed, the spontaneous magnetization appears. As opposed to three-dimensional magnetism, the ground state is a one domain state in this case where the magnetization is in the plane of the layer. Hence it is easy to measure the direction and temperature dependence of the magnetization without applying an external field, for instance with the classical magneto-optic Kerr-effect. It is however not easy to determine the absolute value of this spontaneous magnetization which will be another topic of the present paper. It turns out that one even has to screen the magnetic field of the earth to obtain the true temperature dependence $M_o(T)$ of the spontanous magnetization[6]. At temperatures below the Curie point T_c, $M_o(T)$ decreases quite slowly but drops sharply to zero at T_c which is close to 300 K for both Fe and Co. Back, Würsch and Pescia[6] observe no tail of $M_o(T)$ at $T > T_c$ in contrast to work only a few years ago in which such tailing was considered a hardly avoidable natural feature of 2D-ferromagnets. However, application of magnetic fields as small as the earth field dramatically alters $M_o(T)$ and, at temperatures close to T_c, can reestablish a large magnetization[6]. In fact, these 2D-ferromagnets exhibit a giant magnetic susceptibility at T_c of the order of 10^5. This is 10^4 times larger compared to the critical susceptibility in the 3D crystals of the same elements. The giant critical susceptibility in 2D reflects in a unique manner the relevance of dimensionality at the magnetic phase transition which has been expected by the theory but could so far not be realized due to the experimental problems connected with producing monolayer epitaxial films of sufficient quality.

Magnetism in nonmagnetic elements. Several band-structure theorists have pointed out that metallic magnetism in 2D is not a priori restricted to those elements which exhibit magnetism in 3D [7]. Because of the reduced number of nearest neighbor atoms, the d-band width in 2D is considerably smaller than in 3D. The magnetic instability should then occur for a much wider variety of transition metal elements. The same reasoning has also been applied to small clusters of atoms.

Good candidates for this exotic magnetism in low dimensional systems are the 3d transition elements V and Cr, the 4d elements Tc, Ru, Rh, and Pd, and the 5d-elements Os and Ir deposited on inert metal substrates such as Au(100) or Ag(100). However, none of these predictions of self sustained exotic magnetism has been veryfied beyond doubt so far [2,8].

If however nonmagnetic elements are deposited onto a magnetic surface such as Fe(100), it is clear that a sizeable spin moment might be induced. Spin polarized core level spectroscopy yields evidence on the existence and direction of the magnetic moment at specific atomic sites, but it can so far not determine the absolute size of the induced spin and orbital moment. The occurrence of the magnetic moment in atoms such as Cr, V, Pd or Ru can be described by an exchange field transferred from the magnetic substrate to the adatom. Such transfer of exchange fields occurs also within the atom from one shell to another. In Gd for instance, the intraatomic exchange interaction transferred from the $4f^7$-shell to the 6s5d bands causes an exchange splitting of the valence bands leading to a sizeable spin moment of .55 μ_B (Bohr magnetons) in the 6s5d bands that adds to the total magnetic moment of $7\mu_B$ generated by the 4f-shell. It is difficult even for conventional magnetometry on bulk magnetic materials to determine the size of the induced magnetic moment on top of the background of the primary moment and to distinguish the induced moment from the effect of impurities. For instance, the magnitude of the spontaneous magnetization at T = 0 in Gd was dubious until very clean Gd could be prepared. In 4f-3d-alloys, it is still quite difficult to determine the magnetic moment. The problems with classical magnetometry increase dramatically with adatoms[9]. The overwhelming part of the magnetization is now due to the substrate. The magnetic moment of the adatoms typically are at most in the 1% range of the substrate magnetization. The determination of the number of adsorbed monolayers and defect densities is an additional problem. Furthermore, the coupling between adatom moment and substrate may exhibit either sign. There is no doubt that the induced magnetic moment is an essential ingredient in the analysis of the coupling mediated by nonmagnetic spacer layers between ferromagnets. There is also no doubt that it can be relevant for the giant magnetic resistance occurring in coupled multilayer systems. Therefore, it is an important task for spin polarized electron spectroscopy to develop methods for absolute measurement of magnetic moments in adatoms.

SPECTROSCOPY WITH LOW ENERGY ELECTRONS

Spin Valves and Giant Magnetoresistance

Exchange coupled magnetic multilayers consist at least of 2 magnetic layers coupled by a nonmagnetic spacer layer. Further layers are generally needed to induce favourable growth and/or crystal structure, for support, and for chemical stability in atmospheric environment. However, the active ingredients are the two ferromagnetic layers coupled by the nonmagnetic spacer layer. If the spacer layer material and thickness is chosen such that the coupling is antiparallel in the absence of magnetic fields, an external field of appropriate strength will induce parallel alignment of the ferromagnetic layers. The switching from antiparallel to parallel alignment is accompanied by a large change in the electrical resistance. This giant magnetoresistance has greatly helped to understand the mechanism of electric transport in magnetic materials. Many years ago, Fert and collaborators had proposed that there are in fact two independent electrical currents, one carried by spin up or majority electrons, i^+, and one carried by spin down or minority electrons, i^-. The two current model can readily explain the giant magnetoresistance[10]. Generally, the electrical resistance is determined by the mean free path λ between collision events. In the case of transition metals, λ is governed by the scattering of the conduction electrons into the unoccupied d-states, hence the total scattering cross section σ is given by

$$\sigma = \frac{1}{\lambda} = \text{const} \cdot D(E_F) \qquad (1)$$

where $D(E_F)$ is the density of states at the Fermi energy E_F. With ferromagnetic metals, $D(E_F)$ is different for the two spin states, therefore one now has two different total scattering cross sections σ^+ and σ^-. In the case of Ni and Co, the majority spin band is fully occupied and $\sigma^+ \ll \sigma^-$ whereas for Fe, $\sigma^+ > \sigma^-$. Note that the two currents are noninteracting in this model, that is there are no spin flips.

With magnetic multilayers, it is now possible for the first time to measure the mean free path or total scattering cross section for the two spin states separately[11]. The technique uses a $Ni_{80}Fe_{20}$ ultrathin layer, in which $\sigma^+ \ll \sigma^-$, as a spin filter or spin valve. The attenuation of a current i^+ or i^- of majority or minority electrons on travelling a distance x in such a filter is given by

$$i^{\pm} = i_o^{\pm} e^{-\sigma^{\pm} x} \qquad (2)$$

Eq. (2) shows that in NiFe, mainly the majority spin electrons are transmitted. By switching the magnetization to the other direction, it will be the minority spin electrons with respect to an external independent source. This external source is in fact the other magnetic layer separated by the nonmagnetic spacer from the filter layer.

Surprisingly, exactly the same model holds for the attenuation of low energy electrons emitted from a multilayer structure. We will show below that the model is even much simplified by the fact that such low energy electrons at or slightly above the vacuum level E_∞ can scatter into all the unoccupied states above E_F, including all of the unoccupied d-states. Eq.(1) may therefore be written for the case of emitted electrons

$$\sigma = c_1 \cdot \int_{E_F}^{E_\infty} D_o(E) dE + c_2 \cdot \int_{E_F}^{E_\infty} D_d(E) dE \qquad (3)$$

where $D_o(E)$ is the density of all nonmagnetic states making no contribution to the magnetization and $D_d(E)$ is the density of those states which have an occupancy that depends on spin state. These latter ones are mainly, but not exclusively, the d-states. The difference between σ^+ and σ^- is then given by the Bohr magneton number μ_B:

$$\sigma^- - \sigma^+ = c_2 \int_{E_F}^{E_\infty} (D^-(E) - D^+(E)) dE = c_2 \cdot n_B \qquad (4)$$

This is much simpler than the case of electrical conduction in which $\sigma^- - \sigma^+$ is given by the difference of the density of states at E_F; $D(E_F)$ is a complicated composite of the electronic structure and the occupancy of the various electronic states.

The simplicity of eq.(4) arises because the Bohr magneton number n_B is in fact known independently in many cases from classical magnetometry. It will be shown that $\sigma^- - \sigma^+$ can be determined by measuring the spin polarization of the electrons emitted from an overlayer structure. This leads to the magnetometry with cascade electrons. With the experimental verification of eq.(3) and eq.(4) described below, an entirely new and much simpler concept of low energy electron spectroscopy has arisen as well as a number of new fundamental insights.

Probing Depth of Threshold Electrons in Transition Metals

The probing depth λ is the distance from the surface over which specific electronic properties of a solid can be sampled by spectroscopic analysis of energy, angular distribution, or spin polarization of the emerging electrons. λ can be translated via the group velocity into the lifetime τ of an electron in the excited states within the solid. A large amount of data for λ has been accumulated from the overlayer method in which the attenuation of a prominent substrate feature is measured as a function of the overlayer thickness x and fitted with an exponential decay $\exp(-x/\lambda)$. The data for many materials are often displayed as a "universal curve" which shows λ as a function of energy [12]. With low energy electrons escaping from the solid close to the vacuum level, the overlayer method cannot be applied due to the lack of a distinguishing feature. However, a well defined magnetization depth profile can be constructed at a surface by depositing either magnetic layers consisting, for example, of Fe on a nonmagnetic substrate of vice versa by depositing nonmagnetic layers on a magnetic surface. If now the increase or decrease of the spin polarization P of the emerging electrons is measured as a function of the overlayer thickness x one can obtain astonishingly accurate values of λ even for threshold electrons.

Low energy cascade electrons are excited by a primary unpolarized electron beam having an energy of 2-3 keV. An electron optical system collects most of the cascade electrons emerging from the surface in an energy window of 5-10 eV from the Fermi level and focuses them onto the entrance diaphragm of a spin polarization analyzer. The ferrromagnetic substrate, for instance Fe, is deposited onto a special NiFe alloy (permalloy) or any other magnetic material that can be magnetically saturated in a low external field comparable to the earth magnetic field, hence the electron optics are not disturbed by it. The Fe film taken by itself would require high magnetic fields for magnetic saturation and furthermore would not show square magnetic hysteresis loops. But, because the Fe film is directly deposited onto the permalloy, it is exchange coupled to it and hence the substrate surface has the desirable magnetic properties of permalloy but the electronic structure of pure Fe. The polarization P of the low energy cascade electrons emitted from the Fe upon irradiation with primary electrons of a few keV can be measured with a high accuracy of $\pm < 0.5\%$. If this surface is now covered with increasing amounts of a nonmagnetic transition metal, P decreases because some polarized electrons from the magnetic substrate are removed during passage through the nonmagnetic overlayer, while some unpolarized electrons are added. If the number of electrons from the substrate and overlayer are labelled with subscripts s and o respectively, the polarization as a function of x is given by

$$P = [(J_s^+ - J_s^-)\exp(-x/\lambda)]/J, \qquad (5)$$

where the total intensity J is given by

$$J = (J_s^+ + J_s^-)\exp(-x/\lambda) + (J_o^+ + J_o^-)[1-\exp(-x/\lambda)]$$

and where + and - stands for majority and minority spin electrons respectively.

With the spin polarized overlayer technique, accurate and reproducible values for the total scattering cross section $\sigma = 1/\lambda$ are obtained in the energy range $5 \leq E \leq 15$ eV above E_F. It turns out that the results cannot be described by the "universal curve" [12] but clearly depend on the transition metal. However, there is still a quite simple behavior independent of whether the metal is magnetic or not. Compilation of the data available from many laboratories reveals that the total scattering cross section σ consists of a constant part σ_o and a part $\sigma_d(5-n)$ that is proportional to the number of unoccupied states (5-n) in the d-band. n is the number of occupied and 5 the total number of d-orbitals available to one spin state. Fig. 1 shows that σ_o and σ_d can be obtained from a fit of the data to the equation

$$\sigma = \sigma_o + \sigma_d(5-n) \qquad (6)$$

With magnetic metals, the spin averaged number of holes $n = 1/2(n^+ + n^-)$ must be inserted into eq.(6) to obtain the spin averaged total cross section $\sigma = 1/2 (\sigma^+ + \sigma^-)$.

To interpret the empirical rule, eq.(6) one has to consider that the dominant processes by which electrons at within 10 eV from the Fermi-level lose energy in a transition metal are electron-electron scattering events. Hence σ_o accounts for inelastic scattering other than into the holes of the d-band including, for instance, scattering into the s-p-derived bands, whereas σ_d describes the scattering into one unoccupied d-orbital. According to this interpretation, the noble metals Cu, Ag, and Au should have identical $\sigma = \sigma_o$. This is indeed the case. The elastic scattering of electrons is of minor importance for threshold electrons, because as many electrons are scattered into the escape cone than out of it.

Eq.(6) is the experimental verification of eq.(3). A good description of the results reported from many laboratories is obtained with $\sigma_o = 0.6$ nm^{-1} and $\sigma_d = 0.7$ nm^{-1}. It yields extremely short mean free path approaching the lattice spacing for the early transition metals,

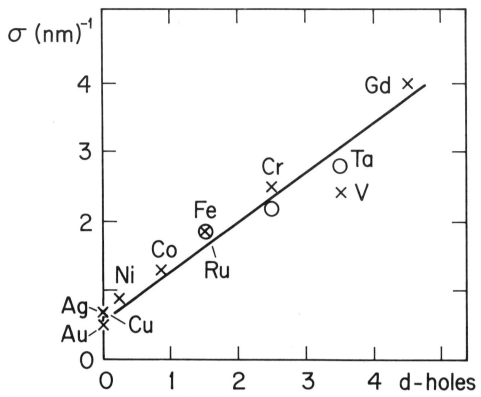

Fig. 1 Total scattering cross section σ in nm^{-1} versus number of unoccupied d-orbitals for the transition metals indicated. Data from various sources according to Ref. [2] are valid for electron energies within 5-10 eV from the Fermi level. The circled points indicate average values from different laboratories. The Co value is from Ref. [20]; the upper value for Cr is on NiFe substrate, the lower on Fe(100). The Cu, Ru, and V data are by courtesy of Karin Totland and Peter Fuchs [34].

in contrast to earlier belief. Previously, it had been assumed that λ should become very large when the energy of the electrons approaches the Fermi energy. However, the overlayer method for electrons close to E_F as described by Gurney et al.[11] yields very short mean free paths even in this case.

It is not expected that eq.(6) describes the fine details of electron-electron scattering in transition metals. It neglects many potentially important phenomena like the varying stability of the d_n-configurations and the scattering into the variable surface or interface states. It will indeed be shown that σ_d as well as the ratio σ_0/σ_d of spin independent to spin dependent scattering may need adjustment to correctly account for the observations in all cases.

Ferromagnetic Layers as Spin Filters

When a transition metal becomes ferromagnetic, the occupancy of the spin up part of the d-shell increases by Δn while the occupancy of the spin down subshell decreases by Δn. It follows from eq.(6) that the total inelastic cross section becomes spin dependent according to

$$\sigma^{(\pm)} = \sigma_0 + \sigma_d [5 - (n(\pm)\Delta n)] \qquad (7)$$

where $\sigma^{(\pm)}$ is the cross section for majority (minority) spins. For the bulk of the material, Δn can be calculated from the Bohr magneton number n_B obtained in conventional magnetometry according to $n_B = n^+ - n^- = 2\Delta n$, because the contribution of s-p-electrons to the magnetization is small.

Eq.(7) allows one to independently test the simple behavior of σ in transition metals stated in eq.(6) because it predicts that electrons travelling through ferromagnetic materials will acquire a transport polarization **a**. If x is the thickness of the ferromagnetic material over which the transport occurs, one has

$$\mathbf{a} = \{[\exp(\sigma^- - \sigma^+)x] - 1\} / \{[\exp(\sigma^- - \sigma^+)x] + 1\} \tag{8}$$

As already stated in eq.(4), $\sigma^- - \sigma^+ = \sigma_d \cdot n_B$. Hence the transport polarization can be calculated from the spin magnetic moment of the ferromagnet. It can also directly be measured either by letting unpolarized electrons pass through the ferromagnetic layer and detecting the acquired transport polarization or by determining the spin dependent attenuation of polarized electrons for instance form a GaAs electron source. Both methods are fully equivalent. In the first case, the ferromagnetic layer acts as polarizer, in the second case as analyzer of the degree P of spin polarization of the incident electrons. The transport polarization $\mathbf{a}(x)$ of the transmitted electrons in the first experiment is also the analyzing power of the spin filter. Namely, the asymmetry of the intensity transmitted when P and the magnetization of the filter are parallel, I^{++}, and the one where they are antiparallel, I^{+-}, is given by

$$(I^{++} - I^{+-})/(I^{++} + I^{+-}) = P \cdot \mathbf{a}(x) \tag{9}$$

The figure of merit or efficiency of this spin filter is given by $\mathbf{a}^2(x) \cdot I$ which is approximately inversely proportional to the time it takes to determine P with a preset accuracy. We see that in both limits $x \to 0$ and $x \to \infty$, the figure of merit tends to zero, while the optimum thickness is approximately reached at $\lambda < x \leq 2\lambda$.

Yet another way of determining the spin filtering characteristics of ferromagnets experimentally relies on the measurement of the spin polarization P_c of the low energy cascade electrons from a thick ferromagnetic sample. This polarization shows an enhancement over the average spin polarization P_0 of the N electrons in the valence bands made up of d-, p-, and s-orbitals. If one assumes that all valence electrons are excited to an escape level with equal probability, it follows that $P_0 = \Delta n/N$ where N now includes the s-p-electrons, but Δn is made up almost entirely by the 3d-electrons alone, hence can be determined with conventional magnetometry on bulk material. It has been proposed right after the first observations of the enhanced cascade polarization P_c that scattering into the unoccupied holes of the d-band could be responsible for it. A summary of these first arguments is given in Ref. 13.

The enhancement of P_c over P_0 can readily be calculated from the transport polarization $\mathbf{a}(x)$ given in eq.(8). One can write for the average transport polarization A of an infinitely thick sample:

$$A = \left(\int_0^\infty \mathbf{a}(x)I(x)dx\right) / \left(\int_0^\infty I(x)dx\right) = (\sigma^- - \sigma^+)/(\sigma^+ + \sigma^-) \tag{10}$$

where the intensity $I(x)$ from depth x was taken to be:

$$I(x) = J_0[\exp(-\sigma^+ x) + \exp(-\sigma^- x)]$$

With eq.(7), eq.(10) yields:

$$A = \Delta n / [\sigma_0/\sigma_d + (5-n)] \tag{11}$$

It has been assumed that $P_c = P_0 + A$. This is valid if $P_0 \ll 1$. The full calculation is given in Ref. 2. Even with Fe, the error introduced by this simplification is only a few %. The enhancement factor $f = P_c/P_0$ is then given by

$$f = 1 + N/[\sigma_0/\sigma_d + (5-n)] \qquad (12)$$

Eq.(12) does not contain Δn, therefore the enhancement f does not depend on the magnetization. Eq.(12) leads to the following statements that can be verified in experiments:

1. The polarization P_c of the low energy cascade electrons is proportional to the spin magnetic moment.
2. The enhancement factor increases with the filling of the d-shell.
3. The enhancement decreases when more empty states are available, because σ_0 increases.

Therefore, $A \to 0$ and $f \to 1$ when the energy of the electrons increases above threshold. This is in contrast to $a(x)$ which changes only when the matrix elements for transitions into the d-states change.

Furthermore, if A is known one can calculate $a(x)$ and vice versa. From eq.(10) one obtains

$$2 A\sigma = \sigma^- - \sigma^+ \qquad (13)$$

where $\sigma = 1/\lambda$ is the spin averaged total scattering cross section.

Experimental Verification of Spinfiltering in Ferromagnets

1. Enhancement of the spin polarization of low energy cascade electrons.
Metal surfaces emit copious amounts of low energy electrons when irradiated with a primary beam of electrons. These low energy secondary or cascade electrons are of great value for magnetometry and their spin polarization P_c has been measured in a number of laboratories. Table 1 gives some data on the ferromagnetic metals Gd, Fe, Co and Ni together with the enhancement A of P_c over P_0 and the enhancement factor $f = P_c/P_0$. The case of Gd requires some special comments. The magnetic moment is 7.55 μ_B/Gd-atom. 0.55 μ_B must be attributed to the 5d-polarization yielding $\Delta n = 0.27$. As opposed to a recent investigation by Tang et al.[14], the scattering into the $4f^7$-state will be neglected here for reasons explained later.

Table 1 Total spin polarization P_0 of s-, p-, and d-electrons in the bulk, estimated number n of occupied d-orbitals in the paramagnetic state and their increase (decrease) Δn due to the spontaneous magnetization at $T = 0$ in Gd, Fe, Co, and Ni; The spin averaged mean free path λ and the cascade polarization $P_c(T \to 0)$ are taken from various authors [2, 31, 38]; A: average transport polarization of the infinitely thick sample.

	P_0	n	Δn	λ(nm)	A	P_c	$f = P_c/P_0$
Gd	0.18	0.5	0.27	0.24	0.04	0.22	1.22
Fe	0.28	3.50	1.10	0.6	0.28	0.53	1.89
Co	0.19	4.10	0.85	0.8	0.25	0.42	2.21
Ni	0.05	4.75	0.25	1.1	0.09	0.14	2.80

All the 3 predictions derived in the previous chapter have been verified with Gd, Fe, Co, Ni and alloys. The first statement on the proportionality of the surface magnetization M_s and the cascade polarization P_c is verified by the T-dependence of P_c according to Ref. 2. Second, the enhancement factor clearly increases with the filling of the d-shell as expected, see Table 1. Third, the decrease of A and f with increasing energy of the cascade electrons is directly evident with amorphous ferromagnets[13], while crystalline surfaces show some structure superimposed on the decay[15] which has been attributed to elastic scattering and spin splittings in the unoccupied density of states.

Calculating A from eq.(11) and taking $\sigma_0/\sigma_d = 0.7$ from Fig. 1, one obtains very high values for $P_c \simeq A + P_0$ that do not agree with the observations. But if one assumes a much larger relative spin independent scattering $\sigma_0/\sigma_d = 2.5$, one calculates P_c values that are quite close to the observations for all ferromagnetic transition metals Gd, Fe, Co, and Ni. The fact that this agrees with the observations on all transition metals adjusting σ_0/σ_d only once is remarkable and would be difficult to explain in any other way. It cannot be expected that the observed cascade polarization agrees perfectly well with the P_c calculated from bulk magnetization data. Since the cascade is formed in the very few last layers at the surface, this would mean that the surface magnetization is not different from the bulk magnetization. This is generally not the case at finite temperatures as the probability density of spin waves, depending on the strength of the exchange interaction in the surface, can be quite different in the surface [2]. Hence it is important to compare to P_c measured at very low temperatures. Even then the surface magnetization can be somewhat different from the bulk magnetization. The fact that the spin independent scattering is much higher than expected could be due at least in part to scattering into the unoccupied surface states. Inverse photoemission experiments provide direct evidence that this scattering can be significant [1]. The proposed simple model connects the very short probing depth with the enhancement of the cascade polarization P_c over the ground state polarization P_0. It can explain a large number of independent experimental observations.

Other models explaining the conspicuously high polarization P_c of the low energy cascade electrons have also been put forward. However, none of them can give a consistent explanation for all the phenomena observed. In one model [16] it is assumed that quasi-elastic scattering occurs which transforms a minority spin into a majority spin electron leaving behind a reversed spin (Stoner excitation). Although Stoner excitations have been experimentally observed [17,18], their rôle in the formation of the cascade is not important. This is clearly proven by the fact that the total cross section eq.(6) does not depend on magnetic order. The situation can be summarized by the statement that the mean free path for conservation of spin is larger than the mean free path for conservation of energy. Furthermore, at electron energies within or below ~1 eV from threshold, the electrons that generated a Stoner excitation cannot escape, hence the spin flip model fails to explain the enhancement of P_0 at very low energies. In yet another explanation of the high value of P_c it is proposed that the spin dependence of the *elastic* scattering of the electrons on the magnetic atoms is responsible for it [19]. While the experiments confirm without doubt that the spin dependence of elastic electron scattering can be quite large at certain electron energies and in some crystal directions, it is impossible to explain the uniformly *positive* sign of the polarization enhancement over a wide range of energies with all materials alike irrespective of whether they are crystalline or amorphous.

2. Spin filtering in Fe/W(110) and Co/W(110). The energy distribution curves of the photoelectrons excited from the valence bands of W(110) exhibit a sharp feature at a binding energy of 2.9 eV below E_F. The attenuation of this well defined feature on depositing overlayers of Fe or Co yields the desired information on the total scattering cross section in Fe and Co, while the dependence of the spin polarization on the layer thickness x is given by the transport polarization $a(x)$[20]. To calculate $a(x)$ from eq.(8), one needs to know $\sigma^- - \sigma^+$. This difference can be obtained from the observed enhancement A of the cascade polarization and the spin integrated total scattering cross section σ according to eq.(13) or from the product of the Bohr magneton number n_B and the partial scattering cross section σ_d determined from Fig.1 (compare eq.(4). The latter yields a larger value for $\sigma^- - \sigma^+$ that does not agree with the observations. Apparently, the assumption of a universal cross section σ_d valid for all transition metals is too crude to predict the delicate difference $\sigma^- - \sigma^+$. However, with the more empirical approach eq.(13) using the observed A, the calculated transport polarizations agree very well with the observations. A similar problem was already encountered when calculating A from σ_0/σ_d eq.(11).

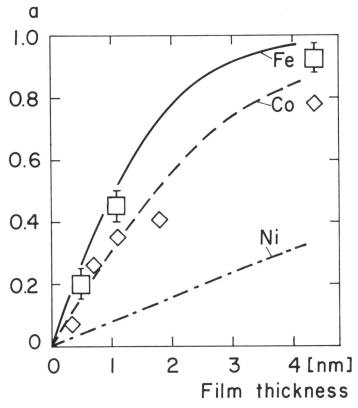

Fig. 2 Transport spin polarization a(x) vs thickness x of the ferromagnetic material for Fe, Co and Ni. The calculated curves are for T = 0 K, whereas the data points were obtained at T = 300 K. Squares denote Fe data, diamonds Co [20].

Fig. 2 shows the a(x) calculated from A at T = 0 together with the experimental data taken at room temperature. There is generally a very good agreement. In the case of Fe, **a** has a tendency to be lower than expected. To discuss this one has to remember that the calculations are for T = 0, and that for a few monolayers of Fe, the Curie-point T_c is not very far from room temperature, whereas in the case of Co ultrathin films, T_c is considerably higher [2]. Yet deviations from the predictions of the simple model larger than the experimental error occur really only with the thickest overlayer of Co were background subtraction is uncertain. The data point obtained by Pappas et al. [21] with Fe on Cu at x = 0.7 nm is **a** = 0.16, also taken at T = 300 K and with approximately the same electron energy above E_F. This is roughly 2 times lower than expected from eq.(8). However, Fe on Cu(100) is known to exhibit the fcc structure for which various magnetic properties up to nonmagnetic have been reported [2]. Taking into account the quoted experimental uncertainties one can say that the simple model eq.(8) accounts well for the observations, provided that the enhancement A of the cascade polarization is used to calculate $\sigma^- - \sigma^+$ according to eq.(13).

Fig. 3 shows the efficiency $a^2(x) \cdot I$ of transport spin filtering for Fe, Co and Ni. We see that the optimum film thickness is 5, 8 and 14 monolayers (ML) of Fe, Co and Ni with figures of merits as high as ~ 0.07, ~ 0.06 and ~ 0.01 respectively, hence at least one order of magnitude more than with detectors based on reflection of electrons, and two orders of magnitude more than with detectors cased on differential spin orbit scattering. Co is particularly attractive as one can use a film as thick as 8 ML and still achieve almost the same figure of merit as with 5 ML of the more delicate Fe. Alloys of Fe and Co would presumably offer the best choice. The reason why Co is so favorable is the fact that compared to Fe, it has much less majority spin holes, yet still a large number of minority spin holes so that the spin independent scattering cannot assume the leading rôle as in Ni. Furthermore, the Co-films generally have a high T_c so that the thermal decrease of the surface magnetization is not sizable even at ambient temperature and, quite importantly, the life time of the Co surface is one order of magnitude higher than the one of Fe. The simple model provides helpful guidance in looking for the best spin filter material which might ultimately be found to be an oxide.

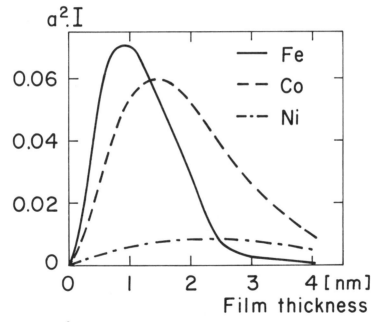

Fig. 3 Figure of merit a^2I for the spin polarization detector based on electron transmission vs thickness of the ferromagnetic material [20].

3. Spin Filtering using a source of spin polarized electrons. Recently, several experiments have been performed in which the ferromagnetic film is the analyzing spin filter. In these experiments, a source of spin polarized electrons is needed, and the change in the transmitted intensity is measured when the relative direction between the magnetization of the filter and the spin polarization of the incident electrons is inverted. J.C. Gröbli et al. [22] excited polarized electrons in GaAs by absorption of circularly polarized photons. These electrons are photoemitted into vacuum after having passed through 5 monolayers of Fe. The work function of the Fe-film is reduced below 1.5 eV so that the highly polarized electrons excited at the band gap of 1.5 eV in GaAs can escape. 5 monolayers of Ag act as a diffusion barrier between the Fe film and the GaAs inhibiting chemical reactions between Fe and GaAs that would otherwise occur and make it impossible to obtain ferromagnetism in only 5 monolayers of Fe on GaAs. In this experiment it is important that the functioning of the two essential elements of the multilayer structure can be tested independently. These elements are the source and the analyzer of the spin polarization. The source is tested by switching off the external magnetic field. The magnetization direction lies then in the plan of the Fe-film, and no spin filtering occurs for the electrons from GaAs polarized perpendicular to the surface. The spin polarization of the electrons emitted through the Ag/Fe/Cs-layers into vacuum is now measured vs the energy of the circularly polarized photons. One observes the typical signature of the GaAs source of electrons: the highest polarization for hv = 1.5 eV, and a sign reversal at hv = 3 eV. From this spectral dependence one can conclude that some of the electrons excited in GaAs have traversed the surface barrier. The reduced degree of spin polarization indicates which percentage of the emitted electrons have originated in the Ag/Fe/Cs overlayers, as these latter ones are unpolarized. The analyzing ferromagnetic spin filter is tested by measuring the spin polarization of the photoelectrons emitted with unpolarized light as it depends on the strength of the external magnetic field applied perpendicular to the film surface. The electrons originating in the Fe as well as the ones having been transported through it will exhibit spin polarization under these conditions. Magnetic saturation perpendicular to the surface is reached at an external field of 1.5 Tesla. If this field is applied and the photons are circularly polarized, one observes circular dichroism in photoemission: The photoelectron intensity changes when the handedness of the photons is inverted. The same change occurs when the direction of the magnetic field is inverted. The experiment yields the analyzing power a(x) = 26% of 5 monolayers of Fe at electron energies $1.5 \lesssim E \lesssim 2$ eV above the Fermi energy. This certainly meets the general expectations, but a closer analysis must await a better characterization of the magnetic properties of the Fe-film. Particularly interesting and novel in this experiment is the fact that it can be done with very low energy electrons. A conceptually quite similar experiment testing the analyzing power of a 5-6 ML Co-film was reported by Y. Lasailly et al. [23]. The spin polarized electron beam is again prepared with a GaAs photocathode, and the intensity transmitted through a spacially separated, free standing metallic target is measured in a Faraday cup. The 3 mm wide metallic target consists of a 1 nm thick Co film sandwiched between 21 and 2 nm thick gold layers. At low electron energy, close to the clean surface vacuum level, the transmission of majority spin electrons is found to be 1.43 times higher than the one of minority spin electrons. This yields an analyzing power of a(x) = 18% for the electron energy $E \cong 5$ eV above E_F. The intensity of the electrons was attenuated by 5 orders of magnitude mainly during passage through the first Au-film of 21 nm, in reasonable agreement with Fig. 1 from which an attenuation of $2 \cdot 10^{-6}$ is expected. Although there is no quantitative agreement with the analyzing power of a(x) = 28% given in Fig. 2, this result is really unique in that it points the way to new and much simpler detectors of spin polarization [20]. The task is to reduce the thickness of the supporting substrate.

A still different version of spin filtering using a source of polarized electrons has been reported by Celotta, Unguris and Pierce [24]. The source of polarized electrons is the (100) surface of an Fe-wisker. The wisker is covered with 0 to 25 ML of Ag, Au or Cr, and on top of this spacer there are 3-12 ML of Fe. The ferromagnetic Fe layer at the surface serves as the spin filter. Depending on the thickness of the spacer layer, it is magnetized parallel or anti-parallel to the substrate. Secondary electrons are excited by a primary electron beam of a few keV energy. The intensity of the low energy cascade electrons is plotted vs the thickness of the spacer layer. One observes that the intensity shows steplike variations as the magnetization of the spin filtering overlayer changes sign with increasing thickness of the spacer material. For quantitative analysis of the effectivity of this spin filter and comparison to the calculated a(x) one would have to extrapolate the amplitude of the variation to zero spacer thickness. The actual spin filtering is reduced to the 1% level as most of the low energy cascade electrons originated in the spin filter itself and in the spacer layer. Electrons from both of these sources produce no intensity variation when the magnetization of the ferromagnetic surface layer reverses. Only those few electrons that reached the surface from the Fe wisker substrate are spin polarized and can produce the observed variation in intensity with relative magnetization direction of substrate and overlayer. Nevertheless, it is found that the most effective thickness of the Fe-overlayer is 6 ML in agreement with the predictions of Fig. 3.

Sirotti, Panaccione and Rossi [25] have shown that it is possible to reduce the background from the spacer layer which was Cr in this case with synchrotron radiation. There is a large increase of the yield of secondary electrons by more than a factor of five when the photon energy is tuned to the $L_{2,3}$ excitation edge of Fe. The decay of the 2p-core hole produces most of the secondary electrons close to the Fe-atom. This is revealed by a sizeable increase of the spin polarization of the emitted secondary electrons. If however the photon energy is tuned to the Cr $L_{2,3}$-edge, there is a corresponding decrease of the secondary electron spin polarization. With this technique, one can perform very effective testing of the spin filtering and also determine its energy dependence.

4. New interpretation of older experiments. The first experiments with spin polarized electrons had to be as simple as possible in order to prove that photoelectrons from ferromagnetic materials were at all spin polarized. Therefore, the spin polarization of photoelectrons excited close to the photoelectric threshold was measured with Gd, Fe, Co, Ni and other model materials like EuO [26]. Almost all types of theories on magnetism predict that the spin polarization of threshold photoelectrons should be negative in Co and Ni because the majority spin states must be very nearly or completely occupied in both cases. Table 1 shows that this statement simply requires knowledge of the total number n of electrons and the Bohr magneton number $n_B = 2\Delta n$. In agreement with this expectation, the spin polarization of the electrons at the Fermi energy E_F is negative in the energy distribution curves with both Co and Ni. However, threshold photoelectrons show negative spin polarization only with Ni [27], but not with Co [28]. We have recently repeated this experiment with the same result [29]. Fig. 4 shows the polarization of the photoelectrons from Co(0001) and Ni(111) vs the photon energy. Cs was deposited to vary the vacuum level from 1.5 to 4.85 eV in order to exclude final state effects as the cause of the absence of negative polarization. These findings, namely the difference between Co and Ni for threshold electrons, and the appearance of negative polarization for electrons emitted from E_F at higher energies, can now be understood by taking into account that the photoelectrons have to travel to the surface, and that the enhancement A has to be added to the initial polarization P_i which is negative for both Ni and Co. According to table 1, A is as large as 25% for Co, but only 9% for Ni. If one shifts the zero line in fig. 4 by A, the negative initial state polarization P_i appears even for Co. The presence of P_i in the case of the energy distribution curves is explained by the higher kinetic energy E_{kin} at which the electrons travel to the surface in this case, and the observation that A decreases rapidly with increasing E_{kin}. Fig. 4 demonstrates also that the polarization enhancement is present even with electrons emitted within less than 100 meV from threshold. This excludes the spin flip model [16,17,18] as a mechanism for the polarization enhancement as discussed above.

Another example for the difficulties generated by the old picture of low energy electron emission in which the threshold electrons had a large mean free path, are the early observations with Gd overlayers on Fe(100) [30]. Gd is the ferromagnetic metal with the largest total scattering cross section $\sigma = 4$ nm^{-1} according to Fig. 1. This is due to the largely unoccupied

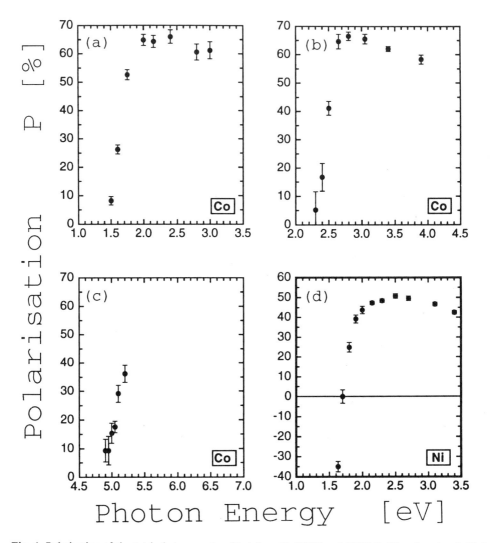

Fig. 4 Polarization of the total photocurrent emitted from Co(0001) and Ni(111). The photothreshold Φ was adjusted by depositing cesium on top of the clean metal surfaces. a) $\Phi = 1.5$ eV; b) $\Phi = 2.3$ eV; c) $\Phi = 4.85$ eV; d) $\Phi = 1.5$ eV. From Ref. 29.

5d-states. Most of the magnetic moment of 7.55 μ_B is produced by the 4f^7-level, but 0.55 μ_B must be attributed to the 6s-5d-conduction bands that are strongly exchange coupled to the 4f-states. This means that the transport polarization is not large, about the same as in Nickel. Fig. 5 shows the spin-polarization of secondary electrons vs energy excited with a primary electron beam of 2.5 keV from x = 0.24 nm of Gd deposited on Fe(100) at T = 150 K. The spin polarization of the Fe $M_{23}M_{45}M_{45}$ Auger electrons is positive showing that the substrate is magnetized in the positive direction, whereas the Gd $N_{45}O_{23}N_{67}$ and $N_{45}N_{67}N_{67}$ Auger electrons exhibit negative polarization showing that the Gd-film is magnetized opposite to the Fe-film. This is expected since it is known that the spin moment of the rare earth moment couples antiparallel to the 3d-ferromagnets. As E = 0 is approached, one observes negative spin polarization. This means that most of the low energy electrons starting with $P_c = +53\%$ in Fe could not penetrate through 0.24 nm of Gd, that is approximately 1 monolayer.

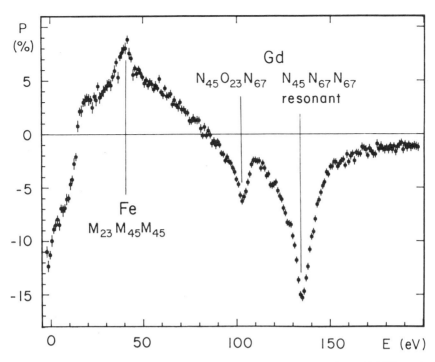

Fig. 5 Spin polarization vs. kinetic energy of secondary electrons from a Gd film on Fe(100), excited with primary electrons of 2500 eV. T = 150 K; the film thickness is ≅ 1 monolayer (2.4 Å).

O. Paul [31] has deposited Gd on Ni confirming the enormous attenuation of low energy electrons in Gd. Historically, the interpretation of this phenomenon has been very difficult as one was not ready to accept such a short attenuation length. F. Meier et al. [32] deposited paramagnetic Gd on Ge and found a very strong depolarization in one monolayer Gd for the spin polarized electrons excited in the Ge substrate. It was proposed that this is due to spin exchange scattering developed for the case of ferromagnetic insulators [33]. Similarly, Taborelli et al. [30] proposed that the electrons from the Fe-substrate invert their spin polarization by exchange scattering on the Gd-$4f^7$-moments. Yet today it is clear that there is simply an enormous attenuation of low energy electrons in early transition metals.

A further observation with the experimental results shown in Fig. 5 concerns the scattering of electrons on the $4f^7$-states. As shown in Ref. 14, the $4f^8$ autoionizing state is just a few eV above vacuum level. If this scattering was important, it would favor the escape of majority electrons from Fe, which are minority electrons in Gd. The observed negative polarization demonstrates therefore that such scattering is not important. For simplicity we have then omitted the scattering on $4f^7$-states in the discussion.

Magnetometry with Low Energy Electrons

Magnetometry is the measurement of the direction and magnitude of the magnetization vector. It is important to perform magnetometry as a function of temperature T and external magnetic field H. Of particular interest with ultrathin films is the spontaneous magnetization, that is the magnetization in the absence of any external field; even the magnetic field of the earth can be a major disturbance in 2D-films[6]. It has been shown that magnetometry using spin polarized electron spectroscopy in its various forms can access a number of new phenomena although the determination of the magnitude of the spin moment has not been possible so far. The most spectacular results using magnetometry with spin polarized electrons include the detection of the metastable magnetization in liquid Fe based on the time resolution at the picosecond level, the imaging of magnetic domains and domain walls at spatial resolution in the nanometer range, the detection of exchange couplings mediated by metallic and insulating spacer layers, and the investigation of magnetism in the various types of surface and quantum well states.

Low energy cascade electrons are emitted in copious amounts from metal surfaces upon irradiation with a primary beam of electrons or photons. The depth of information carried by the cascade electrons and their high degree of spin polarization in the case of ferromagnetic surfaces are intimately connected as shown above. The fact that the hole states in the d-shell are responsible for the magnitude of the total scattering cross section as well as for the spin magnetic moment, compare eq.(4), makes it a worthwhile proposition to test the capabilities of absolute magnetometry using the spin polarization P_c of the low energy cascade electrons. If an overlayer is magnetic, it acts as a spin filter, and the difference $\sigma^- - \sigma^+$ between total scattering cross section for minority and majority spins is proportional to the Bohr magneton number n_B according to eq.(4). The problem is reduced to determining the factor of proportionality, σ_d. There are two ways in which this can be attempted. In the first, one assumes that the partial scattering cross section σ_d of the electrons into one d-state is universal to all the transition metals; $\sigma_d = 0.7$ nm^{-1} is then obtained from Fig. 1. In the second, one takes an empirical approach and determines $\sigma^- - \sigma^+$ from the observed enhancement of the low energy cascade polarization eq.(13), or from the transport polarization $a(x)$ eq.(8), and combines it with the Bohr magneton number as known from conventional magnetometry. The empirical approach leads to a lower value $\sigma_d = 0.42$ nm^{-1} and $\sigma_d = 0.37$ nm^{-1} for Fe and Co respectively. These values are obtained with Table 1, but should be considered preliminary and subject to change. First, authors do not agree on the exact amount of the enhancement A of the cascade polarization P_c at T = 0. This has to do also with the calibration of spin detectors that can vary by up to 5% in different laboratories. Further, the transport polarization $a(x)$ is difficult to determine because it requires high quality films. Additionally the film thickness x has been at variance by up to a factor of 2 in the published literature. Finally, it is not clear what the Bohr magneton number n_B is in thin films, or, for the enhancement, at the surface. It should be noted that classical magnetometry determines the sum of orbital and spin moment. Yet several experimental facts, quite apart from theoretical considerations, exist that point to a quite different orbital contribution in thin films compared to the bulk. Nothing is known about n_B in surfaces as classical magnetometry is excluded. Inspite of these difficulties, it is to be expected that most of the present uncertainties can be reduced with time. The magnetic moment of transition metals generated upon adsorption on a magnetic surface such as Fe(100) lends itself naturally for a first test of absolute magnetometry with cascade electrons. Following Fuchs, Totland, and Landolt[34], the case of V/Fe(100) is discussed as an example. The first questions are whether the adsorption of 1 ML of V will affect the magnetic moment of the Fe(100) surface atoms, and whether the moment, if any, which is induced in V will be directed parallel or antiparallel to the Fe magnetization. These questions are answered by core level spectroscopy. The spin polarization of the electrons emitted in the $M_{23}M_{45}M_{45}$ Auger transition from the Fe-atom is unchanged upon V-adsorption, showing that the magnetic moment localized at Fe-sites is not affected. This is not a priori certain; the adsorption of Ta for instance which is in the same column of the periodic table below V induces a reduction of the magnetic moment under similar conditions[31]. Further, the negative spin polarization of the electrons emitted in the $L_3M_{23}M_{45}$ Auger transition from V proves that V indeed has acquired a moment and that this moment is directed antiparallel to the moment of the substrate. The absolute value of the spin moment cannot be evaluated from the Auger electron spin polarization as the electronic structure in V is modified due to the presence of the

3p core hole. If spin polarized photoemission from core states is used, it is also not clear as discussed in this meeting how the magnitude of the spin moment can be obtained. Note that a

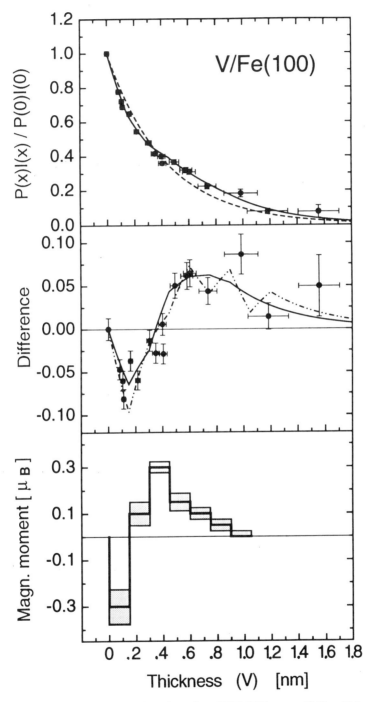

Fig. 6 Secondary-electron polarization times intensity of V/Fe(100) versus V film thickness. Top panel: Raw data. Center panel: After subtraction of an exponential background with an attenuation length of 4.3 Å (dashed line in top panel). Bottom panel: Magnetization depth-profile obtained from best fit (shown as solid lines in the top and center panels). From Karin Totland, Diss. ETH Nr. 10481, Zurich 1994.

closed shell must yield the total photoelectron spin polarization equal to zero if all the multiplets are correctly taken into account even when the atom carries a magnetic moment in an open shell.

The intensity I and polarisation P of the low energy cascade electrons excited in the Fe(100) substrate are attenuated in the V overlayer. If this layer was nonmagnetic, the attenuation is equal to $e^{-\sigma x}$, where x is the thickness and σ the spin averaged attenuation length. The dashed line in Fig. 6 is the best possible fit of an exponential to the data. It yields $\sigma = 2.3$ nm^{-1} for the spin averaged total scattering cross section in V. This is somewhat too small to nicely fit Fig. 1, but agrees within experimental uncertainty with $\sigma = 2.7 \pm 0.4$ nm^{-1} determined for Ta [35,36]. However, in the case of Ta the first adlayer was not included in the fitting as it clearly reduced the magnetization in the Fe(100) surface. This indicates that the evaluation is quite delicate. The oscillations around $e^{-\sigma x}$ are explained naturally by the spin filtering in the magnetic V-adlayers. What is experimentally determined is the difference $\sigma^- - \sigma^+$ of the total scattering cross sections for the two spin states as the number of adlayers increases. With $\sigma_d = 0.7$ nm^{-1} one obtains $n_B = -0.3$ Bohr magnetons for the first adlayer. The spin moment decreases and changes sign in successive layers. Similar studies have been performed on Cr/Fe(100), Ru/Fe(100) and Ag/Fe(100). The latter is an example of a non-magnetic adlayer, while Cr and Ru both show an induced magnetic moment which however ist directed parallel to the one of the substrate in the case of Ru [37].

SUMMARY

Electrons of low kinetic energy are emitted in large quantities from surfaces irradiated with photons or electrons. Despite many important applications based on low energy photoemission or electron multiplication in the process of cascade formation, it is generally not well understood how low energy electron emission occurs. One question of great interest is the depth of information carried by the threshold electrons. Up to very recently the general consensus was that this depth is quite large, of the order of 20 Å and more. But experiments with spin polarized electrons revealed that the mean free path λ of low energy electrons can be as short as an atomic layer with early transition metals and gradually increases to about 10 layers as the d-shell gets fully occupied in the noble metals. Essential for obtaining this new result was the application of magnetic multilayers in electron spectroscopy. With this technique, well defined magnetization profiles can be built at a surface that allow one to apply a magnetic form of the well known overlayer technique with which the depth of information or inelastic mean free path λ can be determined[38]. The results of these experiments are summarized in a simple rule which says that the inelastic scattering cross section $\sigma = 1/\lambda$ is devided into a term that accounts for scattering into unoccupied states other than the d-states, and a term $\sigma_d(5-n)$ that is proportional to the number of holes (5-n) available to one spin state in the d-orbitals[39]. In this picture, elements in the same column of the periodic table such as Cu, Ag and Au, or Fe and Ru, or V and Ta should have identical σ. This is indeed observed within the experimental uncertainties. The model can also explain why the early transition metals with a large number of unoccupied d-states have a very large total scattering cross section resulting in a probing depth λ of as little as ~ 1 monolayer in Gd. An even more interesting consequence of the model is that the total scattering cross section becomes spin dependent with the ferromagnetic transition metals because the number of holes is different for majority (+) and minority (-) spin electrons. In fact, the difference $\sigma^- - \sigma^+ = \sigma_d \cdot n_B$, where n_B is the number of Bohr magnetons per atom. This leads to a transport polarization a(x) when electrons travel a distance x in ferromagnetic material, and to an enhancement A of the spin polarization of low energy electrons emitted from magnetic surfaces. Several recent experiments have been specifically designed to test the predictions of the simple model and satisfactory agreement was found in all cases. The distinguishing feature of this new picture of low energy electron emission is the existence of two independent currents of electrons for the two spin states. Interactions between the currents, that is spin flips, can be neglected. This corresponds exactly to the two current model for conduction electrons with which the giant magnetoresistance in magnetic multilayers can be understood. The spin dependent mean free path of conduction electrons is measured just as in electron spectroscopy, using very thin magnetic films as spin filters or spin valves. It turns out that the mean free path of the conduction electrons is inversely proportional to the unoccupied density of states accessible to these low energy electrons just as in electron spectroscopy. The relevant quantity is then the density of states at the Fermi-energy. $D(E_F)$ is however not generally known a priory from magnetization measurements as is the case with the Bohr magneton number n_B. Therefore, the spin filtering of emitted low energy electrons turns out to be simpler and directly linked to the spin moment. Based on this, absolute magnetometry now also seems to be possible in low energy electron spectroscopy.

The new coherent and simple picture of low energy electron emission can also explain some puzzling results of older experiments such as the absence of negative spin polarization in threshold photoemission from Co. Furthermore, other models such as spin dependent elastic scattering or Stoner spin flip excitations can be excluded on the basis of the existing evidence as an explanation for the transport polarization or the enhanced polarization in low energy electron emission.

ACKNOWLEDGEMENT

I would like to thank E.L. Garwin and F. Meier for helpful discussions and comments.

REFERENCES

*   Permanent address: Swiss Federal Institute of Technology, Zurich, Switzerland
1. M. Donath, Surf. Sci. **287/288** (1993) 722
2. H.C. Siegmann, J. Phys.: Condensed Matter **4** (1992) 8375
3. T. Maruyama, E.L. Garwin, R. Prepost, and G.H. Zapalac, Phys. Rev. **B 46** (1992) 4261 and ref. cited
4. Proc. of the Workshop on Photocathodes for Polarized Electron Sources for Accelerators, 1993, Stanford Linear Accelerator Center, Stanford, Ca.
5. H. Pinkvos, H. Poppa, E. Bauer, and J. Hurst, Ultramicroscopy **47** (1992) 339
6. C.H. Back, C. Würsch, D. Pescia, to be publ. in Z. Physik Condens. Matter
7. S. Blügel, Phys. Rev. Lett. **68** (1992) 851 and ref.
8. A.J. Cox, J.G. Londerback, S.E. Apsel, and L.A. Bloomfield, Phys. Rev. **B 49** (1994) 12295 and ref.
9. C. Turtur and G. Bayreuther, Phys. Rev. Lett. **72** (1994) 1557
10. R. Coehorn, Europhys. News **24** (1993) 43
11. Bruce A. Gurney et al., Phys. Rev. Lett. **71** (1993) 4023
12. M.P. Seah and W.A. Dench, Surf. Interface Anal. **1** (1979) 2
13. D.R. Penn, S.P. Apell, and S. M. Girvin, Phys. Rev. **B 32** (1985) 7753
14. H. Tang, T.G. Walker, H. Hopster, D.P. Pappas, D. Weller, J.C. Scott, Phys. Rev. **B 47** (1993) 5047
15. R. Allenspach, M. Taborelli, and M. Landolt, Phys. Rev. Lett. **55** (1985) 2599
16. J. Glazer and E. Tosatti, Solid State Comm. **52** (1984) 11507
17. D.L. Abraham and H. Hopster, Phys. Rev. Lett. **62** (1989) 1157
18. D. Venus and J. Kirschner, Phys. Rev. **B 37** (1988) 2199
19. M.P. Gokhale and D.L. Mills, Phys. Rev. Lett. **66** (1991) 2251
20. G. Schönhense and H.C. Siegmann, Ann. Phys. **2** (1993) 498 and ref.

21. D.P. Pappas, K.P. Kämper, B.P. Miller, H. Hopster, D.E. Fowler, C.R. Brundle, A.C. Luntz, and Z.-X. Shen, Phys. Rev. Lett. **66** (1991) 504
22. J.C. Gröbli, D. Guarisco, S. Frank, and F. Meier, to appear in Phys. Rev. B, 1994
23. Y. Lasailly, H.-J. Drouhin, A.J. van der Sluijs, G. Lampel, and C. Marlière, Poster Session this meeting and preprint
24. R.J. Celotta, J. Unguris, and D.T. Pierce, J. Appl. Phys. **75** (1994) 6452
25. F. Sirotti, G. Panaccione, and G. Rossi, subm. to J. de Physique
26. H.C. Siegmann, Physics Reports **17** C (1975) 39
27. W. Eib and S.F. Alvarado, Phys. Rev. Lett. **37** (1976) 444
28. G. Busch, M. Campagna, D.T. Pierce, and H.C. Siegmann, Phys. Rev. Lett. **28** (1972) 611
29. J.C. Gröbli, A. Kündig, F. Meier, and H.C. Siegmann, to appear in Physica B
30. M. Taborelli, R. Allenspach, G. Boffa, and M. Landolt, Phys. Rev. Lett. **56** (1986) 2869, and R. Allenspach, Diss. ETH Nr. 7952, Zürich 1985
31. O.M. Paul, Diss. ETH Zürich Nr. 9210, 1990
32. F. Meier, G.L. Bona, and S. Hüfner, Phys. Rev. Lett. **52** (1984) 1152
33. J.S. Helman and H.C. Siegmann, Solid State Commun **13** (1973) 891
34. P. Fuchs, K. Totland, and M. Landolt, to be published, and K. Totland, Diss. ETH Nr. 10481, Zürich 1994
35. M. Donath, D. Scholl, H.C. Siegmann, and E. Kay, Appl. Phys. A **52** (1991) 206
36. O. Paul, S. Toscano, K. Totland, and M. Landolt, Surf. Sci. **251/252** (1991) 27
37. K. Totland, P. Fuchs, J.C. Gröbli, and M. Landolt, Phys. Rev. Lett. **70** (1993) 2487
38. H.C. Siegmann, Surf. Sci. **307-309** (1994) 1076
39. H.C. Siegmann, J. of El. Spectroscopy and Rel. Phenomena **68** (1994) 505

PHOTOEMISSION AND FERROMAGNETISM

P. D. Johnson

Physics Dept.,
Brookhaven National Laboratory,
Upton, NY 11973

Abstract

Photoemission is a well established technique for the study of the electronic structure of atoms and solids. In particular, angle-resolved photoemission has been used extensively to map the band structure of clean and adsorbate covered surfaces, both metal and semiconductor. Extending the technique by measuring the spin of the photoemitted electrons allows the possibility of examining the exchange split band structures characterizing ferromagnetic systems. Here the technique becomes particularly useful in the study of the magnetic properties of surfaces, thin films and associated interfaces.

1. Introduction.

The photoexcitation of an electron from initial state $|i\rangle$ with binding energy E_i to some final state $|f\rangle$ with Energy E_f results in a photocurrent I given by

$$I(h\nu, E_i) = \sum \left|\langle f|\underline{A}\cdot\underline{P}|i\rangle\right|^2 \delta(E_f - h\nu - E_i) \quad (1)$$

where $\underline{A}$ is the classical vector potential of the photon and $\underline{P}$ is the momentum operator. $h\nu$ represents the energy of the incident light and the δ-function describes the energy conservation during the excitation.[1] The summation is over all states at an initial binding energy E_i. The introduction of an angle resolving capability allows, via momentum conservation, the investigation of the electronic structure at discrete k-points in the Brillouin zone.[2] In this way it is possible to map out the band structure of the material under investigation. The matrix element in eqn. (1) also introduces selection rules which allow the determination of the symmetry of the initial state.

A more complete description of the process allows for the many body response of the system to the excitation. Here the δ-function of eqn (1) is replaced by the function

$$A(h\nu, E_i) = \frac{1}{\pi} \frac{\mathrm{Im}\,\Sigma(h\nu, E_i)}{\{[h\nu - E_i - \mathrm{Re}\,\Sigma(h\nu, E_i)]^2 - [\Sigma(h\nu, E_i)]^2\}} \quad (2)$$

where $\Sigma(h\nu, E_i)$ represents the complex self energy of the electron. This correction results in a broadening and shifting of the peaks with respect to the binding energies that would characterize the one electron spectrum. In a ferromagnetic system we may anticipate that the correction will be spin dependent reflecting the different spin densities.

For the present purposes we note from equation (1) that the photon will couple only to the orbital momentum of the electron, not to its spin. The spin of the electron will therefore be conserved during the excitation and by measuring the spin in the final state we determine the spin in the initial state.

2. The Experimental Method.

The valence band spin-polarized photoemission experiments reported in this paper were all carried out on an apparatus that has been described in considerably more detail elsewhere.[3] Shown in fig. 1, spin detection is achieved by mounting a compact low energy spin detector of the type developed at the National Bureau of Standards[4] onto the exit lens of a commercial hemispherical analyzer.[5] This latter analyzer has an angular resolution of the order of +/- 1.5°. The overall energy resolution of the experiment is typically 0.35 eV.[3]

In the spin detectors, the photoemitted electrons are scattered off a polycrystalline gold surface at an energy of 150 eV. Spin-orbit coupling produces a scattering asymmetry which may be measured to determine the spin polarization of the electrons. This polarization P is given by

$$P = \frac{N^\uparrow - N^\downarrow}{N^\uparrow + N^\downarrow} \quad (3)$$

where $N^\uparrow$ and $N^\downarrow$ represent the numbers of electrons with the two different spin components in the beam.

The figure of merit (FOM) used in comparing different polarimeters is defined as [6]

$$FOM = S^2 \frac{I}{I_0} \quad (4)$$

where S, the Sherman function reflects the polarimeter's ability to distinguish the two spins, I is the sum of the current collected by the two opposite detectors and I_0 is the incident beam current. The spin detector used in the present studies has a FOM of approximately 10^{-4}.

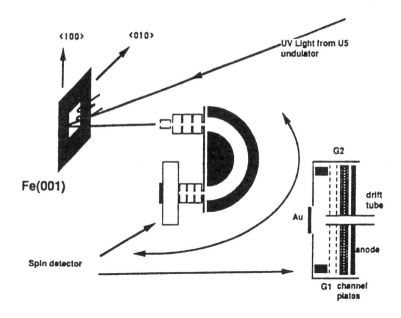

Figure 1. Schematic of the spin polarized photoemission experiment showing the picture frame crystal and the hemispherical analyzer with associated spin analyzer.

In spin polarized photoemission experiments based at synchrotron radiation sources, it is particularly useful to be able to mount the experiment on a beamline having an undulator or insertion device as the source.[3] These sources, consisting of a linear array of magnets, enhance the available photon flux by typically two orders of magnitude per milliradian when compared with the normal bending magnet source of synchrotron radiation.

In order to define a quantization direction for the spin of the electrons, it is necessary to magnetize the sample under investigation. However it is important to minimize the stray fields that will distort the electron trajectories in front of the sample. For elemental metals such as Fe, Co or Ni this is often achieved by shaping the sample in the form of a picture frame as indicated in fig. 1. The magnetizing field is then obtained from a solenoid wrapped around one arm of the frame. In the case of thin films the magnetizing field is obtained by discharging a current pulse through a small coil in close proximity to the film. The photoemission measurements are always made with the sample in a remnant magnetic state.

2. Clean Surfaces.

First principles calculations predict that the clean surfaces of many materials will have enhanced magnetic moments. As an example, the magnetic moment at the surface of an Fe crystal cut in the (001) plane is predicted to be 2.96 μ_B as opposed to the 2.2 μ_B

characteristic of the bulk.[7] This enhancement of the moment reflects the reduction in the number of nearest neighbors as the dimensionality is reduced. These predictions have never been directly verified although several studies of magnetic thin films appear to indicate that the average magnetic moment increases as the film becomes thinner.[8]

Although it is not possible to determine the magnitude of a magnetic moment from angle resolved photoemission studies, it is possible to verify many other aspects of the calculation. In particular it is possible to identify and characterize the spin dependent surface states that are responsible in part for the surface magnetic structure.

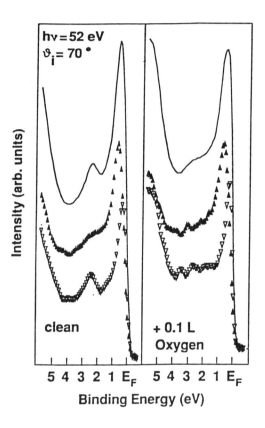

Figure 2. Spin integrated and spin resolved photoemission spectra recorded along the surface normal for photons of incident energy 52 eV. the angle of incidence corresponds to p-polarized light (70°). The spectra are shown before and after exposure to 0.1L of oxygen.

Figure 2 shows an example of a minority spin even symmetry state that has been identified on the (001) surface of Fe.[9,10] The surface character of the state is confirmed through it's sensitivity to oxygen adsorption; it's symmetry is determined from its sensitivity to the angle of incidence of the light. By varying the angle of detection of the photoelectrons, we may follow the dispersion of the surface state as a function of $k_\parallel$ across

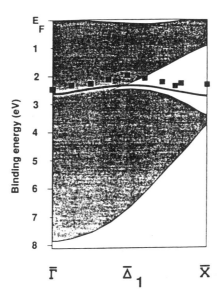

Figure 3. Experimental dispersion of the minority spin surface state (•) shown in fig. 2 compared with the theoretical dispersion (thick solid line) determined from a 13 layer slab calculation. The calculated minority bulk projected bands are shown by the shaded region.

the zone. This is shown in fig. 3 where we compare the experimentally determined dispersion of the state in the $\overline{\Gamma X}$ azimuth with the calculated dispersion. The parallel momentum $k_{\parallel}$ is given by

$$k_{\parallel} = (2m/\hbar^2)^{1/2} (E_f - \phi)^{1/2} \sin\theta \qquad (5)$$

where ϕ is the workfunction of the surface, θ is the angle of emission with respect to the surface normal and E_f is the final state energy above the Fermi level.

3. Thin Films and Interfaces.

Because of the potential technological applications in both the recording and device industries, magnetic multilayers are currently the subject of an enormous research industry.[11] These multilayers, consisting of ferromagnetic layers interlaced with non-magnetic layers, have been shown to exhibit an oscillatory exchange coupling between the adjacent ferromagnetic layers dependent on the thickness of the intervening non-magnetic layer.[12] Example systems showing this property include the Cu/Co[13] and Fe/Ag[14] multilayers both grown in the (001) direction. It is not possible to use photoemission to study the "bulk" electronic structure of these multilayers because the technique is limited by the escape depth of the photoelectrons. However it is possible to study the properties of related thin films and interfaces.

The finite thickness of the individual components within the multilayer imposes boundary conditions which lead to a quantization of the allowed electronic states within the layer. Thus rather than seeing electronic transitions between bands characterizing the continuous bulk electronic structure of the material, we may anticipate seeing discrete states at binding energies dependent on the thickness of the layer. The energy separation ΔE between the different states is given by [15]

$$\Delta E = \frac{dE}{dk}\frac{2\pi}{L}. \qquad (6)$$

where dE/dk reflects the dispersion in the associated bulk bands from which the states are derived and L is the thickness of the film. The discrete quantization is clearly seen in photoemission studies of silver films deposited on an Fe(001) substrate.[10,16] Figure 4 shows the photoemission spectra recorded from along the surface normal from silver films of different thicknesses as indicated. In the region immediately below the Fermi level a series of peaks are observed with binding energies dependent on the thickness of the film. In the case of the ultrathin films these states have been described as interface states, a reflection of their considerable weight in the interface between film and substrate. Inverse photoemission studies indicate that in the thicker films the states evolve into the "quantum well" states.[17]

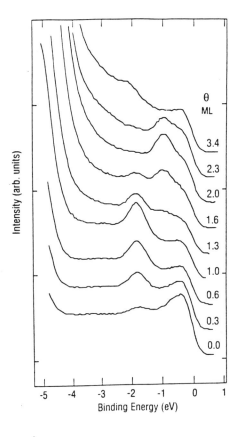

Figure 4. Spin integrated photoemission spectra recorded from different thickness silver films deposited on a Fe(001). The different silver coverages are indicated. The spectra correspond to emission along the surface normal and the incident light is p-polarized with an energy of 52 eV.

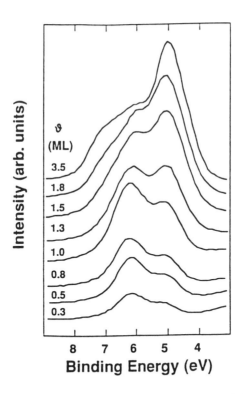

Figure. 5. The coverage dependence of the silver d-bands as a function of silver coverage. The incident light is p-polarized with an energy of 60 eV.

Further information on the interfacial coupling may be obtained by examining the evolution of the silver d-bands as a function of coverage.[10] Figure 5 shows a series of photoemission spectra recorded from the silver d-bands as the silver coverage is varied from sub-monolayer through to coverages exceeding three monolayers. At fixed photon energy the spectra show a continuous coverage change throughout the range. Elsewhere such changes have been discussed in terms of a quantization of the d-bands.[18,19] However in that the spectra present a two peaked structure for both the monolayer and bilayer coverage, it is unclear whether the discrete peaks in fig. 5 represent such a quantization. By varying the incident photon energy we are able to preferentially select emission from either the silver or iron components. This is shown in fig. 6.[10] At a photon energy of 118 eV the cross-section for emission from the silver d-bands is reduced by nearly two orders of magnitude. Thus the spectrum at this photon energy provides a measure of the substrate contribution to the spectra recorded at lower photon energies. The conclusion that the peak at a binding energy of 6.0 eV contains a strong interfacial component is supported by spin polarization analysis which shows this peak to be polarized but only in the interfacial layer.[10]

Quantization of the electronic structure is also observed in studies of thin copper films deposited on a Co(001) substrate.[20,21] Figure 7 shows the spin resolved photoemission spectra recorded from selected copper films.[20] As in the case of the silver

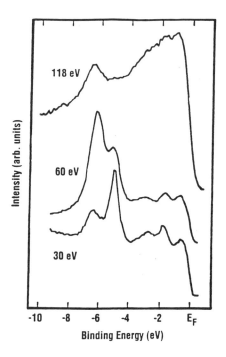

Figure 6. Spin integrated spectra recorded from one monolayer of silver deposited on an Fe(001) substrate at photon energies corresponding to 30, 60 and 118 eV.

films, the states are preferentially polarized with minority spin even in the thicker films where the photoelectron mean free path will restrict the sampling to the copper layers. The observation that these quantum well states are spin polarized and crossing the Fermi level with the appropriate frequency has lead to the suggestion that they are the states that mediate the exchange coupling in the magnetic multilayers.[17]

As shown in fig. 8, with increasing thickness, the s/p derived quantum well states appear to move up to and through the Fermi level with a periodicity or frequency identical to that observed for the long period oscillation in the exchange coupling in the related multilayers.[13] In a simple phase model the states may be viewed as stationary states trapped within a one-dimensional potential well. Within such a model the states will exist at binding energies corresponding to wave-vector k such that

$$\phi_1 + \phi_2 + 2kd = 2\pi n \qquad n = 1,2,3,\ldots \qquad (7)$$

where ϕ_1 and ϕ_2 represent the phase changes at the two interfaces and d is the thickness of the well. Recognizing that with each additional layer another half wavelength is added to the electron's wavefunction it is a simple matter to show that the Fermi surface will be sampled every m layers where[21]

$$m = \frac{k_{BZ}}{k_{BZ} - k_F}. \qquad (8)$$

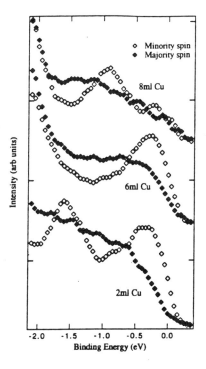

Figure 7. Spin resolved photoemission spectra recorded from 2, 6 and 8 ml thick copper films. The incident photon energy is 24.0 eV and the angle of incidence of the light corresponds to p-polarization.

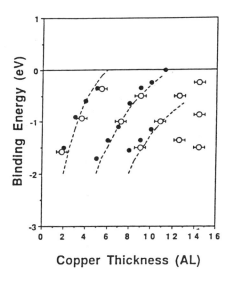

Figure 8. Comparison of the observed binding energies for quantum well states of the Cu/Co(001) from the experiment described in the present paper (open circles) (ref. 20), an earlier photoemission study (solid circles) (ref. 17) and the extrapolation of lines modeling inverse photoemission observations of the same states above the Fermi level (dashed lines) (ref. 17). The different copper thicknesses are indicated.

Here k_{BZ} and k_F define the zone boundary and Fermi wave vectors respectively. In the direction of interest, the difference in these wavevectors $k_{BZ} - k_F$ represents one half the Fermi surface spanning vector.

The quantization and spin polarization of the states may be modeled within the tight-binding formalism.[20] Our approach is based on the use of a Hubbard Hamiltonian of the form

$$H = \sum_k E(k)n_k + \frac{U}{N}\sum_{k',k} n_{k\uparrow}n_{k'\downarrow} \qquad (9)$$

where the first term represents the paramagnetic band structure and the second term represents the modification due to an on-site spin dependent potential, U. Constructing the calculation in a slab formulation, we take the parameters associated with a tight binding fit to the appropriate paramagnetic band structure for each layer,[23] split the on-site spin dependent energies for the d-blocks by an amount Δ_l and then integrate the spin-dependent densities of states up to the Fermi level to obtain layer dependent moments, m_l. A self consistent solution is sought such that for each layer, $\Delta_l = U_l m_l$. The "exchange" potential, U_l, is taken as the effective Stoner parameter from LSD calculations of the susceptibilities for the different elements.[24]

To simulate the photoemission spectra recorded along the surface normal, the calculation of the spin dependent densities of states is restricted to a narrow region around the center of the surface Brillouin zone. The calculated eigenvectors in each layer of the slab are weighted by an escape depth and representative photoionization cross-sections.[25] Figure 9 shows the results of such a spin-dependent calculation for different thickness copper films deposited on a seven layer Co slab.[20] The minority spin states are observed to

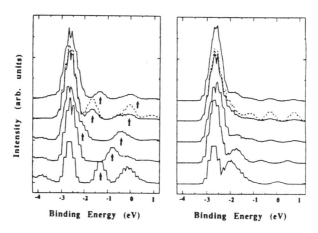

Figure 9. Calculated spin resolved photoemission "spectra" from different thickness copper films. The left panel shows minority spin spectra and the right panel shows majority spin spectra. The copper thicknesses range from two through to six monolayers from bottom to top in each panel. The dashed lines indicate spectra calculated for the five monolayer films with equal weight given to s, p, and d states. The arrows indicate the relevant quantum well states in the minority spin panel.

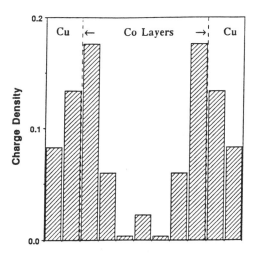

Figure 10. The calculated layer dependent charge densities for the quantum well state of a 2ml Cu film either side of a 7ml Co(001) slab. The vertical dashed lines indicate the Cu/Co interfaces. In these interfaces the orbital character is predominantly d_{z^2}.

move up to and though the Fermi level with a frequency similar to that seen in the experiment. States are also observed in the majority spin channel but with considerably less intensity.

The calculation highlights other interesting characteristics of these states. Aside from the obvious s/p character, the states also have significant copper d-character. This is shown in the figure for the five monolayer case where we compare the spectra obtained by giving equal weight to the s, p and d electrons with the spectra obtained preferentially from the d-electrons. The d-component reflects the strong $sp_z d_{z^2}$ hybridization that occurs in the vicinity of the copper d bands and explains the observation of a d-moment on the copper site in recent magnetic circular dichroism (MCD) studies of the Cu/Co multilayers.[26] There is also a significant interfacial component in these states as indicated in fig. 10 for the 2ML Cu state. Earlier we discussed the interfacial characteristics for the ultrathin Ag films deposited on an Fe(001) substrate.[10] It should be noted that as a result of the d-d hybridization, the charge density peaks in the interface even for the states associated with thicker Cu films and in fact an enhanced moment on the interfacial copper sites was reported in the MCD studies of the system.[26]

As noted earlier, theories invoking the bulk Fermi surface of Cu predict two period lengths for the oscillatory exchange coupling.[27] The long period of oscillation reflecting the spanning vector at the center or belly of the Fermi surface "dog bone", the short period of oscillation reflecting the spanning vector at the neck of the "dog bone". In an isolated copper slab, the latter point on the Fermi surface, approximately 0.75 $\overline{\Gamma X}$, is sampled by states that have dispersed upwards from the center of the zone with a dispersion in the plane of the film, $E = h^2 k_{\parallel}^2 / 2 m_e$, characterized by a free electron like mass, m_e.[28] As a function of thickness, we would expect from eqn. 8, the Fermi surface to be sampled with a periodicity inversely related to the bulk Fermi surface spanning vector at this point in the zone. However in the thin films or multilayers this behavior may be modified by hybridization in the interfaces.

Figure 11 shows the photoemission spectra recorded from a 2 ml thick copper film deposited on a 20ML thick Co substrate, as a function of the angle of emission in the $\overline{\Gamma X}$ azimuth. At the center of the zone the quantum well state characterizing this thickness is again observed at a binding energy of 1.6 eV. However away from the center of the zone the dispersion of this state in the plane of the film is clearly not free-electron like but, rather is characterized by an effective mass of approximately 4.0 m_e where again m_e represents the free electron mass.

Shown in fig. 12, the modification of the in-plane effective mass extends out to copper film thicknesses of the order of 7-8ML.[29] This thickness will determine the pre-

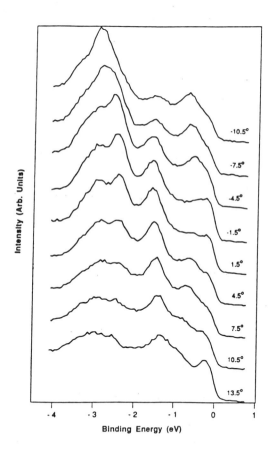

Figure 11. Photoemission spectra recorded from a 2ml thick copper film as a function of the angle of emission as indicated. The incident photon energy is 22.7 eV and the angle of incidence of the light corresponds to p-polarization. The quantum well state has a binding energy of 1.4 eV with respect to the Fermi level near the center of the zone.

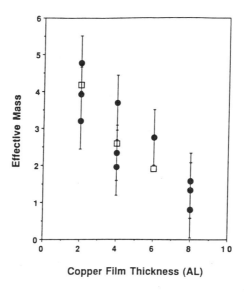

Figure 12. Plot of the fitted effective masses of the quantum well states as a function of the copper film thickness in monolayers (ML). The solid circles indicate the fits to the experimental data; the open squares indicate the results of tight-binding calculations of the effective masses. The dashed line represents the free electron mass.

asymptotic limit below which it will not be valid to describe the properties of the copper layer simply in terms of its bulk electronic structure. The effective mass enhancement may be viewed as a competition between on the one hand the properties of the interface and on the other the properties of bulk copper. In the very thin films the interfacial hybridization dominates.

The modification of the effective mass is an indication that the Fermi surface in the copper film will also be modified, an observation that has implications for theories of the oscillatory exchange coupling that invoke the bulk Fermi surface.[27] Indeed in the preasymptotic regime it is likely that the calculated coupling periods involving spanning vectors away from the center of the zone will be modified as a result of this interfacial hybridization.

4. Core level Photoemission

Photoemission from the more localized core levels provides a means of examining the site specific properties. Indeed core level photoemission is well established as a means of probing site specific valency.[30] The extension of these ideas to the possibility of probing site specific magnetic information has already been demonstrated.[31-35] Here we will review some of the basic ideas. In discussing the process we will concentrate our discussion on the photoemission from the 3s core level of Fe. It is a relatively well studied system[33,34] and does not have the complications associated with spin-orbit effects as is the case of the 3p core level.[31,32]

As in any spectroscopy, the spin polarization carried by the photoemitted core electron will reflect both initial and final state effects. We measure the electron in the final state but are interested in the properties of the initial state, particularly the initial state of the valence bands. The total angular momentum, S, and the z-component of the total angular momentum, S_z, represent good quantum numbers. To model the excitation we assume that the orbital moment in the valence bands is quenched and therefore take the total angular momentum S as the total spin of the system in the initial state. This is simply the net spin of the electrons in the valence bands, S^v. We further assume that the alignment of the spins is determined by the magnetization direction and take the z-component $S_z = S^v$.

If we photoemit an electron with spin $|1/2|$, the ion in the final state will be left either in a low spin state corresponding to total spin $S^v - 1/2$ or in a high spin state corresponding to total spin $S^v + 1/2$. The high spin state $S^v + 1/2$ may have z-components S_z equal to $S^v + 1/2$ or $S^v - 1/2$. We write the wavefunction, Ψ_{HS}, for this state as

$$\Psi_{HS} = A|S^v + 1/2, 1/2 \rangle |S_z^v + 1/2, -1/2\rangle + B|S^v + 1/2, 1/2\rangle|S_z^v - 1/2, +1/2\rangle \qquad (10)$$

where we have retained the subscript z to distinguish the z-component from the total spin. In equation (10) we recognize that to conserve S_z^v we must emit a photoelectron with spin $+1/2$ if the ion final state corresponds to $S^v - 1/2$ and vice versa. Analysis of the Clebsch Gordon coefficients yields

$$A = \sqrt{\frac{2S+1}{2S+2}} \quad \text{and} \quad B = -\sqrt{\frac{1}{2S+2}} \qquad (11)$$

The spin polarization, P, in this high spin state will therefore be given by[36-38]

$$P = \frac{B^2 - A^2}{B^2 + A^2} = -\frac{S}{S+1} \qquad (12)$$

The low spin state may have only one z-component $S_z = S^v - 1/2$ and the wavefunction for this state, Ψ_{LS}, will be given by

$$\Psi_{LS} = C|S^v - 1/2, 1/2\rangle|S_z^v - 1/2, +1/2\rangle \qquad (13)$$

Here the photoelectron carries spin $+1/2$ resulting in a 100% spin polarized peak in the spectrum. The polarization difference between the low spin and high spin peaks has lead to the suggestion that photoionization of the 3s core levels be used as an "internal source" of spin polarized electrons for photoelectron diffraction studies.[38]

A spin polarized photoemission spectrum of the Fe 3s core level is shown in fig. 13.[39] These latter spectra were obtained with an instrument similar to that described in section 2 but with a larger solid angle of collection.[40] In the majority spin channel we are clearly able to observe for the first time the two peaks corresponding to the low spin state at higher binding energy and the high spin state at lower binding energy. In the minority spin channel we observe a single peak corresponding to the high spin state. These spectra represent strong evidence for an atomic like multiplet excitation.

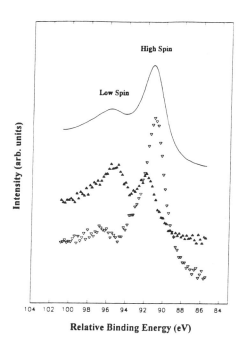

Figure 13. Spin resolved photoemission spectra from the Fe 3s level recorded at a photon energy of 250 eV. The solid line shows the spin integrated spectrum, the filled triangles represent the spin resolved majority spin spectrum and the open triangles the minority spin spectrum. The high spin and low spin ion states are indicated.

The separation between the two peaks in the majority spin channel reflects the exchange interaction between the core electron and the valence electrons in the final state. As has been noted by several authors previously,[41-43] this splitting is considerably smaller than would be expected on the basis of a Van Vleck type model where the splitting would scale with the (3s,3d) exchange integral. Indeed such a model would predict a splitting of the order of 8.0 eV. However models recognizing that the interaction between different final state electronic configurations plays an important role, predict an exchange splitting between the peaks of the order of 2.7 eV,[44] i.e close to our experimental observation of 3.4 eV.

The origin of the energy separation between the two components of the high spin state is unknown at the present time. It may reflect different final state screening involving the 4s electrons or it may represent some small initial state separation. This effect requires further investigation.

Extensions of the technique to provide site specific magnetic information have already been demonstrated.[32,35] Here we will present as an example the results of a study of the oxidation of the Fe(001) surface.[32] Figure 14 shows the characteristic spin resolved photoemission spectra from the clean Fe surface in the 3p core level region. The spectra are characterized by an intense minority spin peak and a less intense majority spin peak displaced to higher binding energy by 0.5 eV.

Figure 15 shows the spin resolved Fe 3p photoemission spectra following exposure of the surface to 16L of O_2 and after annealing the latter surface to 650°C. The main substrate iron emission is represented by the minority spin feature at a binding energy of 52.9 eV. The majority counterpart is again at a binding energy 0.5 eV higher. With the initial oxidation, fig. 15(a), we identify a new majority spin peak shifted to higher binding energy by 4.5 eV with respect to the clean surface peak. The shift in binding energy for this peak suggests that we may associate it with Fe^{3+} valency. This together with the observation that it is spin polarized, suggests that the initial oxide is of the ferrimagnetic form γ-Fe_2O_3. The observation that the substrate Fe emission is dominated by minority spin electrons but the initial oxide is dominated by majority spin emission suggests that the moments on the outer Fe ions associated with the oxide are antiferromagnetically coupled to the moments of the substrate via a superexchange mechanism involving subsurface atoms.

Following annealing, fig. 15(b), it is still possible to identify emission from the substrate but now the spectra are dominated by unpolarized emission shifted approximately 2.0 eV from the main peak. The latter shift in binding energy is indicative of a Fe^{2+} phase. These observations suggest that the oxide following annealing is most probably the antiferromagnetic Fe_xO.

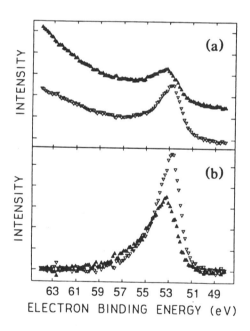

Figure 14. (a) Spin-resolved Fe 3p photoemission spectra recorded from clean Fe(001) with 90-eV photons. (b) The same as in (a) but with the background subtracted. In both figures, upward triangles indicate majority spin, downward triangles indicate minority spin.

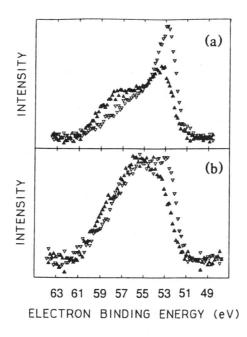

Figure 15. (a) Spin resolved Fe 3p spectra showing the initial oxidation of Fe after exposure to 16L of O_2 with the background subtracted. [1L (Langmuir) = 10^{-6} Torrsec]. (b) The same as in (a) but after annealing to 650°C. The photon energy used in these studies is 90 eV.

4. Summary.

In this paper we have reviewed the use of spin polarized photoemission as a technique for the study of the magnetic properties of surfaces, thin films and interfaces. Valence band studies allow the identification of magnetic states unique to the solid-vacuum and solid-solid interfaces. Studies of noble metal thin films deposited on ferromagnetic substrates reveal spin polarized quantum well states that mediate the oscillatory exchange coupling in the associated multilayers.

Core level photoemission provides site specific magnetic information. However as with all spectroscopies the technique is influenced by both initial and final state effects. The challenge in developing this spectroscopy is whether it will ultimately be able to provide information on the site specific moments in the ground state.

5. Acknowledgments

The author is pleased to acknowledge many stimulating conversations with K. Garrison, Y. Chang, N.V. Smith, B. Sinkovic, N. Brookes, and M. Weinert. This work has been supported in part by the Department of energy under contract DE-AC02-76CH00016.

References

1. "Photoemission in Solids", ed. by M. Cardona and L. Ley (Springer-Verlag, Berlin, Heidelberg, New York, 1978).
2. "Angle-Resolved Photoemission", ed. S.D. Kevan (Elsevier, Amsterdam, 1992).
3. P.D. Johnson, S.L. Hulbert, R. Klaffky, N.B. Brookes, A. Clarke, B. Sinkovic, N.V. Smith, R. Celotta, M.H. Kelly, D.T. Pierce, M.R. Scheinfein, B.J. Waclawski and M.R. Howells, Rev. Sci. Inst. **63**, 1902 (1992).
4. J. Unguris, D.T. Pierce, R.J. Celotta, Rev. Sci. Instrum., **57** (1986) p. 1314 ; M.R. Scheinfein, D.T. Pierce, J. Unguris, J.J. McClelland, R.J. Celotta, and M.H. Kelley, Rev. Sci. Instrum. **60(1)**, 1 (1989).
5. VSW Ltd., HA 50 Analyzer
6. J. Kessler, "Polarized Electrons" (Springer, Berlin, 1985)
7. S. Ohnishi, A.J. Freeman, and M. Weinert, Phys. Rev. B **28**, 6741 (1983).
8. W. Clemens, T. Kachel, O. Rader, E. Vescovo, S. Blugel, C. Carbone and W. Eberhardt, Sol. Stat. Comm. **81**, 739 (1992).
9. N.B. Brookes, A. Clarke, P.D. Johnson and M. Weinert, Phys. Rev. B **41**, 2643 (1990).
10. N.B. Brookes, Y. Chang, and P.D. Johnson, to be published Phys. Rev. B.
11. L.M. Falicov, Physics Today, **45(10)**, 46 (1992).
12. S.S.P. Parkin, Phys. Rev. Lett. **67**, 3598 (1991).
13. Z.Q. Qiu, J. Pearson, and S.D. Bader, Phys. Rev. B **46**, 8659 (1992).
14. J. Unguris, R.J. Celotta, and D.T. Pierce, J. Magn. Magn. Mater. **127**, 205 (1993).
15. P.D. Loly and J.B. Pendry, J. Phys. C **16**, 423 (1983).
16. N. B. Brookes, Y. Chang, and P. D. Johnson, Phys. Rev. Lett. **67**, 354 (1991).
17. J.E. Ortega, F.J. Himpsel, G.E. Mankey, and R.F. Willis, Phys. Rev. B **47**, 1540 (1993).
18. D. Hartmann, W. Weber, A. Rampe, S. Popovic, and G. Guntherodt, Phys. Rev. B**48**, 16837 (1993).
19. D. Li, J. Pearson, J.E. Mattson, S.D. Bader and P.D. Johnson, to be published.
20. K. Garrison, Y. Chang and P.D. Johnson, Phys. Rev. Lett. **71**, 2801 (1993).
21. N.V. Smith, N. B. Brookes, Y. Chang, and P. D. Johnson, Phys. Rev. B **49**, 332 (1994).
22. C. Carbone, E. Vescovo, O. Rader, W. Gudat, and W. Eberhardt, Phys. Rev. Lett. **71**, 2805 (1993).
23. D.A. Papaconstantopoulos, Handbook of the Band Structure of Elemental Solids (Plenum Press 1986).
24. J.F. Janak, Phys. Rev. B **16**, 255 (1977).
25. J.J. Yeh and I. Lindau, Atomic Data and Nuclear Data Tables, **32**, 1 (1985).
26. M.G. Samant, J. Stohr, S.S.P. Parkin, G.A. Held, B.D. Hermsmeier, F. Herman, M. van Schilfgaarde, L.-C. Duda, D.C. Mancini, N. Wassdahl, and R. Nakajima, Phys. Rev. Lett. **72**, 1112, (1994).
27. e.g. P. Bruno and C. Chappert, Phys. Rev. Lett. **67**, 1602 (1991).
28. A. Euceda, D.M. Blylander, L. Kleinman and K. Mednick, Phys. Rev. B **27**, 659 (1983).
29. P.D. Johnson, K. Garrison, Q. Dong, N.V. Smith, D. Li, J.E. Mattson, J. Pearson and S.D. Bader, Phys. Rev. B, to be published.
30. C.R. Brundle, T.J. Chuang and K. Wandelt, Surf. Sci **68**, 459 (1977).

31. C. Carbone, and E. Kisker, Sol. Stat. Comm. **65**, 1107 (1988).
32. B. Sinkovic, P.D. Johnson, N.B. Brookes, A. Clarke, and N.V. Smith, Phys. Rev. Lett. **65**, 1647 (1990).
33. F.U. Hillebrecht, R. Jungblut, and E. Kisker, Phys. Rev. Lett. **65**, 2450 (1990).
34. C. Carbone, T. Kachel, R. Rochow and W. Gudat, Sol. Stat. Comm. **77**, 619 (1988).
35. F.U. Hillebrecht, Ch. Roth, R. Jungblut, and E. Kisker, and A. Bringer, Europhys. Lett. **19**, 711 (1992).
36. S.F. Alvarado and P.S. Bagus, Physics Letters, 67A, 397(1978)
37. G.M. Rothberg, J. Magn Magn Mater. 15-18, 323 (1980).
38. B. Sinkovic, B. Hermsmeier and C.S. Fadley, Phys. Rev. Lett **55**, 1227 (1985); B. Sinkovic, D.J. Friedman and C.S. Fadley, J. Magn Magn Mater. 92, 301 (1991).
39. Z Xu, Y. Liu, P. D. Johnson, B. Itchkawitz, K. Randall, J. Feldhaus and A. Bradshaw, to be published.
40. Z Xu, and P. D. Johnson, to be published.
41. C.S. Fadley, D.A. Shirley, A.J. Freeman, P.S. Bagus and J.W. Mallow, Phys. Rev. Lett. **23**, 1397 (1969).
42. P.S. Bagus, A.J. Freeman, and F. Sasaki, Phys. Rev. Lett. **30**, 850 (1973).
43. D.A. Shirley, Chapter 4 in Vol. 1 of ref. 1.
44. P. Bagus and J. Mallow, to be published in Chem. Phys. Lett.

THE STUDY OF EMPTY ELECTRON STATES OF SOLIDS WITH CORE X-RAY ABSORPTION AND INVERSE PHOTOEMISSION

L. Braicovich

Dipartimento di Fisica del Politecnico,
Piazza Leonardo da Vinci 32, 20133 Milano, Italy.

1. INTRODUCTION

The main goal of these lectures is to discuss the connection (and the differences) between two spectroscopies used to study empty electron states of solids. One is a typical core spectroscopy i.e. XAS (X-Ray Absorption Spectroscopy) and the other is Inverse Photoemission in which one sends electrons onto a sample and on measures a photon spectrum bringing information on the empty states of the system. In what follows we will treat the general concepts in some rather abstract limiting cases (par 2 and 3) and we will conclude with a short presentation of the spectroscopy of Fe oxides (far from being complete) which are of interest for this school; this will show how the general concepts are applied.

The comparison between the two methods will be treated here in a very schematic and pedagogical way. We will consider in an elementary approach two extreme situations having important methodological implications:
(i) the case in which it is possible to neglect all effects coming from the perturbation due to the creation of an excited final state in the solid; in this case the spectral functions reflect directly the ground state properties of the system
(ii) the case in which the creation of the final excited state contributes in an essential way to the spectral functions which cannot be interpreted directly in terms of ground state properties; thus the problem is how to recover these properties from spectra reflecting the excitation process.

In rather crude and qualitative terms let us consider a solid having delocalized valence electron states for which the local electron density is varying not fast enough to prevent the application of the local spin density approximation (LSD approximation[1]). In such a case the perturbation due to the addition of an electron in IPES or due to the presence of an excited electron coming from core excitation is expected to be small and possibly negligible in practical terms (we forget for the moment the attractive potential of the hole). Thus there is good chance to interpret the spectral functions directly in terms of ground state properties while this is definitely impossible in the opposite situation where the local correlation is very important. As a consequence it is not unreasonable to call case (i) "weakly correlated limit" and case (ii) "strongly correlated limit". As a matter of fact the situation is much more subtle and the detailed discussion of this point is beyond the limits of the present lectures; nevertheless the simple arguments given below should help to introduce some important concepts on this delicate point.

2. THE WEAKLY CORRELATED LIMIT

a. Inverse Photoemission and Direct Valence Photoemission

Let us consider a threedimensional crystal where **k** is a good observable and the band model works satisfactorily; we want to show the information which can be obtained in an inverse photoemission experiment in which all excitations are described as transitions between unperturbed states. The answer comes from an obvious extension of the direct valence photoemission formalism which is briefly summarized here following the treatment by Koyama and Smith of ref [2] (for a general introduction to photoemission see also[3]); for the convenience of the reader we prefer to treat as a first step the direct photoemission since it is more intuitive.

Let us make the assumption of constant matrix elements and let us consider the optical absorption which is proportional to the sum of all transitions conserving the electron momentum **k** since the photon momentum is in general negligible (we neglect transitions assisted by other elementary excitations; for example phonons). In this case the optical absorption is proportional to the so called Joint Density of States (JDOS)

$$I(h\nu) = (2\pi)^{-3} \sum_{i,f} \int d^3k \, \delta(E_f(\mathbf{k})-E_i(\mathbf{k})-h\nu) =$$
$$=(2\pi)^{-3} \sum_{i,f} \int (dS_{fi}/|\nabla_\mathbf{k} E_f(\mathbf{k}) - \nabla_\mathbf{k} E_i(\mathbf{k})|) \qquad (1)$$

where the sum is over all initial occupied (i) and final empty (f) states having energies $E_i(\mathbf{k})$ and $E_f(\mathbf{k})$ and the second integral is carried over the surface S_{fi} defined in **k** space by the condition $\delta(E_f(\mathbf{k})-E_i(\mathbf{k})-h\nu)=0$ i.e. the surface of constant interband energy.

Note that by definition the total density of states in the ground states is given by

$$DOS(E) = (2\pi)^{-3} \int d^3k \, \delta(E-E(\mathbf{k})) =$$
$$= (2\pi)^{-3} \int (dS/|\nabla_\mathbf{k} E(\mathbf{k})|) \qquad (2)$$

where the integration is over the surface S defined by the condition $\delta(E-E(\mathbf{k})) = 0$.

The comparison between the two formulas shows clearly that the critical points (i.e. the features) in the DOS and in the optical spectrum are not the same being related to different topologies in **k**-space[4].

On the other hand the photoemission experiments do not measure the total number of transitions but measure also how the optically excited electrons are distributed in kinetic energy. Since one can reconstruct from a photoemission spectrum the initial energy of the electron, the spectrum is proportional to an integral containing two constraints represented by two delta functions (we consider here an angle integrated experiment i.e. without constraints on the **k** direction); this integral called Energy Distribution of the Density of States (EDJDOS) is given by

$$D(E,h\nu) = (2\pi)^{-3} \sum_{i,f} \int d^3k \, \delta(E_f(\mathbf{k})-E_i(\mathbf{k})-h\nu) \, \delta(E-E_i(\mathbf{k})) =$$
$$= (2\pi)^{-3} \sum_{i,f} \int (dl_{fi}/|\nabla_\mathbf{k} E_f(\mathbf{k}) \times \nabla_\mathbf{k} E_i(\mathbf{k})|) \qquad (3)$$

A crucial aspect is that the critical points of the DOS are also critical points of the valence photoemission spectrum as seen by comparing formula (2) with formula (3); moreover there are other critical points coming from the **k** points where the two gradients are parallel and which are typical of photoemission. In the EDJDOS the integration is over the line l_{fi} of intersection between the two surfaces defined by the two delta functions. Due to this topology[5] it is possible to recover in a valence photoemission experiment an information which is much closer to the ground state DOS than in an optical absorption experiment.

A further important possibility is offered by valence photoemission when also angular distributions of photoelectrons are measured. If one can account for refraction at the surface one can reconstruct the direction of the **k** vector within the solid; in this case one has to add a third constraint to the integral i.e. a third delta function. Thus the integral collapses to a sum over a discrete set of points related by symmetry. In other words it is possible to establish directly a connection between **k** and the energy values i.e. one measures the dispersion relations (band mapping); this is particularly easy in bidimensional systems where the the component **k**// parallel to the surface is a good observable and is conserved during the emission.

Another important point is the behavior of the first delta function in the EDJDOS at increasing photon energies; it is intuitive that this delta function is relaxed in the limit of high excitation energies since it is always possible to accommodate the excited electron independently of the initial **k**-point; also non vertical transitions have to be included since the photon momentum is no more negligible with respect to the electron momentum. This is basically the situation in X-Ray Photoemission Spectroscopy (XPS) so that the spectra are a good image of the initial occupied DOS.

Of course one has to keep in mind that the present treatment assumes constant matrix elements i.e. constant photoemission cross sections; in the real case one recovers information weighted on the photoemission cross sections. In fact the cross sections have an important energy dependence (for the atomic case see[6]) which can be exploited to pick up contributions coming from states which locally project onto different angular momentum components; the systematic use of this photon energy dependence is one of the very important outcomes of the tunability of Synchrotron Radiation applied to photoemission.

In inverse photoemission the situation is analogous and the extension of the above treatment is straightforward once the nature of the experiment is specified.

First of all it is necessary to consider that inverse photoemission experiments are very often intrinsically **k**-resolved; in inverse photoemission the solid angle occupied by the incoming electron beam is in general better defined than the input solid angle of a photoelectron spectrometer used in direct photoemission unless an accurate angle resolved direct photoemission experiment is considered. For this reason a lot of band mapping of crystal empty states has been done in the UV region. On the other hand in the inverse XPS regime one recovers **k**-integration in analogy with direct photoemission. In the UV, if one wants to measure a **k**-integrated information, one must use either diffuse incidence of the electrons (with crystalline samples) or polycrystalline samples.

More precisely the inverse photoemission experiments can be carried out in two distinct approaches as far as the scanning variable is concerned. One can use some kind of bandpass to detect photons having all the same energy (the same "color") by measuring the intensity vs. the kinetic energy of the incoming electrons; this mode is called with an obvious name "isochromat spectroscopy". On the other hand one can keep the electron kinetic energy fixed and one can measure the spectrum as a function of the photon energy ("dispersive mode"). For the details of the two modes the reader is referred to the literature[7]; here we mention only a few facts:
(i) in the inverse XPS regime one works in the isochromat mode by using in the inverted mode the fixed monochromator used to monochromatize the conventional X-ray sources used in XPS (typically Al-Ka at 1486.7 eV); this operation mode is called BIS = Brehmstrahlung Isochromat Spectroscopy
(ii) in the UV the dispersive approach which gives the maximum of flexibility in band mapping is technically the most difficult requiring a grating and a position sensitive detector having a known dependence of the efficiency on the photon energy [8]
(iii) in the UV most of the isochromat work is done around 10 eV by using a typical bandpass obtained by combining a low pass (typically an ionic crystal) and a high pass (a Geiger with Iodine filling or a particular photocathode as CuBe); in such a case the efficiency is very high and one can use large solid angles. For this reason spin resolved inverse photoemission which is of interest for this school is performed in this operation mode [9,10,11]
(iv) in the UV a compromise between tunability and simplicity is to use a grating to measure simultaneously a set of isochromat spectra at different photon energies [12,13].

In the isochromat mode the application of the above formalism shows that the sum of all transitions giving a photon hn is a joint density of states between states which are normally empty; the selection of the initial electron energy which is the sweeping variable defines a subset through another delta function so that the spectral function in the isochromat mode is analogous to that of formula (3) with the condition that both E_f and E_i are empty states. The treatment for the dispersive inverse photoemission spectra is obvious and it is left to the reader.

Up to now we have treated the excitation in the strict limit of single particle processes; an important step forward is to consider the interaction between different excitation channels. In this connection the resonant processes are extremely important; the basic theory is due to Fano[14]. With reference to direct photoemission from a conductor let us consider an excitation of a valence electron with the same energy of the transition from a core level to the Fermi level; in this case the virtual excitation at the Fermi level can accompany the photoemission transition giving rise to an interference between the two channels with an antiresonance before threshold and a resonant peak above threshold. A typical example is given by the comparison between a closed shells and a shell with one hole; in the former case there is no excitation generating resonance while this can be very strong in the second case. For example Yb metal has the f^{14} configuration so that there is no 4d-4f transition and no resonance in 4f emission; as soon as the surface is oxidized some Yb atoms go to the f^{13} configuration opening the 4d-4f channel and generating a strong Fano lineshape at the crossing of the core 4d level[15]. The resonant photoemission projects onto the core level involved in the virtual excitation thus giving some degree of site sensitivity to valence spectroscopy. The resonance effect has been seen also in inverse photoemission [16,17] and it has been used recently in Ce compounds across the 3d-4f threshold to pick up locally projected information[18]. We have mentioned in this subparagraph the resonant processes but this does not imply that the process is limited to weakly correlated systems; in fact the states involved in the resonance can be either weakly or strongly correlated. As a matter of fact the above examples on rare earths refer to atomiclike states which are strongly correlated.

b. XAS vs. IPES and Examples

In what follows we give a brief description of core absorption by stressing in the weak correlation limit the differences in empty state spectroscopy in comparison with inverse photoemission. In the weak correlation limit it is expected that the core absorption is simply the sum of the possible transitions between unperturbed states connecting the core wave function with the empty states. Due to this fact the following points are relevant:
(i) the information is intrinsically site selective i.e. it is related to the the local contribution to the empty DOS projected onto the site where the core absorption takes place. Thus the information is sensitive to the chemical species and sometimes to the particular chemical state of the excited species.
(ii) the selection rules are very stringent; thus the sampling of the final states in the more intense features is due to the dipole selection rules. Thus one measures some of the contributions coming from the different orbital characters in a development of the empty DOS. Taking into account the well known dominance of the (l +1) channel this gives a great selectivity and in the meantime makes it difficult to measure some component as the s-like term in empty states. Note that in order to make the argument elementary we have neglected the effect of the crystal field.

In the above points we have stressed the connection with the unperturbed empty bands in a rather extreme case in which any perturbation due to the excitation is negligible; in this sense it is legitimate to call the absorption 'bandlike". As a matter of fact we will see that this is rather naive since the perturbation due to the core hole can be severe; however there is a variety of problems in which this perturbation can be represented to a good approximation by a rigid shift due to the attraction to the core hole. In all these cases the expression "bandlike absorption" is still appropriate.

In a comparison with IPES the main difference is in the much greater selectivity of core absorption which makes the two approaches complementary even when the two spectroscopies are both bandlike. In some cases each of the two spectroscopies give unique information as it is the case of **k**-mapping in IPES. Before explaining these points with some examples it is important to stress other features of the two spectroscopies coming from the fact that IPES has always an electron in the ingoing channel while XAS can be carried out in a variety of modes.

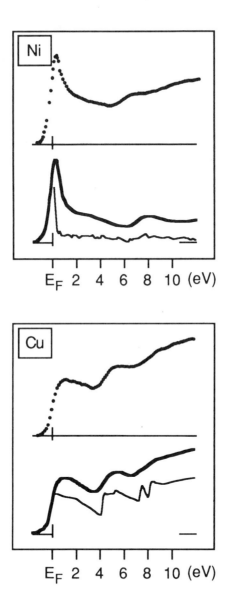

Fig. 1. Measured BIS (dots) and DOS curves for Ni and Cu. The thin curves are the unbroadened DOS and the bold lines are the broadened DOS to simulate lifetime and resolution effects (after ref [20])

In IPES the surface to bulk sensitivity is determined by the penetration depth of the electrons in close analogy to the situation of direct photoemission where the electron escape depth determines the depth scale below the surface. Thus in general the surface sensitivity is rather high in all inverse photoemission spectroscopy although less pronounced in the BIS regime due to the use of higher energy electrons. Moreover in UV inverse photoemission one has to keep in mind that the penetration depth is rapidly varying with the energy so that the sensitivity to the surface is much higher at 30 eV that at 10 eV[19].

On the other hand XAS is in general much more sensitive to the bulk; if the experiment is of the type photon in-photon out (for example in transmission) XAS is a bulk spectroscopy. If the XAS spectra are obtained by detecting the total electron yield the experiment is controlled by the electron escape depth just above the vacuum level giving some sensitivity to the surface which however is much lower than in IPES. Finally, if a particular subset of electrons are detected in the outgoing channel of XAS (for example a particular Auger electron), one can carry out the measurements in a surface sensitive mode; in this last case the limitation comes in general from the counting rate but with modern Synchrotron Radiation sources it is possible to obtain extraordinary good results.

Having in mind these precautions on the surface to bulk sensitivity we present some specific applications of IPES and XAS.

(i) An overview of the total density of states is typically obtained with BIS i.e. inverse valence XPS. In this respect the work of ref.[20] is remarkable since it makes a successful comparison with the ground state calculations of the DOS, as seen in Fig 1 for Ni and Cu. The empty d-state peaks are well seen in the spectra and correspond to the so called "white lines" seen in optical absorption from the p-states (see for example Ni spectra in [21]). For finer effects showing that the incoming electron also in this case induces some perturbation of the system see .

(ii) The effect of the preferential orientation of the spin in magnetic systems is seen in empty state spectroscopy through spin resolved IPES; in this case one uses an electron beam with a preferential orientation of the electron spin and one measures the spectra with electron polarization parallel and antiparallel to the magnetization of the sample. A typical result on Ni d-holes is given in Fig 2 taken from Dose[9]. The minority nature of these holes is immediately apparent; the information on the unbalance of minority and majority carriers in core absorption comes from dichroism as explained in other chapters of this book (in Ni see[21]). For details on spin resolved IPES and on the experimental technique see [9,10].

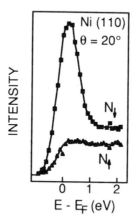

Fig. 2. Spin resolved inverse photoemission from Ni showing the minority and majority components (after ref [9])

(iii) Band mapping is a typical application dealing with **k** as a good observable; by definition this requires a spectroscopy able to see the dispersion of the empty states as IPES in the weakly correlated limit. In this school on magnetic materials it is appropriate to give examples on transition metals by presenting in Fig 3 a compilation of band mapping taken from G.J. Mankey et al [22].

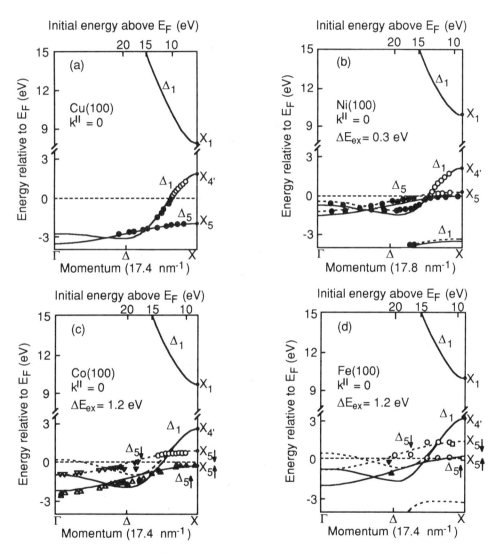

Fig. 3. Band mapping of Cu(100) and of the magnetic fcc pseudomorphs Ni(100), Co(100) and Fe(100) (after ref. 33)

(iv) As we will show and it is discussed in others chapters of this book the transition metals (TM) oxides are strongly correlated compounds so that the final state effects are important in electron spectroscopy. However there is a good evidence that the XAS spectra of the Oxygen K level are to a great extent "bandlike" as opposite to the $L_{2,3}$ edges of the TMs having a much stronger coupling of the core hole with the valence states due to the non zero angular momentum of the hole. For this reason there is a remarkable general agreement between the shape of the oxygen K-edge spectra and the angle integrated inverse photoemission spectra in about the first 4-5 eV above the bottom of the conduction band as shown in Fig 4 on Fe_xO taken from ref [23] (the Oxygen K-edge is taken from[24]). The same happens in Fe_2O_3 (the XAS is given in[25] and IPES in[26]) and in an even more evident case in Fe_3O_4 (the IPES is compared in[27] with the XAS taken from[26]). In all these cases, due to cross sections reasons, IPES contains a strong contribution from the d-holes while XAS probes the p-holes in the Oxygen site. The general agreement between the two spectroscopies is remarkable since it suggests directly some degree of covalent O(p)-Fe(d) mixing ruling out the extreme ionic picture in which the p-shell of Oxygen should be completely filled and the Oxygen K-edge spectrum should not show any peak just above threshold.

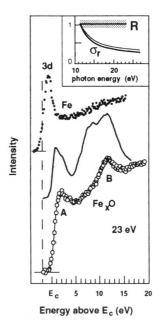

Fig. 4. Comparison of inverse photoemission (lower curve) and Oxygen K-edge in Fe monoxide (middle curves) (after ref [24] and ref. quoted therein); for further details see the original paper.

2. PROBLEMS ON THE STRONGLY CORRELATED LIMIT

In this case the perturbation due to the excitation process is extremely important so that the two spectral functions (IPES and XAS) differ substantially from the ground state and are very different one from the other. Thus the main problem is to try to reconstruct the ground state properties starting from the measured spectral function; this requires models as it will be discussed below in some particular cases. In principle this can be done starting from one particular spectroscopy; however, since any model is necessarily a simplification of the reality, it is much more convenient to try to obtain the same information on the ground state simultaneously from more than one spectroscopy. In fact this strongly reduces the ambiguities in the analysis; here we refer for pedagogical reasons to IPES and XAS but a wider set of spectroscopies is even better.

Before entering this subject we want to give the feeling on the importance of the correlation by showing in Fig 5 the valence spectra of a typical Rare Earth i.e. Gd covering both direct and inverse photoemission[28]. In direct photoemission and in inverse photoemission the peaks coming from 4f states are found at completely different energies whose separation is by definition the Hubbard energy U connected with the fluctuation (f^n-f^n) $\rightarrow$ (f^{n-1}-f^{n+1}). This typical experimental situation motivates the name "electron subtraction spectroscopy" or "ionization spectroscopy" given to photoemission and the name "electron addition spectroscopy" or "affinity spectroscopy" given to inverse photoemission.

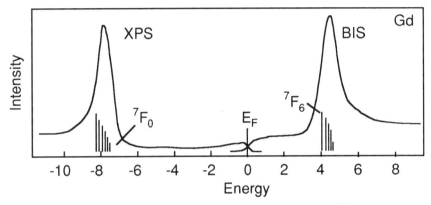

Fig. 5. Valence XPS and BIS from elemental Gd (after ref [29]).

In what follows we select a very particular case which can give a glimpse on the approach to the problem. We consider a system containing a transition metal bound to a ligand and we assume that the first excitation has a finite energy and involves an electron transfer from the ligand to the TM atom. Thus we choose a charge transfer insulator in the sense clarified by Zaanen Sawatsky and Allen in ref [29]. This problem will be treated in the simplest case in which only two configurations are important; the two configurations are

$$|1\rangle = |d^n\rangle \text{ and } |2\rangle = |d^{n+1}\bar{L}\rangle \tag{4}$$

where $\bar{L}$ denotes a hole on the ligand atom. This is a very convenient pedagogical choice and does not imply that it can give an exhaustive description of a real system. Moreover we will assume for simplicity that both configurations are not degenerate (also this is an oversimplification; see for example[30])

In this case the states of the system come from the diagonalization of the following 2x2 matrix being the two configurations mixed by the hybridization matrix element T (assumed real).

$$\begin{vmatrix} E_1 & T \\ T & E_1 + \Delta \end{vmatrix} \tag{5}$$

By measuring all elements in unit of the energy separation Δ between the two configurations (not hybridized) the energetics is that shown in Fig.6 (upper part) where δ is the hybridization shift obviously identical for the two levels being the representation bidimensional.

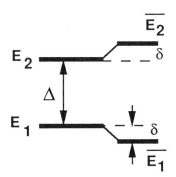

Weights of the two components

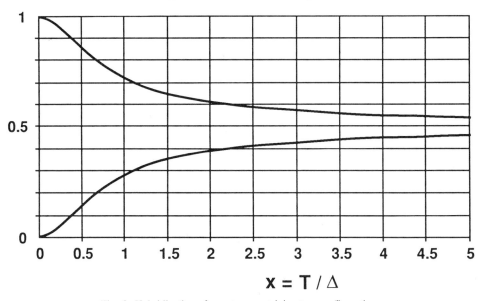

$x = T / \Delta$

Fig. 6. Hybridization of a system containing two configurations.

The diagonalization brings to two states containing the two base states with the weights given in Fig 6 (lower part); the hybridized states can be easily expressed as

$$\psi_1 = (\cos\theta)|1\rangle + (\sin\theta)|2\rangle$$
$$\psi_2 = -(\sin\theta)|1\rangle + (\cos\theta)|2\rangle \qquad (6)$$

where θ is the so called hybridization angle

$$tg|\theta| = |\overline{E}_1|/T \qquad (7)$$

Note that the hybridized states are represented in very useful vectorial diagram as in the Fig 7 where the axis are taken on the non hybridized base. Although these are very elementary notions it is useful to remember that the energy eigenvalues have a dependence on $x=T/\Delta$ which is parabolic at the small x values and is asymptotically linear in the case of large x (appropriate to nearly degenerate interacting configurations) where the relative weights of the two configurations tend to become equal.

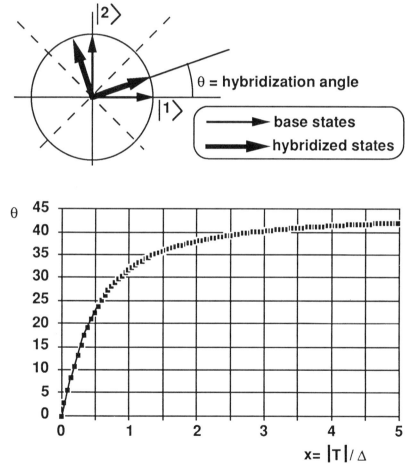

Fig. 7. Hybridization angle in a system containing two configurations.

Starting from the above description of the ground state we treat the case of inverse photoemission i.e. we calculate the transitions to a system having an extra electron added. In this case as seen at the right in Fig 8 the separation between the two non hybridized

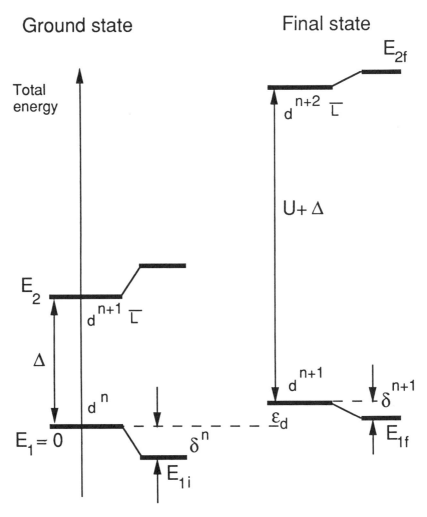

Fig. 8. Total energy diagram in charge transfer insulator containing two configurations. At the left the ground state and at the right the excited state created in inverse photoemission due to electron addition.

configurations in the final state is U+Δ since U is the price to pay to increase the d-occupancy by one. Being the system a charge transfer insulator (Δ < U) the gap between the two configurations in the final state is more than twice the initial gap so that the system is much less hybridized than in the initial state. This dehybridization due to the electron addition is represented by the reduction of the hybridization angle shown in the vectorial diagram of Fig. 9; the heavy arrow represents the ground state and the two light arrows correspond to the final state with θ_f smaller than θ_i being $\Delta\theta=\theta_i-\theta_f$ the decrease of the hybridization angle. In the sudden approximation two peaks are found in the spectra due to the projection of the initial state onto the two final states, the one without and the other with the ligand hole. The square of the projections of these vectors are the weights of the two components so that the ratio of the intensities of the two components is the square of the tangent of $\Delta\theta$ (see also ref [31]); with some degree of arbitrariness one component is called main line and the other the "satellite". The energy positions of the two components come from the differences between the energies of the initial and final states. Note that in the above figures the levels represent the energy situation of the system as a whole and are not to be thought of as single particle levels.

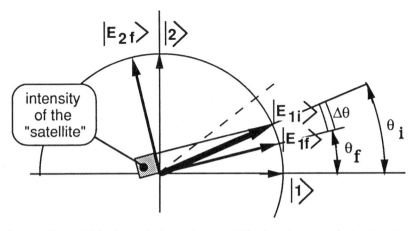

Fig. 9. Change of the hybridization angle due to electron addition in a charge transfer insulator; the figure refer to the energies represented in fig 8. The two components in the spectrum come from the projection of the initial state (heavy arrow) onto the two final state axes (light arrows).

The above example is the simplest formulation of the so called cluster calculations and suggests several interesting considerations.
(i) If one thinks of the problem in terms of the straightforward approach typical of the weakly correlated limit the perturbation in the final state due to electron addition might seem a severe inconvenience. However this perturbation is far from being dangerous; in fact it is just the variation of the hybridization in the final state which creates a multiplet of lines in the spectrum putting in evidence the configuration interaction in the ground state which otherwise would not be seen.
(ii) It is not a paradox to say that a severe perturbation in the inverse photoemission final state is very useful since the final state eigenvectors become very close to the unhybridized initial state configurations (asymptotically they are coincident in the limit U → ∞) so that the intensities of the lines give an immediate insight on the hybridization angle in the ground state which is the interesting quantity.
(iii) Also in this very simple hypothetical system the interplay between the parameters is rather subtle and one has to be cautious in drawing conclusions. A typical case is the intensity of the satellite. In the ground state the amount of the ligand hole configuration is stronger at the higher values of the hybridization T/Δ; however this does not imply that the satellite in the spectra is necessarily stronger than in less hybridized systems. In fact a strongly hybridized system requires a considerable perturbation in the final state to counteract the strong effect of T/Δ and to develop a strong satellite; in realistic situations a strongly hybridized system with its real U could be far from this situation. On the other hand a system with a smaller T/Δ is more sensitive to the electron addition perturbation so that it can

develop, at small U, a satellite stronger than in a more hybridized system although asymptotically (U → ∞) the satellite is smaller than in the other case. In other words the normalized intensity (I_2/I_{tot}) of the satellite increases monotonically with U but the curves pertaining to different hybridizations have crossings as in the case shown in the Fig. 10.

In the real cases the situation is more involved but in general the above arguments remain significant and help to understand qualitatively the basic physics.

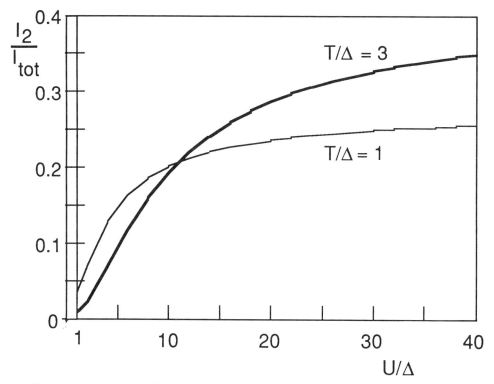

Fig.10. Intensity of the "satellite" in a charge transfer insulator containing two configurations. The intensity is plotted as the function of the Mott-Hubbard energy U (in units of the charge transfer energy Δ) using as a parameter the hybridization matrix element T (in units of the charge transfer energy Δ).

It is noteworthy that the reaction of the system to the addition of an extra electron is different in a Mott-Hubbard insulator where the first excitation is $(d^n-d^n) \rightarrow (d^{n+1}-d^{n-1})$ involving the energy U. In this case the separation between the two configurations in the ground state is U and becomes 2U in the final state due to the extra electron addition. Thus the change of hybridization in the final state with respect to the initial state is fixed whereas in a charge transfer insulator it is dependent on the relative intensities of the two parameters Δ and U. In the limit of the small T/U values the perturbation can have important effects since the hybridization angle varies rapidly at low T/U values; on the other hand for very high T/U the relative weights af the two configurations do not differ too much in the initial and final state so that the intensity of the satellite drops to small values; this generates a nonmonotonic dependence of the satellite intensity on T/U as in the Fig. 11 showing also that the maximum intensity of the satellite is rather small in this case (Mott-Hubbard insulator) as compared with the charge transfer insulator treated above.

In XAS the final state is quite different with respect to inverse photoemission because a core hole is created while an extra electron is deposited in the first empty states. The repulsion due to this electron partially counteracts the attraction to the core hole Q so that this situation is less perturbed with respect to inverse photoemission where the number of electrons changes. In this sense XAS is sometimes referred to as a "well screened spectroscopy" as opposite to electron addition and electron subtraction spectroscopies which

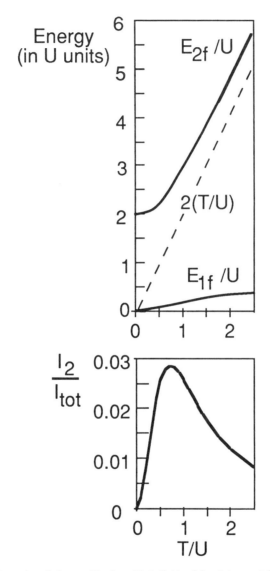

Fig.11. Intensity of the satellite in a Mott-Hubbard insulator containing two configurations as a function of the hybridization matrix element (in units of the Mott-Hubbard energy U)

are "poorly screened". In a charge transfer insulator the energetics is shown in Fig.12 where for completeness we have also shown the configurations due to core photoemission. The separations of the configurations in the final state of XAS is Δ-Q+U in comparison with Δ+U in inverse photoemission. This originates a smaller perturbation; moreover the ordering of the different configurations along the energy scale can be different depending on the numerical values of the parameters; the detailed application of the above algebra to this problem is left to the reader. The important point is that the joint interpretation of a variety of spectroscopies adds strong constraints to the parameters and makes the application of the model more reliable in order to recover the properties of the ground state. To quote an example of this procedure applied to a realistic case we mention CoO from ref [32].

Before concluding the paragraph we mention that the core hole attraction often has to be treated in a more sophisticated approach i.e. not only with the attraction Q but also by accounting for the deformation of the spectral function due to the attraction to the core hole; an example is the Oxygen K-edge in Cu_2O treated in[33]. This deformation is rather small and this confirms the bandlike behavior of the Oxygen K-edge while in the same paper the great importance of correlation in the Cu $L_{2,3}$ edges is shown; this last point is important since in this case the Cu atom is formally in the d^{10} configuration.

3. A SHORT AND SIMPLIFIED CASE HISTORY ON Fe OXIDES

As an example of the use of valence and core spectroscopies introduced in the previous paragraphs we present some results on Fe oxides; the data are taken from materials published before April 94 and the reader is referred to the original papers for a detailed discussion. We stress that we do not aim to be comprehensive and conclusive also because some points are still open to controversies.

An interesting case is hematite i.e. Fe_2O_3. The direct photoemission spectroscopy has been treated by Fujimori et al[34] with a cluster model. On the basis of the shape of the Fe 3p-3d resonance as seen in photoemission the authors concluded that the system behaves as a charge transfer insulator and assumed as basis in their octahedral cluster (with the appropriate symmetry and degeneration) the following states:
- the Fe state with 5 electrons $t_2^3e^2$ (where t_2 and e denotes the E_g and T_{2g} symmetries)
- the states with a ligand hole and 6 electrons on Fe having respectively E_g and T_{2g} symmetries.

On this basis the authors obtained a good fitting of the direct valence photoemission features explaining the presence of a broad distribution of Fe-d character in a very wide energy range extending down to about 16 eV below the top of the valence band. Later on, this fitting has been confirmed in detail with more sophisticated measurements by Lad and Henrich[35]. All this work has been done in the absence of information on the inverse photoemission so that the constraints to the parameters coming from electron addition spectroscopy could not be taken into account. As a matter of fact the successive measurements of ref[27] have shown that the correlation in the previous work has been probably overestimated; in fact Fujimori et al[35] predicted an onset of the d-hole distribution which is at too high energies as shown in[27]. This is a typical example of the usefulness of more than one spectroscopy in the use of cluster models.

On the other hand the choice of the relevant configurations is strongly influenced by the nature of the experiment. In this connection hematite gives a further interesting example connected with dichroism at the $L_{2,3}$ edges of Fe. This system is antiferromagnetic and thus gives linear dichroism; a very interesting point is that hematite has a phase transition (known as Morin transition) at about -10 °C in which the angle between the direction of the magnetic moments and the c axis of the crystal changes by 90° without change of the antiferromagnetic ordering. Thus one can get rid of all the anysotropies coming from the orientation of the light polarization with respect to the crystal by studying the effect of the crossing of the Morin transition. This has been done very successfully by Kuiper et al.[36] who have also shown that the main experimental results can be interpreted by calculating of the transition $2p^63d^5 \rightarrow 2p^53d^6$ (in the presence of the crystal field) and by disregarding the states $2p^53d^7\underline{L}$ having a ligand hole.

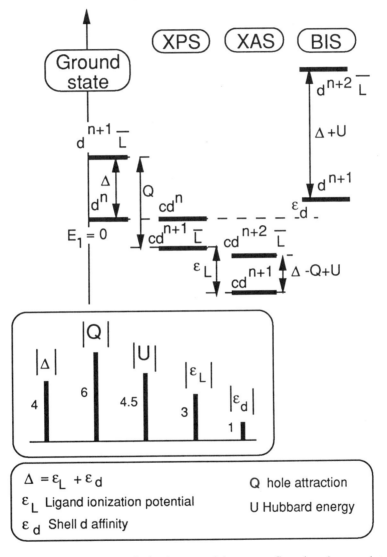

Fig.12. Total energies in a charge transfer insulator containing two configuration: the ground state (at the left) and the final state reached in core XPS in core absorption and in BIS. The drawing is to scale with the values of the parameter shown by the bars in the inset.

Another very interesting case is the monoxide Fe_xO showing once again the importance of the choice of an appropriate basis to treat the correlation in atomiclike calculations. In the pioneering direct photoemission works by Eastman and Freeouf [37] and Alvarado et al.[38] the d^5 final state configuration was considered on the basis of crystal field theory; this originated a too narrow distribution of the d-states along the energy scale as demonstrated successively by Bagus et al.[39] who showed the importance of a better description of the Fe atom in a FeO_6 octahedral cluster. This has been done by including configuration interaction within the final d^5 states, by considering not only the $t_{2g}^3 e_g^2$ and $t_{2g}^4 e_g^1$ final states reached via emission of a t_{2g} or of a e_g electron from the $t_{2g}^4 e_g^2$ ground state but also the $t_{2g}^2 e_g^3$ final state obtained via ionization and d-electron rearrangement. Later on Fujimory et al[40] stressed the importance of including also the ligand hole i.e. of describing the final states not only as d^{n-1} but also as $d^n \underline{L}$ states. As a matter of fact a general consensus on the ordering of the configurations and on the dimensionality of the minimum basis is not completely reached.

In the meantime the monoxide has been one of the basic cases to test the merits of the recently introduced Self Interaction Corrected Local Spin Density approximation (SIC-LSD)[41]. As it is well known, in traditional LSD the use of the electron density creates an artificial interaction of the electron with itself which is explicitly removed in SIC-LSD. In this case one obtains directly from Kohn and Sham eigenvalues a gap in many oxides; in particular one solves the problems connected with the impossibility of obtaining a gap in conventional SIC (for Fe monoxide see[42]; for orbital polarization correction in Fe monoxide see[43]). The SIC-LSD formalism has been applied[24] to the discussion of Fe_xO by studying jointly the d-distributions from direct photoemission (taken from ref[36]) and from inverse photoemission measured to this purpose. In this description the states involving a ligand hole cannot be obviously described; however it has been shown in[24] that the SIC-LSD description can be used to evaluate theoretically U and Δ values which can be used as a good starting point to discuss the experimental results with a cluster model; moreover it is noteworthy that the values of the parameters compare reasonably well with those estimated empirically from core XPS[44].

REFERENCES

[1] R. O. Jones, and O. Gunnarsson, Rev. Mod. Phys. 61, 689 (1989)

[2] R. Y. Koyama and N.V. Smith, Phys. Rev. B2, 3049 (1970)

[3] C.N. Berglund and W.E. Spicer, Phys. Rev. 136,A1030 (1964); 136, A1044 (1964)

[4] J.C. Phillips, in Solid State Physics, ed. by F. Seitz and D. Turnbull (Academic Press, New York, 1966), Vol.18, p.55.

[5] E. O. Kane, Phys. Rev. 175,1039 (1968)

[6] J. J. Yeh and I. Lindau, At. Data Nucl. Data Tables 32, 1 (1985)

[7] N. V. Smith, Rep. Progr. Phys. 51, 1227 (1988)

[8] Th. Fauster, F.J.Himpsel, J.J. Donelon, A. Marx, Rev. Sci. Instrum. 54, 68 (1983); Th. Fauster, D. Straub, J.J. Donelon,D. Grimm, A. Marx, F.J.Himpsel, Rev. Sci. Instrum. 56, 1212 (1985)

[9] U. Kolac, M. Donath, K. Ertl, H. Liebl, and V. Dose, Rev. Sci. Instrum. 59,1933 (1988)

[10] M. Donath, Surf. Sci. Rep. 20, 251 (1994)

[11] F. Ciccacci, E. Vescovo, G. Chiaia, S. De Rossi, and M. Tosca, Rev. Sci. Instrum. 63, 3333 (1992)

[12] M. Sancrotti, L. Braicovich, C. Chemelli, F. Ciccacci, E. Puppin, G. Trezzi, and E. Vescovo, Rev. Sci. Instrum. 62, 639 (1991)

[13] P.T. Andrews, Vacuum 38, 257 (1988)

[14] U. Fano, Phys. Rev. 124, 1866 (1961)

[15] L.I. Johansson, J.W. Allen, I. Lindau, M.H. Hecht, and S.B.M. Hagström, Phys. Rev. B21, 1408 (1980)
[16] Yongjun Hu, T.J. Wagener, Y. Gao, and J. H. Weaver, Phys. Rev. B38, 12708 (1988)
[17] L. Duo', M. Finazzi, and L. Braicovich, Phys. Rev. B48, 10728 (1993)
[18] P. Weibel, M. Grioni, D. Malterre, B. Dardel, and Y. Baer, Phys. Rev. Lett. 72, 1252 (1994)
[19] M.P. Seah, and W.A. Dench, Surf. and Interf. Anal. 1, 2 (1979)
[20] W. Speier, J.C. Fuggle, R. Zeller, B. Ackermann, K. Szot, F.U. Hillebrecht, and M. Campagna, Phys. Rev. B30, 6921 (1984)
[21] C.T. Chen, F. Sette, Y. MA, and S. Modesti, Phys. Rev. B42, 7262 (1990)
[22] G.J. Mankey, R.F. Willis, and F.J. Himpsel, Phys. Rev. B48, 10284 (1993)
[23] L. Braicovich, F. Ciccacci, E. Puppin, A. Svane, and O. Gunnarsson, Phys. Rev. 46, 12165 (1992)
[24] Shun-ichi Nakai, T. Mitsuishi, H. Sugawara, H. Maezawa, T. Matsukawa, S. Mitani, K. Yamasaki, and T. Fujikawa, Phys. Rev. B36, 9241 (1987)
[25] F. M. F. de Groot, M. Grioni, J. Fuggle, J. Ghijesen, G.A. Sawatsky, and H. Petersen, Phys. Rev. B40, 4886(1989)
[26] F. Ciccacci, L. Braicovich, E. Puppin, and E. Vescovo, Phys. Rev. B44, 10444 (1991)
[27] M. Sancrotti, F. Ciccacci, M. Finazzi, E. Vescovo, and S.F. Alvarado, Z. Phys. B84, 243 (1991)
[28] J.K. Lang, Y.Baer, and P.A. Cox J. Phys. F: Metal Phys. 11,121 (1981)
[29] J. Zaanen, G.A. Sawatsky, and J. W. Allen, Phys. Rev. Lett. 55, 418 (1985)
[30] J. Zaanen, G.A. Sawatsky,, J. Fink, W. Speier, and J. Fuggle, Phys. rev. B32, 4905 (1985)
[31] G. van der Laan, C. Westra, C. Haas, and G.A. Sawatsky, Phys. Rev. B23,4369 (1981)
[32] J. van Elp, J.L. Wieland, H. eskes, P. Kuiper, G.A. Sawatsky, F. M. F. de Groot, and T.S. Turner, Phys. Rev. B44, 6090 (1991)
[33] M. Grioni, J.F. van Acker, M.T. Czyzyk, and J.C. Fuggle, Phys. Rev. 45, 3309 (1992)
[34] A. Fujimori, M. Saeki, N. Kimizuka, M. Taniguchi, and S. Suga, Phys. Rev. B34, 7318 (1986)
[35] R.J. Lad and V.E. Henrich, Phys. Rev. B39, 13478 (1989)
[36] P. Kuiper, B. G. Searle, P. Rudolf, L.H. Tjeng, and C.T. Chen, Phys. rev. Lett. 70, 1549 (1993)
[37] D.E. Eeastman, and J. L. Freeouf, Phys. Rev. Lett. 34, 395 (1975)
[38] S. F. Alvarado, M. Erbudak, and P. Munz, Phys. Rev. B14, 2740 (1976)
[39] P. S. Bagus, C.R. Brundle, T.J. Chuang, and K. Wandelt, Phys. Rev. Lett. 39, 1229 (1977)
[40] A. Fujimori, N. Kimizuka, M. Taniguchi, and S. Suga, Phys. Rev. B36, 6691 (1987)
[41] A. Svane, and O. Gunnarsson, Phys. Rev. Lett. 65, 1148 (1990)
[42] E. Terakura, A.R. Williams, T. Oguchi, and J. Kübler, Phys. Rev.Lett. 52, 1830 (1984), and Phys. Rev. B30, 4734 (1984)
[43] M.R. Norman, Phys. Rev. Lett. 64, 1162 (1990); 64, 2466(E)(1990) and Phys. Rev. B44, 1364 (1991)
[44] Geunseop Lee, and S.-J. Oh, Phys. Rev. B43 ,14674 (1991)

THE SPIN DEPENDENCE OF ELASTIC AND INELASTIC SCATTERING OF LOW ENERGY ELECTRONS FROM MAGNETIC SUBSTRATES

D. L. Mills

Department of Physics and
The Institute for Surface and Interface Science
University of California, Irvine
Irvine, California 92717-4575 USA

We review the information that may be obtained from the analysis of the spin dependence of the scattering of electrons from the surface of magnetic materials. Two sources of spin dependencies are explored: those introduced by the spin orbit effect, and by the exchange coupling between the beam electron spin, and the unpaired spins in the substrate. Each of these can be used to obtain information about the near surface environment of magnetic materials, through the elastic scattering of electrons from the material. We also describe recent theoretical analyses of the inelastic scattering of electrons by Stoner excitations and spin waves in ferromagnets. The implications of these results are explored.

I. INTRODUCTION

Much of our current knowledge of both the geometrical and electronic structure of the outermost atomic layers of crystals comes from the various forms of electron spectroscopy, in which an electron with kinetic energy in the range 50 - 300eV is scattered from the surface, emitted from it in response to an external probe, or absorbed by a transition to an unfilled state. Electrons in this energy range have mean free paths of two or three interatomic spacings, with the consequence that the information generated from the experiment concerns the properties of the very outermost two or three layers.

The data can be interpreted in a detailed quantitative manner in many cases, because presently multiple scattering theory provides very accurate description of the propagation of the electron through the crystal. Such calculations required large state-of-the-art computers a decade ago, but are now carried out routinely on workstations of modest cost. In the energy range mentioned above, the scattering of the electron from an atom embedded in the solid is controlled primarily by the potential

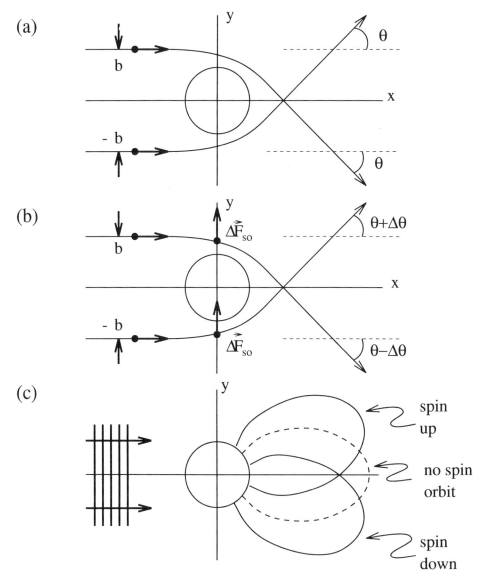

Figure 1. A sketch which illustrates the origin of the spin orbit asymmetry for an electron scattering from a single atom. We have (a) two trajectories followed by a classical charge scattering from a potential, in the absence of spin orbit coupling, (b) the influence of the spin orbit force on the same trajectories, and (c) the spin orbit asymmetry induced in the quantum mechanical cross section, by means of the mechanism illustrated in (b).

trajectory, the orbital angular momentum ℓ_z has the opposite sign from that where the impact parameter is $+b$. It follows from Eq. (II.4) that for this case also, $\Delta \vec{F}_{SO}$ points in the $+\hat{y}$ direction. Hence, the electron with impact parameter $-b$ suffers a greater deflection than that whose impact parameter is $+b$. To first order in the spin orbit effect, the deflection angle is $\theta - \Delta\theta_{SO}$ for the first case and for the second, it may be written $\theta + \Delta\theta_{SO}$.

This means that the scattering cross section for the magnetic moment bearing particle is no longer an even function of θ; if a beam strikes the target, with uniformly distributed order parameters, more electrons will be deflected in the upper half plane $y > 0$, and less to the lower half plane $y < 0$. This feature survives in a full quantum theoretic situation, of course, to produce the result illustrated in Fig. (2c), where we compare the differential cross section for up spin electrons, with the symmetric form appropriate to the case where the spin orbit effect is absent. A reversal of the spin of the beam particle reverses the sign of the asymmetry, as illustrated in Fig. (2c).

We have assumed nowhere that the atom from which the electron is deflected has a magnetic moment itself. Thus, the spin orbit induced asymmetries are operative in scattering from non magnetic atoms and substrates.

The effect just described has been known for many decades,[3] and forms the basis for a device referred to as a Mott detector that allows one to measure any polarization present in an electron beam. If an unpolarized beam, which consists of an equal number of up and down spin electrons, the total scattered current $I_{scatt.}(\theta)$ will be an even function of θ, if the total scattered current is measured. If the beam is partially polarized, say with net polarization upward, then there will be an asymmetry in the total scattered current, with more electrons in the half plane $y > 0$, and fewer in the half plane $y < 0$. The difference $I_{scatt.}(\theta) - I_{scatt.}(-\theta)$ serves as a measure of the degree of polarization in the beam.

A second source of spin dependence in electron scattering has its origin in the quantum mechanical exchange interaction between the beam electron and the substrate magnetic moment; in contrast to the spin orbit symmetry, this requires the atom in question to possess a net magnetic moment. Let the magnetic moment point in the direction $\hat{n}$. An electron with spin $\vec{\sigma}$ parallel to $\hat{n}$ will see an effective potential $V_\uparrow(r)$ different than that $V_\downarrow(r)$ experienced by a down spin electron. In the calculations reported below, we shall employ model potentials generated by spin polarized versions of density functional calculations, carried out within the local density approximation (LDA). These calculations are phrased in terms of local, spin dependent potentials. A quantum theory of spin bearing fermions involves the two component wave function described in standard texts. For general orientation of the beam electron spin $\vec{\sigma}$ with respect to that $\hat{n}$ of the magnetic moment in the substrate, if $B(r) = V_\uparrow(r) - V_\downarrow(r)$ we have the potential

$$\mathbf{v}(r) = V(r)\mathbf{I} + B(r)\sigma \cdot \hat{n} \qquad (II.5)$$

In the next subsection, we discuss the means of calculating the electron-substrate interaction, with the effects above incorporated into the analysis. In the end, the scattering amplitude $\mathbf{f}$ which describes scattering of the electron from the state with wave vector $\vec{k}_I$ to that with wave vector $\vec{k}_S$ can be written as an operator in the 2×2 spin space of the electron. In the limit of interest to us, both exchange and spin orbit effects will not be large. For the purposes of discussion, noting that $\vec{k}_I \times \vec{k}_S$ is a pseudo vector normal to the scattering plane defined by the two vectors $\vec{k}_I$ and $\vec{k}_S$,

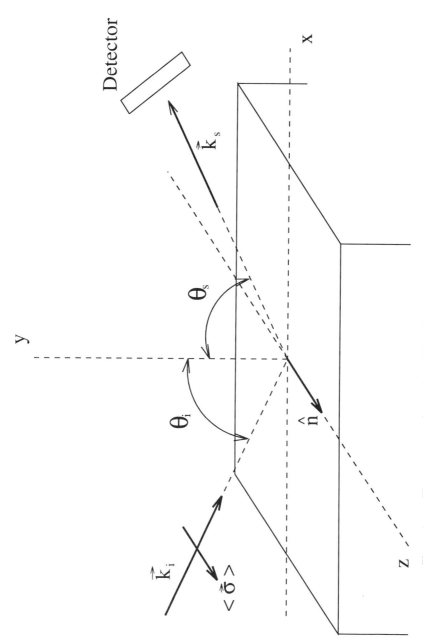

Figure 2. The geometry of a typical SPLEED experiment. The magnetic moment $\hat{n}$ of the substrate is parallel to the surface, and normal to the scattering plane. The beam polarization $\langle\vec{\sigma}\rangle$ (shown parallel to $\hat{n}$) is parallel or anti-parallel to $\hat{n}$. If one explores the specular beam, $\theta_I = \theta_S$. The detector measures the total scattered current.

we may write

$$\mathbf{f}(\vec{k}_I \to \vec{k}_S) = f_o(\vec{k}_I \to \vec{k}_S)\mathbf{I} + \Delta f_{SO}(\hat{k}_I \times \hat{k}_S) \cdot \vec{\sigma} + \Delta f_{ex}\hat{n} \cdot \vec{\sigma}. \qquad (II.6)$$

A typical spin polarized electron diffraction (SPLEED) experiment is carried out with a polarized beam incident on a ferromagnetic sample. The direction $\hat{n}$ of the magnetic moment of the substrate is parallel to the surface, as illustrated in Fig. (2). The scattering plane is perpendicular to $\hat{n}$, while the beam polarization $\langle\vec{\sigma}\rangle$ is either parallel or antiparallel to $\hat{n}$. If one examines the specular beam, $\theta_I = \theta_S$, though Bragg beams are studied as well. The detector measures the <u>total</u> current (spin up, plus spin down).

Both $\hat{n}$ and $\langle\vec{\sigma}\rangle$ are parallel or antiparallel to the z axis, in Fig. (2). There are then four geometries we may denote symbolically by $(+1,+1),(+1,-1),(-1,+1)$ and $(-1,-1)$. The first digit refers to the direction $\langle\sigma_z\rangle$ of the beam polarization, and the second the direction $\hat{n}_z$ of the substrate magnetization. One sees from Eq. (II.6) that <u>difference</u> between $I(+1,+1)$ and $I(-1,-1)$ is controlled by the spin orbit asymmetry Δf_{SO}, while the difference between $I(+1,+1)$ and $I(+1,-1)$ is controlled by the exchange asymmetry. If P_o is the beam polarization, it is conventional[1] to form from the four intensities $I(\pm 1,\pm 1)$ the two ratios

$$A_{SO} = \frac{I(+1,+1) + I(+1,-1) - I(-1,-1) - I(-1,+1)}{P_o\left\{\sum_{\sigma_z}\sum_{n_z}I(\sigma_z,n_z)\right\}}, \qquad (II.7a)$$

and

$$A_{ex} = \frac{I(+1,+1) + I(-1,-1) - I(+1,-1) - I(-1,+1)}{P_o\left\{\sum_{\sigma_z}\sum_{n_z}I(\sigma_z,n_z)\right\}}. \qquad (II.7b)$$

If Δf_{SO} and Δf_{ex} are both small, then to first order in Δf_{ex}, A_{SO} is independent of exchange and controlled by only the spin orbit effect. Similarly, A_{ex} is controlled by only the exchange contribution to the scattering amplitude in this limit. Thus two ratios thus are referred to as the spin orbit and the exchange asymmetries as a consequence.

It is the case that A_{SO} is independent of exchange, and A_{ex} is insensitive to the spin orbit asymmetry only in the limit that these two effects are both small. One may test this question by calculating a third intensity ratio A_u discussed by Feder,[1] which vanishes to first order in both Δf_{SO} and Δf_{ex}. When $A_u \ll A_{SO}$ and A_{ex}, one is justified in asserting that the spin dependencies introduced by the spin orbit and exchange interactions may be separated by forming the combination described in Eqs. (II.6).

This completes our general discussion of the origin of spin dependence in the scattering of electrons from magnetic materials. In the next subsection, we discuss the procedures used in our work, along with a philosophy we wish to implement.

B. Multiple Scattering Descriptions of SPLEED; General Remarks

As noted earlier, Feder[1] has developed a formalism with which one can calculate the elastic scattering of electrons from magnetic materials, in a manner which allows one to generate theoretical descriptions of both A_{ex} and A_{SO}. This uses the standard "muffin tin" representation of the solid. There are two basic analyses that must be performed in such a study.

Within each muffin tin, one imbeds a potential $V(r)\mathbf{I} + B(r)\hat{n}$ taken to be spherically symmetric in our work, described below. It is necessary to generate the single site t matrix, which describes the scattering of a spin bearing electron from such a "solid state atom". This is done through use of the full Dirac equation in our work (following Feder), so a full and complete description of relativistic effects such as spin orbit coupling is incorporated in the analysis. In our work, we use bulk ground state potentials generated from *ab initio* LDA studies of the magnetically ordered material. For example, our work on Fe utilizes potentials generated in a study of an Fe(100) slab by Fu and Freeman.[4] These authors employ a seven layer slab; we utilize the potential from the central layer. However, we wish to simulate magnetic moments in the surface region whose magnitude may differ from those in the bulk material. This we do by simply increasing the magnitude of $B(r)\hat{n}$ for atoms in and near the surface. That is, for the near surface layers, we employ the potentials $V(r)\mathbf{I} + r_\ell B(r)\hat{n}$ where r_ℓ is viewed as the ratio between the moment in layer ℓ, and that in the bulk.

Given the single site t matrix, we then calculate the scattering amplitude by means of the appropriate summation of the multiple scattering series. Of course, the Dirac wave function is a four component object, while the non-relativistic Schrödinger equation to which this is matched outside the muffin tin is a two component object. In the outer portion of the muffin tin, the potential is quite weak, and at the energies of interest here, the third and fourth components of the Dirac spinors are quite small. The matching is thus achieved straightforwardly.

For the magnetic scatterings of interest to us, we wish to proceed in a manner analogous to current descriptions of low energy electron diffraction (LEED). Here, bulk potentials are employed in the surface region, since as noted earlier at energies appropriate to LEED experiments, the electron is scattered primarily from the core region of the ion, where the potential is little affected by rearrangements of the valence electrons induced by the surface environment. The nature of the complex inner potential is also well understood for many solids at this stage, so in a modern LEED analysis the only adjustable parameters are those which describe the geometry of the surface environment. In our analyses of magnetic materials, we wish to proceed within a framework where the potentials contain no adjustable parameters, save for the constants r_ℓ used to simulate the magnitude of near surface magnetic moments. In some instances, as we shall see below, the near surface geometry is unknown as well.

One issue does arise, if one wishes to proceed in the manner just described. This is the nature of the inner potential. It has been argued that in the ferromagnetic materials, the inelastic mean free path of the beam electron may depend on whether its spin is oriented parallel or antiparallel to the magnetic moments in the substrate.[1] A description of this possibility requires the introduction of new phenomenological parameters; at the moment, we have no *ab initio* studies in hand of the self energy of excited electrons in ferromagnetic materials. One can incorporate this effect by introducing a spin dependence in the complex inner potential, adjusting the parameters that describe this through some appropriate procedure. Tamura and co-workers[5] argue one should supplement the ground state potential $V(r)\mathbf{I} + B(r)\hat{n}$ in each muffin tin by an empirically constructed complex supplement, to simulate the spin dependent excitation of the substrate atoms.

A very clever experiment has been interpreted as providing a direct measure of a spin dependent electron mean free path in Fe, for electrons with energy above the vacuum level.[6] The sample probed consisted of a Cu(100) surface, upon which four monolayers of Fe had been grown. Recall that the Cu 3d bands lie well below those of Fe, so in such a sample to good approximation the Cu and Fe 3d levels interact only weakly. Electrons are then photoexcited from the Cu 3d bands, to an energy above the vacuum level. A certain fraction then propagate through the Fe film to the vacuum outside, where they are detected. The experiment examines the intensities $I_\uparrow(E)$, $I_\downarrow(E)$, of electrons emitted along the normal, with a spin detector. This is done for three distinct energies above the vacuum level. The measurements show that $I_\uparrow(E) > I_\downarrow(E)$. The authors of ref. (6) argue that the spin dependence of the emission intensities arise because up spin electrons have a mean free path $\lambda_\uparrow$ in Fe longer than that $\lambda_\downarrow$ appropriate to down spin electrons. An analysis provides explicit values for $\lambda_\uparrow$ and $\lambda_\downarrow$, presumed to be the inelastic mean free path.

However, Gokhale and the present author have noted that an alternative interpretation of the data is possible.[7] We have seen already that by virtue of the exchange interaction between an excited electron, and a substrate magnetic moment, the cross section for purely <u>elastic</u> scattering from an Fe atom depends on the relative orientation of the spin of the excited electron and the magnetic moments of the Fe atoms. Thus, an electron incident on the Fe film will be scattered by the array of magnetic atoms; the transmission coefficient, regarded now as the square of the forward amplitude for <u>purely elastic</u> scattering, will display a spin dependence. Gokhale and Mills carried out explicit calculations of the spin dependence of the forward scattering amplitude, for an electron transmitted through four layers of Fe, to find very good accord with the data, <u>without</u> the need of introducing spin dependence in the inelastic damping of the electron wave.

The present author argues that the above analysis establishes that for Fe, and very likely for the ferromagnetic transition metals Ni and Co, there is no need to introduce new spin dependent features in the complex inner potential used in multiple scattering analyses. It should be quite sufficient to describe such a substrate or film by embedding them in the same inner potential used in LEED analyses. We have proceeded within such a framework.

We now turn to a summary of our studies of the spin dependence of electron scattering from magnetic substrates.

C. Explicit Calculations

In this section, we summarize our studies of the elastic scattering of spin polarized electrons for two systems, the (110) surface of an Fe crystal sufficiently thick to be regarded as semi-infinite, and a bilayer of Fe on the W(100) surface. In the first case, previous LEED studies provide[8] information on the surface geometry, while for the Fe bilayer on W(100), both the nature of the magnetism and the geometrical structure is unknown. We shall argue below that the spin orbit asymmetry is a most useful probe for such a system.

(i) **The Case of Fe(110)** A most extensive series of spin polarized electron scattering studies of the Fe(110) surface have been carried out by Waller and Gradmann. The samples consisted of thick Fe(110) films grown on the W(110) surface. The data set is remarkably extensive, in that for a large number of beam energies from 50eV to over 120eV, the exchange and spin orbit asymmetries have been measured as a function of angle of incidence, for the specular beam. Data is available for the

($\bar{1}$1) beam as well. These data thus provide a critical test of our ability to generate a quantitative description of spin dependent effects in the scattering of electrons from a well known ferromagnet, over a broad range of energies, within the framework of the scheme described in the previous sections.

We have carried out detailed studies of the energy and angle variation of A_{ex} and A_{SO},[10] and compared the results with the data of Waller and Gradmann. On the whole, as the reader will appreciate, the agreement between theory and experiment is very good indeed, over the entire energy range.

In Fig. (3), we show a detailed comparison between theory and experiment for a beam energy of 62eV, a case studied in detail also in ref. (3). The only adjustable parameters are the r_ℓ, described earlier and used to simulate magnetic moments in the surface whose magnitude may differ from those in bulk Fe. The data are represented by squares in Fig. (3), while the solid line is the theoretical calculation based on the picture which emerges from the *ab initio* calculations of Fu and Freeman.[4] This analysis suggests the moment is enhanced over that in bulk Fe by 19.4%, the second layer moments are enhanced by 6.8%, and smaller enhancements are present in the third and fourth layer. Clearly, the theory and experiment are in impressive accord. Note, for example, the comparison between theory and experiment here and elsewhere in this paper is absolute; both A_{SO} and A_{ex} are intensity ratios. The dotted line in Fig. (3) is calculated for a picture in which the magnetic moments equal those of bulk Fe everywhere, while the dashed line is a picture set forth by Tamura *et al.* in reference (5). This has the moment in the surface layer enhanced by 35%, while that in the second layer is diminished by 15%, with all other moments assuming their bulk values. The model of Fu and Freeman would seem to provide the most satisfactory account of the data.

In Fig. (4), we show comparison between the data (dotted line) and theory (solid line) for the model of Fu and Freeman, for three other beam energies. Again the theory does a very adequate job of reproducing the data. Keep in mind that the detector used in the experiments subtends an angle of roughly 2°. Thus, some features in the theoretical curves are sharper than those in the data.

We have also calculated A_{ex} for the two other models used in Fig. (3), for the beam energies used in Fig. (4). The differences between the curves calculated for the three models are rather small at these energies. Indeed, in Fig. (3), the theoretical results display a sensitivity to surface moment only in the near vicinity of the prominent peak near 40 degrees angle of incidence. We comment further on this point below.

In Fig. (5), we show the comparison between theory and experiment for the spin orbit asymmetry, for the four beam energies described above. Again the theory provides a very good account of the data. In reference (10), we also present a comparison with data taken for the ($\bar{1}$1) beam, to find the theory reproduces all features found in the data.

Our conclusion is then that the scheme outlined in the previous subsection, utilizing ground state potentials and a simple spin independent inner potential, can provide an excellent account of the data on Fe(110), over a wide range of beam energy. The only adjustable parameters in the analysis, as discussed above, are the ratios r_ℓ used to simulate the enhancement or reduction of moments in the near surface region, relative to those in bulk Fe. It is our view that the ground state potentials generated by the *ab initio* methods of Fu and Freeman are excellent, at least for the present purpose.

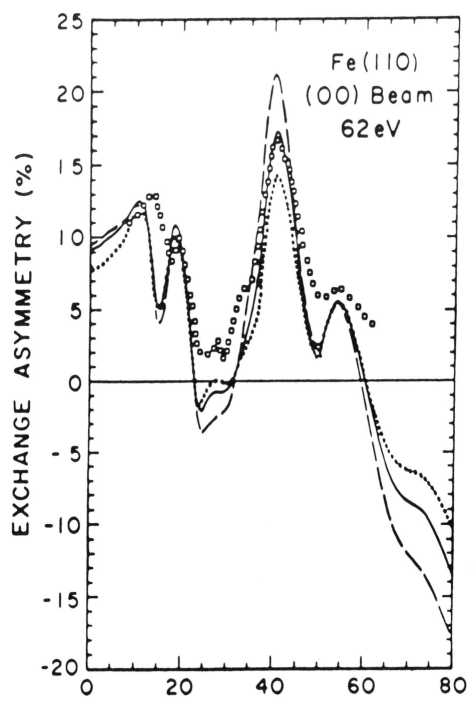

Figure 3. For a beam energy of 62eV, we show a comparison between theory and experiment, for the exchange asymmetry for scattering from Fe(110). The squares are data, taken from ref. (5). The solid line is a calculation which assumes enhanced moments in the surface region, as proposed in ref. (4). The dotted line assumes all near surface moments are the same as those of bulk Fe, while the dashed line is the model set forth in ref. (5), as calculated with our procedures.

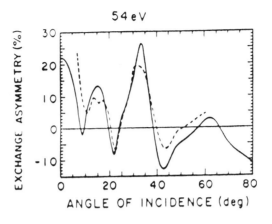
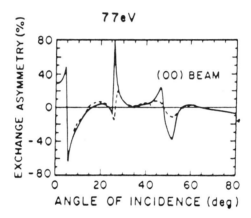
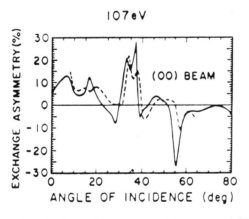

Figure 4. A comparison for the exchange asymmetries measured (dashed lines) on Fe(110) by Waller and Gradmann, and theory (solid lines) for the model of surface moments by Fu and Freeman.

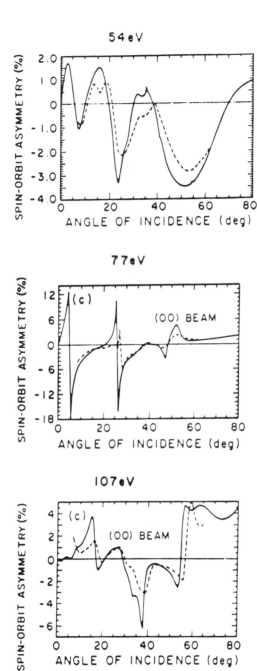

Figure 5. A comparison between theory (solid line) and data (dashed line) for the spin orbit asymmetry, for the four energies used in Fig. (5).

We are concerned, however, that the exchange asymmetries are less sensitive than we desire to the magnitude of the surface moment, save for the feature near 40° angle of incidence at 62eV beam energy. This is the case for the energies selected for study in ref. (10). We thus decided to explore this issue in more detail, since the quantitative analysis of SPLEED data offers our best opportunity for extracting the magnetic moments in the near surface region of magnetically ordered materials.

We proceeded as follows. One would like to have the means of locating particular scattering conditions where A_{ex} is particularly sensitive to the surface magnetic moment. One can find such regimes, of course, by calculating the energy and angle variation of A_{ex}, for a sequence of models of the magnetic moment profile of the surface. This is a tedious and time consuming task, unfortunately. Our procedure[11] was to note that A_{ex} is a function of the set of numbers $\{r_\ell\}$ that define the surface magnetic moment profile. It is then possible to calculate directly, through use of multiple scattering theory, the derivative $(\partial A_{ex}/\partial r_\ell)$, for some choice of r_ℓ of interest. The derivative may be evaluated for the case where all $\{r_\ell\} = 1$. We argue one should search for maxima in A_{ex} where the logarithmic derivative $A_{ex}^{-1}(\partial A_{ex}/dr_\ell)$ is large.

We applied such an analysis to the Fe(110) surface. Not surprisingly, when the derivative logarithmic derivative was calculated for the surface layer moment, the logarithmic derivative was substantial near the prominent peak in the 62eV data. A scan of the range of energy and angle explored by Waller and Gradmann showed that the only other energy regime where A_{ex} displayed high sensitivity to surface moment is the 50-52eV range, and for angles of incidence in the 30° range. The theory reproduces the data at 52eV quite nicely; the data favors an enhanced surface moment, but it is difficult to discriminate between the models put forth in ref. (4) and ref. (5) from the comparison. The data at 50eV beam energy is strikingly different than that at 52eV, and this difference is reproduced very nicely by the scattering calculations. That portion of the 50eV data sensitive to surface moment is accounted for adequately by a picture with less surface enhancement.

On balance, the analyses in ref. (10) and ref. (11) suggest that the surface moment in Fe(110) is enhanced in the 20-30% range. It is the view of the present author that it is questionable to base the assignment of surface moment on comparison between theory and experiment at a single beam energy. A picture developed in such a way should be confirmed by exploring at least one other beam energy. The calculations in ref. (10) and ref. (11), free of adjustable parameters save for those used to simulate enhanced (or reduced) magnetic moments near the surface, are highly successful in that they give a very good account of data taken over a wide energy range. However, for reasons outlined above, it is difficult to make a clear and unambiguous assignment of the magnitude of the enhanced moments on the surface.

The reason why this is so is not hard to appreciate by considering a simple kinematical description of the origin of the exchange asymmetry. If r_s describes the ratio of the surface to the bulk moment, and if the electron is supposed to sample only the outermost atomic layer, the backscattered signal is proportional to

$$\left|\langle \vec{k}_s|V|\vec{k}_I\rangle \pm r_s\langle \vec{k}_s|B|\vec{k}_I\rangle\right|^2$$
$$\cong \left|\langle \vec{k}_s|V|\vec{k}_I\rangle\right|^2 \pm 2r_s Re\{\langle \vec{k}_s|V|\vec{k}_I\rangle\langle \vec{k}_I|B|\vec{k}_s\rangle\}, \qquad (II.8)$$

where the plus sign applies to the case where the beam polarization is parallel to the local moment, and the minus sign is appropriate to the antiparallel case. Quite clearly, if we wish to discriminate between $r_s = 1.2$ and $r_s = 1.4$, the change in the calculated exchange asymmetry will be somewhat less than 20%, on average. This is for the extremely favorable case where the electron samples only the outermost layer of moments before backscattering. In fact, the electron will penetrate two or three layers, so the effect will on average be smaller by perhaps a factor of two.

The above argument suggests that to reliably extract the surface moment from exchange asymmetry data, one must be able to calculate A_{ex} to much better than ten percent accuracy. It is possible, through methods such as those used in ref. (11) to isolate scattering conditions where A_{ex} is particularly sensitive to surface moment. The calculations show that under such favorable conditions, changes in surface moment from 1.0 to 1.4 may lead to 25% alterations in A_{ex}. The data surely has uncertainties in the 10% range. Under these conditions, it remains difficult to make clear statements with confidence, regarding the precise magnitude of the enhanced surface moments, to judge from our experience with Fe(110).

In the theoretical literature, it has been argued that there are very large moments on the Cr(100) surface.[12] Spin polarized electron loss data has been interpreted to suggest the presence of greatly enhanced moments, on the (100) surface of very thin Cr films grown on Fe(100).[13] It would be of very great interest to see a SPLEED study of this surface. Such a large contrast between surface and bulk moment should be readily detected by this method; the unambiguous detection of more modest enhancements will prove a challenge to the theorist.

(ii) The Fe Bilayer on W(100) Recently Elmers and Gradmann have produced bilayers of Fe on the W(100) surface which appear to be of very high quality.[14] We have been engaged in an analysis of spin orbit and exchange asymmetry on this system. In this section, we describe our results, along with the principal conclusions.

For this system, the geometry has not been determined, so we have addressed both the nature of the geometry, and in addition the magnetism in the system. We note, however, that Wu and Freeman[15] have carried out theoretical studies of both the magnetism and geometry of this system, by *ab initio* methods.

Our calculations demonstrate that for this system, the spin orbit asymmetry A_{SO} is quite sensitive to geometrical structure. We argue this quantity should serve as a useful structural probe when a light material such as a 3d metal film is present on a heavy material used as a substrate. First note from the calculations displayed in the previous subsection, that A_{SO} for a pure Fe surface is in the range 1-3% for the energies used. For the Fe bilayer on W(100), A_{SO} is much larger, in the 10% range. Evidently the electrons penetrate the two Fe overlayers, and sample the W substrate within which the spin orbit coupling is very strong, before backscattering out through the Fe bilayers once again. The incoming electron wave is diffracted through the Fe bilayer before striking the W substrate; a change of either angle of incidence or beam energy will thus alter the amplitude of the electron wave that reaches the substrate. Similarly, the electron wave is diffracted by the two Fe layers once again upon exiting from the structure. We may say that the Fe bilayer acts as a "quantum filter", and the energy or angle variation of these inference effects is influenced strongly by the precise position of the Fe layers relative to each other, and to the substrate.

We illustrate this point in Fig. (6), where we give values for A_{SO} measure for the Fe bilayer, for a beam energy (relative to the vacuum level) of 85eV. A series of theoretical curves are shown for comparison. Each is calculated, as a function of angle of incidence, for different locations of the Fe bilayer. We place the atoms in the first bilayer in the fourfold hollow sites, and second layer of Fe atoms reside in the

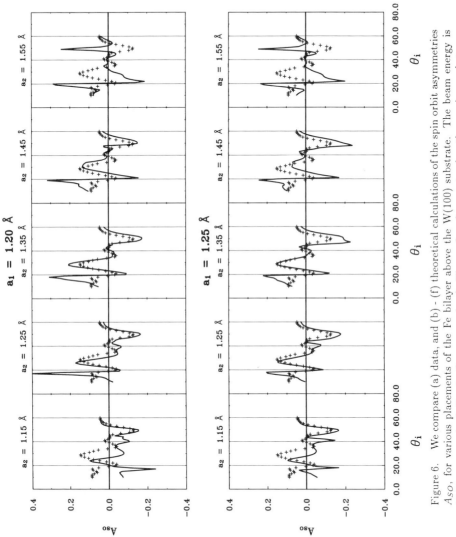

Figure 6. We compare (a) data, and (b) - (f) theoretical calculations of the spin orbit asymmetries A_{SO}, for various placements of the Fe bilayer above the W(100) substrate. The beam energy is 85eV, measured relative to the vacuum level. The preferred geometry has the spacing a_1 between the outermost layer of W nuclei, and the first Fe layer equal to 1.2Å, and that a_2 between the two Fe layers equal to 1.35Å.

fourfold hollow sites of the first layer (BCC stacking). A very good account of the data is found by placing the first Fe layer 1.20A above the outermost layer of W nuclei, and the outermost Fe layer 1.35A above the inner layer. It is evident that if either layer is displaced by ±0.1A from our preferred position, there are substantial changes in A_{SO}. Clearly, the spin orbit asymmetry can serve as a sensitive structural probe, for a system such as this, where the overlayer acts as a "quantum filter" that controls the spin orbit asymmetry induced through interaction of the electron wave with the substrate. Wu and Freeman find 1.08 and 1.34A for the two spacings. Thus, they place the innermost layer closer to the W substrate than our determination suggests.

With a geometrical structure in hand, we may turn our attention to the magnetism, as examined through the exchange asymmetry A_{ex}. In their study of Fe overlayers of W(100), Wu and Freeman find no magnetic moment for the monolayer, a result in agreement with conclusions reached from SPLEED studies of the monolayer.[14] The exchange asymmetry is absent in the data, a result compatible with either the absence of Fe moments for the monolayer, or data taken above the ordering temperature. The theory gives, for the bilayer a magnetic moment of the innermost layer equal to roughly 0.76 that of bulk Fe, while the outer layer has a moment 1.1 that of bulk Fe.

We have computed the exchange asymmetry A_{ex} for the Fe bilayer, and several models of its magnetism. The results for the picture set forth in the paper by Wu and Freeman are rather similar to the data, though there are some discrepancies. Clearly their picture is close to that realized in the actual system. We saw that for the Fe(110) surface that the calculated exchange symmetries are less sensitive than desired to the magnitude of the magnetic moment in the surface, save for selected regimes of angle and energy. We find this to be true for the bilayer as well. A detailed comparison between theory and experiment will be published elsewhere.

III. SPIN DEPENDENT INELASTIC SCATTERING OF ELECTRONS FROM MAGNETIC SOLIDS

The previous section explored the information one may obtain through the spin dependence of the scattering of electrons from magnetically ordered materials. We now turn to the case where the scattering event is inelastic: the electron strikes the surface, and creates an excitation of energy ΔE, then leaves after suffering the energy loss ΔE. The basic experiment then consists of scanning the energy loss spectrum of the scattered electron.

Electron energy loss spectroscopy has been applied to the study of a diverse array of excitations localized on the crystal surface, from intramolecular vibrations of adsorbates,[16] to surface phonons and their dispersion relations,[16] and plasmons or electron-hole excitations on either semiconductors or metals.[17]

Electron energy loss spectroscopy is a most powerful technique, when applied to magnetic surfaces. It is the view of the author that the technique has great potential for future development, in the area of magnetic materials.

Through use of a spin polarized electron beam, combined with detection of the spin orientation of the scattered electron, it is now possible to isolate from the loss spectrum the contributions from various selected "channels". Suppose we use the symbol ↑ to describe a beam electron whose spin is parallel to the magnetization of the substrate, assumed ferromagnetic, and ↓ to denote a beam or scattered electron with spin antiparallel to the substrate magnetization. Then, in what some refer to as a "complete" experiment, it is possible to isolate contributions to the electron loss spectrum from the four distinct combinations $\uparrow\to\uparrow, \uparrow\to\downarrow, \downarrow\to\uparrow$, and $\downarrow\to\downarrow$. This has been done by Kirschner and his colleagues,[18] and by the Hopster group,[19] for the transition metal ferromagnets Fe and Ni, for example. Also, Idzerda and his collaborators have studied ferromagnetic cobalt.[20]

These experiments focus their attention on the spectrum of excitations referred to as Stoner excitations. These are interband particle-hole excitations, in which the spin of an electron in the substrate flips in the process. The physical origin of the excitation process lies in quantum mechanical exchange scattering. We illustrate the process responsible for this contribution to the loss cross section in Fig. (7), with the case of Ni in mind. In Ni, in the ferromagnetic state, the Fermi level lies above the top of the majority spin 3d bands, so these levels are filled completely. There are holes in the minority spin band. A beam electron with spin down can drop into an empty hole in the minority spin band, and then excite an electron from the majority spin band via the Coulomb interaction V_c in the process. The result, at least in the very simple picture sketched in Fig. (7), is an energy loss feature in the spin down channel $\downarrow\to\uparrow$ centered at the exchange splitting ΔE_x between the bands.

Notice, for the process illustrated in the figure, there should be no feature at ΔE_x in the scattering channel $\uparrow\to\downarrow$. This is so, because the beam electron, now with spin up, has no empty state to which to make a transition. Thus, at least for a material such as Ni, where the majority spin band is full, one should see a feature in the $\downarrow\to\uparrow$ channel centered at ΔE_x, while there should be no loss feature in $\downarrow\to\uparrow$. This picture agrees nicely with the data reported by Abraham and Hopster, for scattering from the NI(100) surface.[19]

The exchange splitting ΔE_x scales roughly with the strength of the magnetic moment, as one moves across the 3d series of ferromagnetic metals. Thus, spin polarized electron energy loss (SPEELS) studies should provide information on the magnitude of moments in ultrathin films, and on magnetic surfaces. By this means, the existence of giant magnetic moments on the Cr(100) surface have been inferred; enhanced moments on this surface have been predicted theoretically,[12] and a very large value of ΔE_x is indeed found in SPEELS data.[13]

The experiments cited above show clearly that through use of SPEELS, one may study magnetic excitations on crystal surfaces, and in ultrathin films of magnetic materials. One may inquire if the technique can be used as well to study spin waves in such media. At the time of this writing, virtually no data exists on the dispersion relation of spin waves in ultrathin films, or on surfaces, under conditions where their excitation energy is dominated by exchange. Spin waves have been studied both by ferromagnetic resonance, and by inelastic light scattering (Brillouin scattering), but the modes studied here have wavelengths so long compared to the lattice constant that exchange has very little influence on their frequency. The excitation energy of the modes studied by these methods is controlled by the anisotropies present, by the dipolar fields generated by the spins, and by the external Zeeman fields that may be present. The experiments provide a large fraction of the information we have in hand on the fascinating topic of anisotropy in films and at surfaces, but as remarked earlier, we have no knowledge of the strength of the exchange interactions from the direct study of spin waves. By means of SPEELS experiments carried out in the appropriate geometry, one may explore the entire surface Brillouin zone, just as one does in the surface phonon studies mentioned earlier.

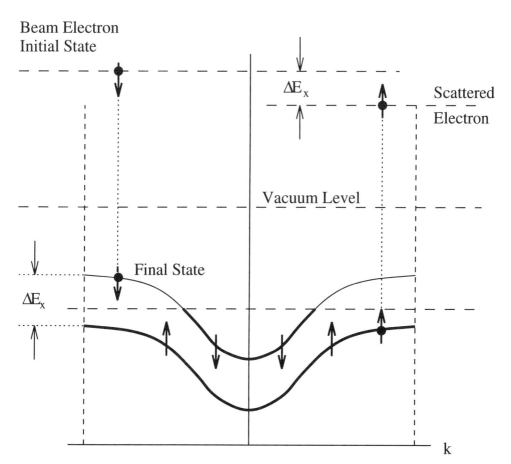

Figure 7. A schematic illustration of the means by which a down spin beam electron creates a Stoner excitation for a material such as Ni, where the majority spin band is filled.

We have initiated a series of theoretical studies of the inelastic scattering of electrons by spin flip processes, with the aim of clarifying the interpretation of studies of Stoner excitations, and also to outline the conditions under which excitation of spin waves may be possible. We comment on the calculations, and on the results we have at the time of this writing.

The theoretical description of the scattering event proceeds as follows. We see from Eq. (II.5) that the beam electron "senses" the magnetic moment on site $\vec{\ell}$ through the exchange term in the crystal potential, which we write

$$B(r)\vec{\sigma}\cdot\hat{n}(\vec{\ell}) = \frac{1}{S}B(r)\vec{\sigma}\cdot\vec{S}(\vec{\ell}) \equiv \frac{1}{S}B(r)[\sigma_z S_z(\vec{\ell}) + \frac{1}{2}\sigma_+ S_-(\vec{\ell}) + \frac{1}{2}\sigma_- S_+(\vec{\ell})]. \quad (III.1)$$

The term proportional to $\sigma_z S_z(\vec{\ell})$, with $S_z(\vec{\ell})$ replaced by its thermal average $\langle S_z(\vec{\ell})\rangle$, describes the spin dependent portion of the <u>elastic</u> scattering of the beam electron off site $\vec{\ell}$. The terms $\sigma_+ S_-(\vec{\ell})$ and $\sigma_- S_+(\vec{\ell})$ describe events in which the beam electron spin flips upon scattering from site $\vec{\ell}$, simultaneously with excitation of the magnetic moment on site $\vec{\ell}$.

We proceed by constructing,[21] within the distorted wave Born approximation, a general expression for the spin flip contribution to the SPEELS cross section. The physical picture is that the beam electron enters the crystal, and propagates to a selected site $\vec{\ell}$, engaging in a multiple scattering sequence when it does so. When is reaches site $\vec{\ell}$, it creates a spin excitation through the $\sigma_+ S_-(\vec{\ell})$ term (assume the beam electron has spin down). It then exits from the crystal, engaging in a multiple scattering sequence as it does so. One then seems the scattering amplitude over all sites $\vec{\ell}$, to form the total amplitude for creating a spin excitation.

In the end,[21] one arrives at a general expression for the SPEELS contribution to the loss cross section rather similar to the well known Van Hove formula which describes neutron scattering from the magnetic degrees of freedom of crystals.[22] The difference between the two cases is first that the neutron interacts very weakly with the crystal, so its wave function in both the initial and the final state is described by a plane wave, to excellent approximation, whereas the electron is scattered strongly as it approaches or exits from the site where its spin flips. Secondly, the neutron interacts with the magnetic moments in the substrate via the magnetic dipole interaction, while in the present case it is exchange between the beam electron and the substrate moments responsible for the dominant coupling.

The general expression, for its implementation, requires also a model of the spin excitations in the substrate. At the moment, to achieve a full description of the spin excitations of a realistic model of either a semi-infinite transition metal (and hence itinerant) ferromagnet, or an ultra thin film is a challenging task, to which we shall devote effort in the near future.

We have, for a first attempt, utilized the simple one band Hubbard model. A mean field treatment of the ground state of the film leads to ferromagnetic configurations, in the appropriate parameter regime, and these are locally stable, according to our analysis. We may describe the spin excitations of such a finite film within the random phase approximation (RPA), to obtain the collective spin wave modes, and the Stoner excitations as well. When this information has been combined with the multiple scattering description of the excitation process, we may obtain a full account of the spin flip contribution to the SPEELS spectrum, for the model. Such an analysis has been completed by Gokhale and the present author.[23]

In Fig. (8), we show for a twelve layer Hubbard model film the spin flip contribution to the SPEELS spectrum. The calculation is for a BCC film with (100) surface and as noted earlier, the (mean field) ground state is ferromagnetic in character. The Fermi level has been placed so the majority spin band is filled, so one may regard this model as a crude representation of Ni metal.

The energy loss ω is given in dimensionless units in the figure; the bandwidth in these units 16 for each of the bands (minority and majority spin), and the average exchange splitting is roughly seven units.

There are two distinct sets of features present in Fig. (8). From $\omega \cong 2$ to $\omega \cong 4.5$, we have a band of losses, with well defined structure. These are the Stoner excitations. The structure has its origin in the fact that the finite film has discrete standing wave "quantum well" states, and the transitions are between these levels. It is the case that the data shows the spectrum of Stoner excitations to be very broad, and not just a peak centered near the exchange splitting ΔE_x.[18-20] The physical reason for this is that wave vector components normal to the surface are not conserved in the single particle transitions, by virtue of the fact that the electron wave in the crystal is attenuated rather strongly; this feature is incorporated into our multiple scattering analysis through introduction of an appropriate complex inner potential. There is no evidence for structure in the existing data, however, even for data taken in rather thin films. Most likely the incorporation of orbital degeneracy into the analysis, and use of a more realistic band structure, would lead to a much smoother spectrum. The "center of gravity" of the Stoner band is in the vicinity of $\omega \cong 3$ while, as remarked earlier, the exchange splitting between minority and majority spin bands is, for the bulk model, $\Delta E_x \cong 7$. Thus, the peak in the Stoner spectrum is downshifted from ΔE_x in this example. Final state interactions between the particle and hole are included in our theory, and are responsible for this shift, in our view.

We have also a cluster of loss peaks at rather low energy, below $\omega \cong 1$. These are produced by spin flip scattering of the electron off the standing spin waves of the ferromagnetic film. It is apparent that an appreciable fraction of the oscillator strength of the total spectrum is contained in these modes.

The results just presented suggest that one should be able to observe spin wave loss features in the SPEELS spectrum of a ferromagnetic surface or film. We have seen from the discussion given above that the Stoner excitations show clearly in several experiments. One needs to extend such studies down into lower loss energies, where the spin waves reside.

Such an extension will prove challenging, however. Many of the studies of the Stoner excitations use rather coarse resolution, in the range of 500meV. This will have to be improved considerably, to resolve the spin wave losses. To improve the resolution, one is required to employ lower beam currents, unfortunately. Then one faces the difficulty that the spin flip cross sections are rather small; the signal will degrade substantially with improved resolution. We have compared the spin flip cross sections for Fe, with those for exciting surface phonons, to conclude the former is weaker by three orders of magnitude.[22] One loses another order of magnitude for Ni, by virtue of its small moment. There is, in our opinion, no need to use a spin polarized beam to see the spin wave loss features, since they lie well above the phonon features, and reside in their own characteristic loss regime. Elimination of the spin detector should compensate, at least in part, for the small excitation cross section.

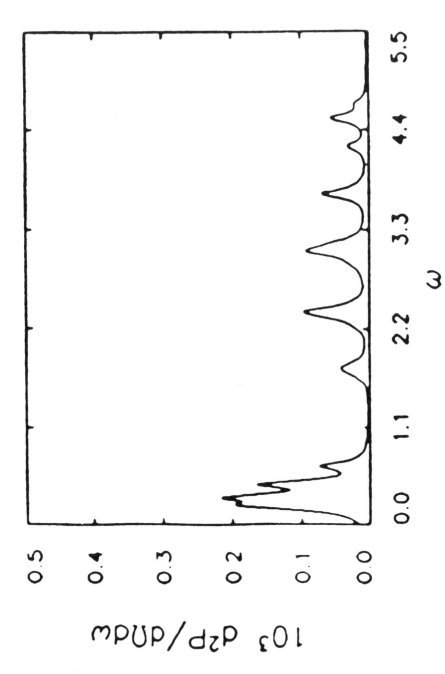

Figure 8. The spin flip contribution to the SPEELS spectrum, calculated for a twelve layer Hubbard model film, with ferromagnetic ground state. The wave vector transfer is 40% of the way to the boundary of the surface Brillouin zone, in the (100) direction.

The remarks above suggest experimental detection of the spin wave losses is possible, but an experiment explicitly directed toward this purpose is required. If the study of short wavelength spin waves on magnetic surfaces and in thin films is successfully implemented by the electron energy loss method, there will be a qualitative impact on our understanding of the magnetism of such systems. We shall be eager to hear of developments in this area.

IV. ACKNOWLEDGMENTS

The research of the author and his colleagues in the area covered by this paper has been supported by the U. S. Department of Energy, through Grant No. DE-FG03-84ER45083.

V. REFERENCES

1. For a review of earlier work, see the chapter by R. Feder in *Polarized Electrons in Surface Physics*, edited by R. Feder (World Scientific, Singapore, 1985).

2. See page 364 of J. D. Jackson, *Classical Electrodynamics* (First Edition), (J. Wiley & Sons, New York, 1962).

3. An excellent general discussion of spin orbit asymmetries and their applications has been given by J. Kessler, *Polarized Electrons* (Springer-Verlag, Heidelberg, 1985).

4. C. L. Fu and A. J. Freeman, J. Magn. Magn. Mater. **69**, L1 (1987).

5. E. Tamura, R. Feder, G. Waller and U. Gradmann, Phys. Status Solidi **B157**, 627 (1990).

6. D. P. Pappas, K. P. Kämper, B. P. Miller, H. Hopster, D. E. Fowler, C. R. Brundle, A. C. Luntz and Z. X. Shen, Phys. Rev. Letters **66**, 504 (1991).

7. M. P. Gokhale and D. L. Mills, Physical Review Letters **66**, 2251 (1991).

8. H. Shih, F. Jona, N. Bardi and P. Marcus, J. Phys. (Paris) Colloq. **C13**, 3801 (1988).

9. The complete set of data may be found in Gerhard Waller, Ph. D. thesis, University of Clausthal, 1986 (unpublished).

10. Alim Ormeci, Burl M. Hall and D. L. Mills, Phys. Rev. **B42**, 4524 (1990).

11. Alim Ormeci, Burl M. Hall and D. L. Mills, Phys. Rev. **B44**, 12369 (1991).

12. R. H. Victoria and L. Falicov, Phys. Rev. **B31**, 7335 (1985).

13. T. G. Walker, A. W. Pang, H. Hopster and S. F. Alvarado, Phys. Rev. Letters **69**, 1121 (1992).

14. H. J. Elmers and U. Gradmann, private communication.

15. Riqian Wu and A. J. Freeman, Phys. Rev. B**45**, 7532 (1992).

16. H. Ibach and D. L. Mills, *Electron Energy Loss Spectroscopy and Surface Vibrations* (Academic Press, San Francisco, 1982).

17. For example, see Burl M. Hall, D. L. Mills, Mohamed H. Mohamed and L. L. Kesmodel, Phys. Rev. B**38**, 5856 (1988).

18. J. Kirschner, Phys. Rev. Letters **55**, 973 (1985); A. Venus and J. Kirschner, Phys. Rev. B**37**, 2199 (1988).

19. D. L. Abraham and H. Hopster, Phys. Rev. Letters **62**, 1157 (1989).

20. Y. U. Idzerda, D. M. Lind, D. A. Papaconstantopoulos, G. A. Prinz, B. T. Jonker, and J. J. Krebe, Phys. Rev. Letters **61**, 1222 (1988).

21. M. P. Gokhale, A. Ormeci and D. L. Mills, Phys. Rev. B**46**, 8978 (1992).

22. Chapter 19, C. Kittel, *Quantum Theory of Solids* (Academic Press, New York, 1963).

23. M. P. Gokhale and D. L. Mills, Phys. Rev. B**49**, 3880 (1994).

SPIN-RESOLVED CORE LEVEL PHOTOEMISSION SPECTROSCOPY

F.U. Hillebrecht, Ch. Roth, H.B. Rose, and E. Kisker

Institut für Angewandte Physik
Heinrich-Heine-Universität Düsseldorf
40225 Düsseldorf, Germany

INTRODUCTION

Core level photoelectron spectroscopy has been used extensively during the past decades not only with the aim of understanding the fundamental effects governing the spectra, but also for its possibilities of helping to solve more technical materials' problems. These aspects are covered by various lectures within this summer school. The analytical capabilities are based on analyzing changes of intensity, binding energy, lineshape, and occurence of satellites in core level photoemission spectra in terms of composition, chemical state of constituents, and local electronic structure.

It had been suggested as early as 1970 that core level spectra may also yield information on magnetic properties. The splitting of the Fe 3s core level spectrum into two components as observed by Fadley and Shirley (1970) was interpreted as an exchange splitting. Generalizing this concept, ns core level photoemission lineshapes were used widely to investigate the magnetic moments of alloys containing Fe and of other materials. Photoelectron diffraction on Mn-compounds was interpreted in terms of the spin polarization of the photoelectrons which is inherently caused by the exchange interaction, although the spin polarization had not been established experimentally (Sinkovic et al, 1985a, b). The spin-resolved study of core levels was initiated by the experiment of Carbone and Kisker (1988) on the Fe 3p level. It demonstrated that the exchange interaction between the core hole and the spin-polarized valence electrons is indeed sufficient to cause spin polarization of the core level photoelectrons. A puzzling result of that experiment was the finite overall spin polarization, despite the fact that the initial core state certainly does not have a spin polarization, being a full (sub-) shell.

The study of core levels by spin-resolved photoemission is more difficult than the study of valence levels, since the cross sections are generally lower, and the available photon fluxes at the energies required are usually lower than at lower energies. Two developments contributed to the growth of this field during the last five years. One was of course the progress in insertion devices generating high photon fluxes at synchrotron radiation sources. The other was the development of a new type of spin polarimeter, which achieves a much higher efficiency than the previously employed polarimeters. Although to date the group in Düsseldorf is the only one using it, this polarimeter will be described in some detail. This is justified by its superior performance, which is all the more remarkable as the apparatus is moved several times per year from the home laboratory to the synchrotron radiation sources in Hamburg and Berlin.

SPIN POLARIMETRY BY *VLEED* ON Fe(100)

The spin polarization of an ensemble of electrons is defined as the vector P= tr (σρ), where σ is the Pauli spin operator, and ρ is the density matrix of the ensemble (Kessler 1985). For particles with spin 1/2, and for a certain quantization axis, this reduces to

$$P = (N_+ - N_-) / (N_+ + N_-)$$

where N_+ and N_- are the numbers of electrons with spin parallel or antiparallel to the quantization axis. Although the spin polarization is a vectorial quantity, for magnetic systems one usually restricts the experimental analysis to the magnetization axis of the sample. This requires a fully magnetized sample. Ideally, one should use an independent method to verify the magnetization state of the sample, such as Kerr microscopy. However, in many cases one may use the polarization of certain well-known spectral features as a criterion for sample magnetization.

Traditionally, spin analysis in photoemission and electron spectroscopy in general is carried out by Mott scattering. Here, the spin-dependent interaction is the spin-orbit interaction. The spin sensitivity is governed by the strength of this interaction, which in turn depends on the energy of the incoming electrons. Various scattering targets have been investigated. Naturally, materials with high nuclear charge are preferred, since in this case the relativistic effect will be large. Nevertheless, in general the spin-orbit interaction in unbound continuum is rather weak, so that the difference in scattering cross section for electron beams of different spin polarization is small. Consequently, a sizeable scattering asymmetry is only encountered when the average cross section is small. This is of course very undesirable for electron spin analysis. Conventional Mott detectors use gold as scattering target, and the scattering is performed at energies around 100 keV. Some improvements can be made by using uranium as scattering target (Pappas and Hopster 1989). In principle the spin-orbit interaction matrix elements scale with the fourth power of the nuclear charge. Since the primary energy affects how far the electrons can penetrate the atom, the scattering occurs with an effective nuclear charge depending on energy and possibly other details. The fourth power law suggests a possible gain of 84% by using U instead of Au, compared to an observed increase of about 50 %. Other possible schemes making use of spin-orbit interaction for spin analysis have been summarized by Unguris et al. (1986).

A completely different approach makes use of another spin-dependent interaction, i.e., the exchange interaction. In a ferromagnet, the electronic states for electrons with spin up or down with respect to some quantization axis (the magnetization axis) are energetically split by the exchange interaction. For the ferromagnets of the 3d transition series, the exchange splitting is most prominent for the 3d states, and is the cause for uneven occupancies of states with spin up or down. Although the magnetism is carried by these states, all other states are also affected both below and above the Fermi level. In general, however, this will not lead to significantly different densities of states for electrons with different spin, because the sp-derived electronic states form wide bands, leading to a fairly uniform density of states away from the d bands. The situation is different if one considers not the density of states integrated over the whole Brillouin zone, but the momentum-resolved band structure, in particular at a critical point. The example of interest in this context is shown in fig. 1 (Tamura et al., 1985). The calculated band structure of Fe along the (100)-direction shows a wide gap between 5 and 10 eV above the vacuum level. The 4 sp-derived band starting at H is, like all the other bands in ferromagnetic Fe, split by the exchange interaction. In other words, at a particular energy and for the corresponding k-vector there are only states for majority spin electrons. If one uses the surface of magnetized Fe as scattering target, such that the incoming electrons can couple to the 4sp-band *only* if they have the right spin polarization, one has a spin detector (Dodt et al., 1988; Tillmann et al., 1989). For this purpose a single crystalline Fe surface has to be used because otherwise the momentum cannot be defined.

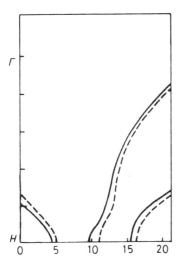

Figure 1. Band structure of ferromagnetic bcc Fe along Γ-H for the range up to 20 eV above the vacuum level as calculated by Tamura et al. (1985). Full lines correspond to majority, dashed lines to minority bands.

The design of such a spin polarimeter was shown schematically by Jungblut et al. (1991). A commercial hemispherical analyzer with 50 mm mean radius is complemented behind the exit slit by a three-element electrostatic lens. It transfers and images the electrons, whose energy corresponds to the pass energy of the spectrometer, towards the scattering target. The small size of the polarimeter (about 10 cm) allows to mount it with the spectrometer as a unit on a goniometer. The target is an Fe film of about 100 Å thickness grown epitaxially on Ag(100). It can be magnetized by current pulses through two pairs of coils along the two in-plane (100) directions of the Fe film, which are parallel to the (110) direction of the Ag substrate. One advantage of using thin films as a scattering target is that the stray field of such a film is negligible. The drift chamber, scattering target and the magnetization coils are all at the same potential. For magnetizing, the coils are disconnected from the electron optics supplies, and connected to the current pulse generator instead. The spin polarization is measured by taking measurements for two opposite target magnetizations sequentially. If the the beam leaving the spectrometer is spin-polarized with the respect to the quantization axis parallel to the axis of target magnetization, then the count rates are different. By magnetizing the target along the other in-plane (100)-direction, the orthogonal spin polarization component can be measured. It has been reported (Ballentine et al., 1989; Qiu et al., 1993) that for thin Fe films on Ag, in the range up to 1 nm, the easy magnetization axis is normal to the film plane. This suggests that it would even be possible to measure the longitudinal spin polarization with this polarimeter, by using a thin Fe film.

Compared to other spin polarimeters, the one described here puts the highest demands on the preparation of the scattering target. Deposition of the Fe film has to be performed at a pressure in the 10^{-10} mbar range onto a clean and well-annealed Ag(100) surface. The lifetime of such a polarimeter is much shorter than that of a gold foil in a Mott detector. In practical operation on a synchrotron beamline the surface of the target has been prepared freshly every day, although operation over two days is possible.

For characterization of the Fe film one may use the standard surface science techniques. A very convenient alternative is to measure the reflectivity of the scattering target as function of incident electron energy. Fig. 2 shows such a measurement without correction for the variation of lens transmission with retardation or accelerating potential. The potential between the exit slit of the spectrometer and the scattering target is varied from $-E_P$, the pass energy, to a ten Volts. The voltages of the lens are varied accordingly, in order to ensure good transmission at all energies. At low energies, the number of backscattered electrons is very low, and rises to a maximum at 10 eV. The high reflectivity is a result of the gap in the band structure of bcc iron along the 100-direction shown in fig. 1. When the energy of the electrons increases further, they can couple to the 4sp-band, and the reflectivity goes down. If the electrons are polarized, those electrons with spin parallel to the majority spin direction in the target can enter at lower energy, leading to different energetic position of the downward shoulder. Consequently, the reflection of electrons with 12.5 eV energy is spin-dependent. For the example shown, the combined asymmetry is 22 %. This is the product of the finite spin polarization of the primary beam, and the scattering asymmetry of the polarimeter. With a primary spin polarization of 50 % (Rose et al., 1994), one obtains a scattering asymmetry of 44 %. This is significantly higher than the analogous figures for any other spin detector reported so far.

The characterization of the Fe film and also the Ag substrate by taking reflectivity data is in principle a LEED experiment where the energy dependence of the specular beam is measured. In the energy range of interest here, the energy of the beam is too low for any other diffracted beam except the specular one to be present (i.e. very low energy electron diffraction, VLEED). Consequently, the relative intensity of the specular beam is very high. At the energy where the maximum spin sensitivity occurs, it is usually about 5%, but also values of 10 % have been observed. This is again higher by an order of magnitude that the figures for Mott or other spin polarimeters based on the spin-orbit interaction. Taking it all together, the

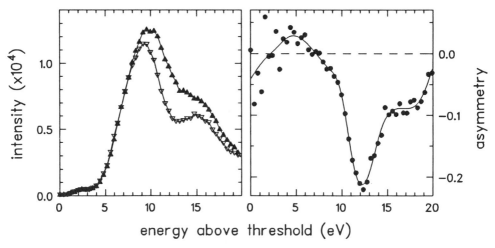

Figure 2. Spin-resolved specular reflection of polarized electrons from Fe(100) as function of incident electron energy. The energy variation of the electron flux incident on the target has not been removed. The primary electrons are photoelectrons excited by 90 eV radiation out of the Fe valence band states at 3 eV binding energy, where the spin polarizationis known to be of majority type. The curve marked ++ corresponds to sample and polarimeter magnetization parallel, the one marked -- to antiparallel magnetization. The right panel shows the asymmetry $A=(I_{++} - I_{+-})/(I_{++} + I_{+-})$, which is the product of the primary spin polarization P (i.e. the spin poalrization of the photoelectrons) and the spin-related scattering asymmetry S (the analogue of the Sherman function of a Mott polarimeter) of the polarimeter. The maximum of 22 % indicates a value of S of 44 %, as the primary spin polarization is about 50% (Rose et al., 1994).

spin polarimeter described here offers a much better performance than conventional polarimeters, at a cost of a more demanding preparation of the scattering target. The investment for such polarimeter can be very low, if, as was done initially, one mounts and prepares sample and polarimeter surface on the same manipulator. In this case the available surface characterization techniques can be used for both sample and polarimeter. The good performance of the Fe(100) polarimeter suggests that one can improve the ultimate energy resolution in spin-resolved electron spectroscopy by using this technique.

All the experimental data presented in this article were obtained from thin films prepared by molecular beam epitaxy. The Fe films were grown on Ag(100), Au(100) or W(110) single crystal surfaces, whereas Co was grown on a Cu(100) substrate. Growth was performed in a vacuum of 10^{-10} mbar range, with substrates either at room temperature (for Fe on noble metals), or at slightly elevated temperature during the initial phase for Fe on W (Gradmann and Waller, 1982) and Co on Cu (Schneider et al., 1990). If not mentioned otherwise, photoelectrons are collected in normal emission with about 8° full acceptance. The energy resolution is ususally 0.3 to 0.4 eV. Film thicknesses were 5-10 nm for Fe films, whereas the Co films were about 2 nm thick.

SPIN RESOLVED CORE LEVEL PHOTOEMISSION: EXCHANGE INDUCED SPIN POLARIZATION

For filled inner shells, there is no net spin or orbital moment in the ground state. This is in contrast to the valence shell of a magnetic solid, where the uneven occupancy of states with spin parallel and antiparallel to the magnetization axis is the source of the macroscopic magnetization, and is associated with an overall spin polarization in the ground state. After removal of an electron from a filled inner (sub-) shell there is of course a spin and possibly an orbital moment.

The Iron 3s core level

The core levels of iron have been studied most thouroughly. Of particular interest was the 3s level, because it is known since 1970 that it shows a distinct doublet structure in spin-averaged photoemission. This doublet structure can be recognized in fig. 3. It was attributed to the exchange interaction between the remaining 3s eletcron and the magnetic moment of the 3d valence shell (Fadley and Shirley 1970). This interpretation implies that the photoelectrons in the main peak and the satellite should carry finite spin polarization, with the signs opposed to each other. The lowest energy final state should be the one where the spin of the remaining 3s electron is parallel to the the 3d spin, consequently the photoelectrons in this peak should have minority spin polarization. The spin-resolved photoemission measurements shown in fig. 3 (Hillebrecht et al., 1990; Carbone et al., 1990) confirm this expectation. The peak of the majority spin electrons is at higher binding energy. The lineshape of the minority peak does not look unusual, except for the rather long tail extending to high binding energy. The majority line shows a feature at the peak of the minority line, and all the structures in the majority line appear to be strongly broadened.

The Fe 3s spectrum to be expected if the 3d electrons were localized was calculated by Thole and v.d. Laan (1991b). Since the core has only spin, but no orbital angular momentum, the spectrum is particularly simple, since the only degree of freedom is how the spin of the 3s electron couples to 3d the spin. The result agrees qualitatively with experiment, with respect to the sign of spin polarization in the main peak and satellite. Also the majority structure at the position of the minority peak is reproduced. However, the lineshapes, in particlar for the majority peak, are not reproduced. The solid state surrounding is of course not taken into account in the calculation, and this may be the source for the more complex lineshapes. From theory, one expects equal integrated intensities in the two spin-resolved subspectra. This is not quite fulfilled by the experimental result, the ratio of the intensities being 1.16, corresponding an overall polarization of 8%. The main features of the spectrum are certainly explained by an atomic-like picture, however, the lineshapes and the finite polarization are beyond such a description.

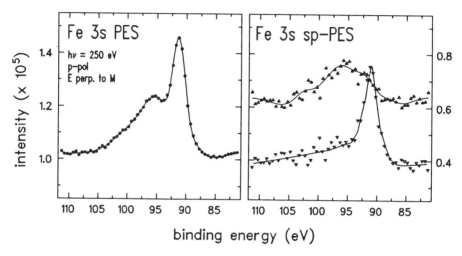

Figure 3. Spin-resolved photoemission spectrum of the Fe 3s level, taken with 250 eV photons (Hillebrecht et al. 1990). Left panel shows spin-averaged spectrum, right panel shows spin-resolved spectra. Full and empty triangles correspond to majority and minority spin spectra.

In atomic theory, the splitting between the two components is governed by the exchange interaction between the single final state 3s electron and the 3d electrons. For systems with non-localized d electrons a similar situation still holds. It was shown by Kakehashi (1985) that the first moment of the spin-averaged spectrum is related to the average d occupancy. This rule is useful for materials which show satellites associated with different d occupancies, like Ni (Hüfner and Wertheim 1975). For Fe this mechanism is not important, since unlike Ni only the 3s spectrum shows such a pronounced satellite. The second sum rule says that the first moment of the polarized spectrum ω_s is given by the negative product of the exchange interaction times the average z-component of the 3d spin. This is similar to the result obtained from atomic theory for localized systems, but generalizing it to systems with delocalized d electrons. From our data we obtain $\omega_s = 1.7$ eV. Assuming a moment of 2.2 μ_B, one finds an exchange coupling constant of 4 eV. Analysis in terms of an atomic model also indicates a reduction by 50 % from the atomic exchange interaction to the one in the solid, so this result is not very surprising. An interesting possibility is offered by the sum rules for the second moments: The second moment of the spin-averaged spectrum should be governed by charge fluctuations and the change of the total spin moment. The early measurements of Fadley and Shirley (1970) showed no change of the Fe 3s spectrum when approaching the Curie temperature. Since the temperature dependence of the charge fluctuations is thought to be negligible over the range up to the Curie temperture, this result indicates that the total moment is essentially unaffected over this temperature range. The same type of experiment has been performed on Cr. It was found that the Cr 3s lineshape is unaffected up to a temperature about three times the Néel temperature. This was taken to imply that the local Cr moment does not change over this temperature range (van Kampen et al., 1991).

THE 3p SPECTRA OF THE 3d FERROMAGNETS

The 3p spectra of the 3d ferromagnets are much more easily measurable, because of the higher cross sections and lower binding energies. Fig. 4 shows a representative Fe 3p spectrum taken with 140 eV photons at normal incidence and normal emission. The data are in qualitative agreement with a number of measurements by other groups (Carbone and Kisker, 1988; Sinkovic et al., 1990; Mulhollan et al., 1990; Kachel et al., 1993; van Kampen et al., 1993). The spin-resolved peaks differ strongly in intensity, yielding a sizeable total spin polarization when one integrates over both peaks, and there is a 0.5 eV energy difference of the peak positions.

Qualitatively similar results have also been obtained for the other 3d ferromagnets (Kachel et al., 1993). Since the effective moment is largest for Fe, Fe shows the strongest polarization features. For Ni, the spectrum is complicated by the occurence of the satellite related to the d^9-d^{10} charge fluctuation (Hüfner and Wertheim, 1975; Hillebrecht et al., 1982). Kachel et al. (1993) suggested on the basis of the observed large spin polarizations that the 3p spectra of the 3d ferromagnets are dominated by the high-spin final states. Their polarizations are given by $P=-(2S_f-1)/(2S_f+1)$, where S_f is the spin of the high spin final state. Although it is difficult to explain why the low spin states should not contribute to the spectra, the trend of spin polarizations observed for Fe, Co, and Ni 3p spectra can be reproduced in this way.

Review of the available spin polarized Fe 3p photoemission data suggests that there may be a photon energy dependence of the total Fe 3p spin polarization. However, for the experiments in the literature the geometries are not always fully specified, so there may be influence of magnetic dichroism. Also, there may be uncertainties in the spin sensitivities of the polarimeters when comparing experiments perfomed on different spectrometers. As will be discussed later, magnetic dichroism cannot occur for the geometry used for the experiment shown in fig. 4, if the sample magnetization is parallel to the electric field E of the radiation.

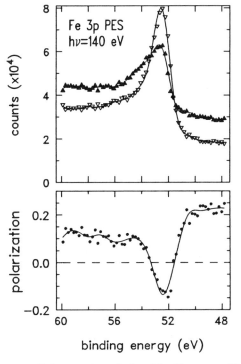

Figure 4. Spin-resolved Fe 3p photo-emission spectrum, excited by s-polarized light of 140 eV (normal incidence, normal emission, M parallel to E). Full and empty triangles show majority and minority spin spectra, respectively.

The magnetic properties of ultrathin films are of interest because they can be drastically different from those of the bulk material. Because of the high surface sensitivity, spin-resolved core level spectroscopy is well suited to study ultrathin films. Due to the distinct binding energies of core electrons, one can in principle distinguish between the magnetic properties of different chemical species. An example for this type of experiment is given in fig. 5. A monolayer of Cr on Fe(100) was predicted to order ferromagnetically, with antiferromagnetic coupling of the Cr and Fe moments (Victora and Falicov, 1985). The magnetic moment per Cr atom in the monolayer on Fe was calculated to be enhanced by a factor of 5 over the bulk Cr magnetic moment (Victora and Falicov, 1985; Fu et al., 1985). Such a large Cr magnetic moment is plausible since for Cr the 3d shell is half-filled, and the surface band narrowing also favours a large moment. This should also facilitate the detection of the Cr moment in spin-resolved core level photoemission. Fig. 4a shows the spin-averaged data for a monolayer of Cr on Fe. The rise of the spectrum on the left is due to the low binding energy edge of the Fe 3p spectrum. The lineshape of the spin-averaged spectrum does not change with coverage, in contrast to the situation found for Cr 3s on Ag (Newstead et al., 1988). The spin-resolved lineshapes differ in intensity, peak energy, and width. This shows first of all that there is a long range magnetic order with non-vanishing magnetization in the Cr overlayer. We find again, as in the case of Fe, that the majority and minrity spin peaks have different total intensity. For Fe and the other 3d ferromagnets, it has been established that the 3p minority spectrum occurs at lower binding energy, has higher intensity, and smaller width than the majority spectrum. The binding energy difference can be understood qualitatively as an extension of Hund's rule to systems with more than one incompletely filled shells, here the 3d and 3p shells. Using the same criteria to identify the Cr 3p minority spin channel, we find that it occurs in the majority spin channel of Fe. This shows that the moments in the Cr overlayer are opposed to the moments in the Fe substrate. As to the *size* of the moment, one may use the splitting between the two components as a guideline. This splitting is of similar size as in Fe, so one may regard the Fe moment as an upper limit to the moment in Cr. The intensity difference, however, is much more pronounced than the energetic splitting. The origin of the intensity difference, i.e. the non-vanishing integral spin polarization, is not understood until now. Atomic calculations yield spectra much wider than the apparent peak widths in the experimental data of solids.

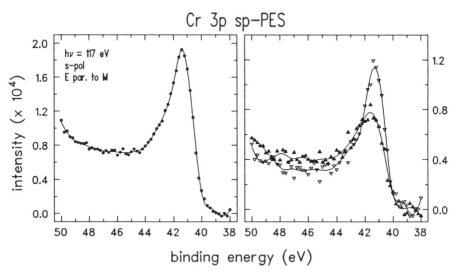

Figure 5. Spin-resolved Cr 3p photoemission spectrum for 1 monolayer of Cr on Fe(100), which in turn is grown on Ag(100). The spin polarization of the secondary background was suppressed. Full and empty triangles correspond to majority and minority spin spectra with respect to the Cr 3d moment.

Nevertheless, the total polarization in such calculations is always zero, although the leading features invariably show minority spin polarization. If one phenomenoligically suggests that in some way the total polarization is influenced by the ground state moment, one again finds that the polarization is not significantly different from that of the Fe 3p level, again pointing to a moment not larger than that of Fe. Applying the picture of Kachel et al. (1993) to the case of Cr yields 0.7 μ_B for the final state spin moment. Recent magnetometry measurements have, on the other hand, shown a drastically enhanced Cr moment for monolayer film on Fe (Turtur and Bayreuther, 1994). At present, the understanding of spin-resolved core level spectra is not sufficiently advanced to extract quantitative information from such spectra, although qualitatively magnetic order and the sign of the magnetic coupling between constituents in multi-component systems can be determined.

MAGNETIC LINEAR DICHROISM IN THE ANGULAR DISTRIBUTION OF PHOTOELECTRONS (MLDAD)

The existence of two new magnetic dichroism effects in photoemission with *linearly* polarized radiation has recently been revealed by a systematic investigation of the dependence of the Fe 3p core level photoemission lineshape on the polarization of the incident light and on the sample magnetization (Roth 1994). The first effect manifests itself in a dependence of the lineshape on the *angle* Φ of the light electric field vector **E** (the polarization vector) with the sample magnetization vector **M**; the spectra are different for $\Phi=0°$ and $\Phi=90°$ (Roth et. al., 1993a). This kind of effect is termed Magnetic Linear Dichroism (MLD) (Thole et. al., 1991) in analogy to similar effects known in photoabsorption spectroscopy under the same name in the visible and also in the x-ray range (van der Laan et. al., 1986).

The other, and potentially even more important new dichroism effect is the dependence of photoemission lineshapes on the *sign* (or *sense*) of the magnetization at fixed angle of **E** with **M**. This dependence is observed in special geometries with non-coplanar (chiral) orientation of the three vectors **E**, **M** and **k**, were **k** is the photoelectron wave vector. Since, in an atomic model, symmetry requires that this effect should only be observable in *angular resolved* photoemission, it was termed Magnetic Linear Dichroism in the Angular Distribution of photoelectrons (MLDAD). This dichroism was first observed in Fe 3p core level photoemission (Roth et. al., 1993b), where asymmetries of up to 25 % (peak-peak) are observed depending on photon energy, and in valence band photoemission from Fe (Rose et. al., 1994). The results obtained in the pioneering experiments on Fe 3p photoemission, which also include spin-polarization measurements, are described in this section.

Experimental Results

The first MLDAD experiments were performed with linearly polarized soft x-rays from the crossed undulator U2 at the dedicated synchrotron radiation source BESSY in Berlin. This undulator was originally designed to produce circularly polarized radiation. For this experiment we took advantage of the fact, that due to the design of the crossed undulator it is also possible to obtain linearly polarized radiation with the polarization vector either in the ring plane (p-polarized) or normal to the ring plane (s-polarized). The change of polarization can be performed relatively quickly (within about 15 min) without any change of the experimental geometry by closing the gap of one of the two undulators, while deactivating the other by opening its gap. The geometry of the experiment is shown in fig. 6. Linearly polarized light impinges on the sample under an incidence angle of 16° with respect to the sample surface. The sample can be remanently magnetized up or down, parallel or antiparallel to the y-axis of the laboratory coordinate system. This means that the sample magnetization is normal to the reaction plane defined by the directions of light incidence and electron emission. The y-axis is also the quantization axis of the spin polarimeter for the supplementary spin-resolved measurements.

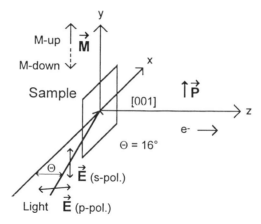

Figure 6. Experimental Geometry. The sample is irradiated under oblique incidence (incidence angle 16°), and the normally emitted photoelectrons are detected. "Magnetic Linear Dichroism in the Angular Distribution" (MLDAD) is only observed with *p-polarized* light. Then, the three vectors **M**, **E** and **k** (the photoelectron wave vector) define a non-coplanar (chiral) geometry.

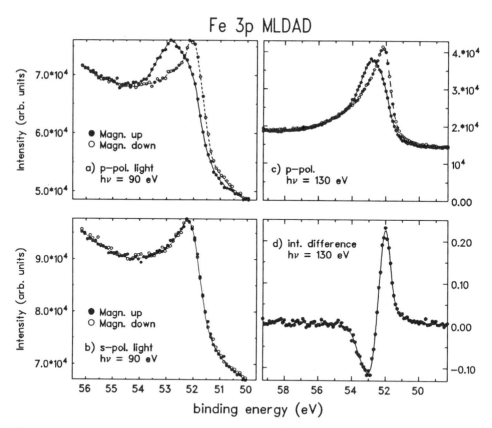

Figure 7. Spin-integrated Fe 3p energy distribution curves (EDCs), taken in normal emission with linearly polarized radiation (90 eV photon energy) at oblique incidence; sample magnetization normal to the reaction plane. a) Spectra excited by p-polarized light; b) same, but excited by s-polarized light. These spectra do *not* depend on the sign of the magnetization. c) same as a, but for 130 eV photon energy. For p-polarized light (a and c) the spectra show a dependence on the sign (or sense) of the magnetization (MLDAD).

Figs. 7a and c show spin-integrated energy distribution curves (EDCs) of the Fe 3p core level obtained for the two directions of the magnetization (M-up, M-down) with *p-polarized* light of 90 and 130 eV photon energy. The EDC's clearly differ drastically for the two opposite magnetization directions. The Fe 3p emission is superimposed on a background of inelastically scattered electrons which for the lower photon energy increases towards larger binding energies. The integrated intensities of the two EDC's above the background are equal within 0.35 %. The EDC for M-down shows a single asymmetric peak with a smaller structure on the high-binding energy side. For M-up, the peak is at 0.8 eV higher binding energy. The line is much wider and its appearance suggests that it is composed of at least two overlapping contributions. These spectra constitute the first evidence for a multiplet structure of the Fe 3p emission line, which is even more clearly seen from our spin-resolved data which are shown below. The energy range in which the spectrum depends on magnetization direction is restricted to 4 eV.

Fig. 7b shows Fe 3p EDCs obtained for the same sample with *s-polarized* light under otherwise identical conditions. These spectra do *not* depend on the sign of the magnetization. This shows that the effect observed with p-polarized light is not produced or noticably influenced by any apparatus effects, e.g. magnetic stray fields in the experimental chamber. The shape of the EDC obtained with s-polarized light differs significantly from both EDC's measured with p-polarized light, and also from their sum spectrum. It is consistent with lineshapes reported previously for the Fe 3p level with s-polarized light. The difference in lineshape between the spectra taken with s-polarized and with p-polarized light (disregarding the additional magnetization *sign* dependence occuring with p-polarized light, i.e. MLDAD) is a manifestation of the first effect mentioned in the introduction, i.e. Magnetic Linear Dichroism (MLD): With s-polarized light, **E** and **M** are collinear ($\Phi=0$), while with p-polarized light they are perpendicular to each other ($\Phi=90°$) (Roth et al., 1993a).

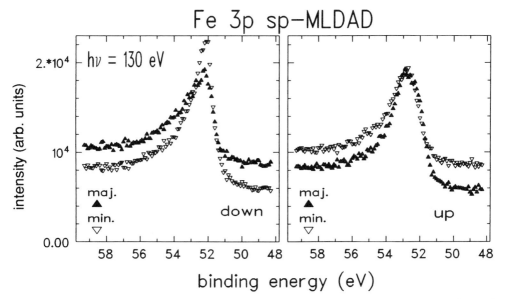

Figure 8. Spin-resolved Fe 3p energy distribution curves (EDCs) for p-polarized light, corresponding to the two spin-integrated EDC's in fig. 7c. Each of the two EDC's in fig. 7c splits into two EDC's, with spin direction either majority or minority with respect to sample magentization. The EDC's are also labeled as Maj. and Min., indicating the spin character with respect to the sample magnetization (majority, minority) as usual.

To elucidate the magnetization dependence (MLDAD) of the spin-integrated spectra taken with p-polarized light (fig. 7a, c), we have measured these spectra also spin-resolved. In fig. 8, each of the two spin-integrated EDCs of fig. 7c is decomposed into two spin-resolved EDCs with defined spin orientation with respect to the y-axis of the laboratory frame of reference as shown on fig. 6. This yields four different spectra for the four possible combinations of sample magnetization and electron spin orientation. The labels "Spin-up" and "spin-down" in fig. 8 refer to electron spin projection parallel or antiparallel to the y-axis, *independent* of the sample magnetization. With respect to earlier work, where the spin orientation was only measured with respect to the sample magnetization direction, the EDC's may also be classified as usual, i.e. according to the majority or minority spin character of the electrons.

In the majority spin channel an increase of intensity is observed when the magnetization is switched from down to up, and correspondingly the spin orientation of the detected electrons is switched from up to down. In the minority spin channel intensity increases when the magnetization is switched from up to down, which again means that electrons with spin-down are measured instead of spin-up electrons. Obviously in both spin channels (majority, minority) the *total* emission increases when electrons with spin-down orientation, as measured with respect to the space-fixed laboratory coordinate system, are detected instead of spin-up electrons. But in detail a mutually *opposite* behaviour is observed in the low- and in the high-energy part of the spectrum, especially in the minority spin channel, as is evident from fig. 8. This is a consequence of spin-orbit interaction in the Fe 3p core states and is a direct evidence of the spin-orbit splitting of the Fe 3p spectrum, as will become clearer from the next figure.

Fig. 9 shows spin-resolved EDC's derived from the data of fig. 8 by averaging the spin-resolved EDC's with the same spin orientation (as referred to the y axis) over the two opposite magnetization directions. These spectra would be obtained from a demagnetized sample with equal number of up and down magnetic domains. At 52 eV binding energy, a small but finite spin-polarization remains after the magnetization has been averaged out. This means that this spin polarization can not be a result of the 'magnetic' exchange interaction, but must be a result of *spin-orbit interaction*. For comparison we also show the result obtained at 90 eV photon energy, which is qualitatitvely similar to the result at 130 eV.

Photoemission studies on atoms have shown, that spin-orbit interaction can lead to spin polarization when linearly or even unpolarized light is used for excitation. The effect was theoretically predicted by Cherepkov (1973) and by Lee (1974). Experimentally, the first evidence for this type of phenomenon was observed in photoionization of outer shells of free atoms (Heinzmann et. al., 1979; Schönhense 1980; Kessler 1985). A related effect was later found in valence band photoemission from metallic Pt and Au (Tamura et. al., 1991; Schmiedeskamp et. al., 1991). According to theoretical treatment, the spin polarization vector **P** is collinear to the vector product [**k** x **E**] of the light polarization vector **E** and the photoelectron momentum **k**. Consequently, for p-polarized light, **P** is normal to the reaction plane defined above. It vanishes after integration over all emission directions, and also for **E** || **k** and **E** ⊥ **k**. The spin polarization is ascribed to the influence of spin-orbit interaction in the bound state combined with quantum mechanical interference of the continuum final states describing the photoelectron.

A similar effect is obviously observed in the Fe 3p spectrum in fig. 9. Here, the angle of **E** with **k** is $\Theta = 16°$ (see fig. 6), and according to the preceding discussion a spin polarization in the y direction due to spin-orbit interaction is possible. In this model, for emission out of a closed p-shell with spin-orbit split final states with total angular momenta j=1/2 and j=3/2, the spin polarization should change sign in the spectrum according to $P^{3/2} / P^{1/2} = -1/2$. Thus we attribute the energy seperation of the main peaks in the spin-resolved EDC's of fig. 9 to the $3p_{1/2}/3p_{3/2}$ spin-orbit splitting, which has not been observed previously, in this way obtaining a value of about 1 eV. Contrary to expectations, the difference spectrum in fig. 9 does not integrate to zero; this may be due to spin-orbit interaction in the photoelectron continuum states (in analogy to the Fano effect) and/or a modification of the core wavefunctions and the corresponding transition probabilitys by the exchange interaction with the polarized valence electrons (Roth, 1994).

The magnetization sign dependence of the spin-integrated EDC's in fig. 7 (MLDAD) can now be explained qualitatively in analogy to magnetic *circular* dichroism (Baumgarten et. al., 1990, Roth, 1994, Hillebrecht et. al., 1994) in the following model: In angle resolved photoemission excited by linearly polarized light, spin-orbit interaction in the bound state and the quantum-mechanical interference of the accessible photoelectron partial waves leads to emission of electrons with a preferred spin orientation with respect to the quantization axis [**E** x **k**]. If the sample is ferromagnetic and magnetized along the same axis, the exchange interaction additionally leads to an energetic splitting of the m_j-sublevels according to their spin character with respect to the same quantization axis. Thus in each of the fine structure components emission of electrons with either minority or majority spin character is favoured, depending on the sign of the magnetization, which in turn have different spectral distributions. In this way an intensity asymmetry must result in the spin-integrated spectrum upon magnetization reversal, as observed experimentally.

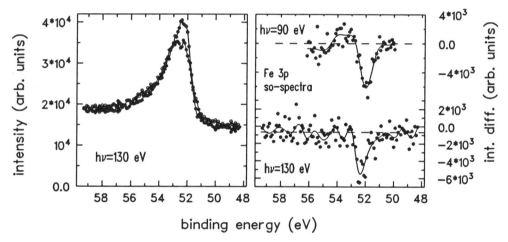

Figure 9. Spin-resolved Fe 3p spectra for p-polarized light showing the *spin-orbit* induced spin polarization. (a) Spin-resolved EDC's for spin directions up and down with respect to the spatially fixed y-axis after averaging over the two opposite sample magnetization directions. (b) Difference spectrum of the EDC's in (a), with same vertical scale units. The result for 90 eV photon energy is also given.

With s-polarized light, however, no spin-orbit induced spin polarization can occur in the geometry employed in our experiment (see fig. 6), since **E** then is perpendicular to **k**, and also the y-axis is coplanar to **E** and **k**. This also follows from symmetry considerations: Parity conservation forbids the existence of spin polarization in a mirror plane of the system for an unpolarized target (Kessler 1985). For s-polarized light the (y,z)-plane is a mirror plane, when in electrical dipole approximation only the electric field of the incident radiation is taken into account, and the photon momentum is neglected in the nonrelativistic limit. Accordingly, no dichroism should be observable with s-polarized light, and indeed there is no dichroism in this case, as seen in fig. 7b.

Within the atomic picture presented here, the spin-orbit induced spin polarization with p-polarized light should vanish after integration over all emission angles, and this should also be true for the dichroism (MLDAD). Consequently, this type of dichroism should not be observable in *photoabsorption* measurements, if one considers photoabsorption as an angle-integrated photoemission experiment at threshold. However, the sample surface may modify the angular distribution of the photoelectrons in such a way that a dichroism persists even after angle integration or in photoabsorption. Preliminary results of absorption measurements at the Fe 3p edge indeed indicate that the effect survives under certain circumstances in photoabsorption.

Our results on MLDAD in Fe 3p photoemission have recently been nicely confirmed (Rossi et. al., 1994), although no measurements of the accompanying spin polarization, apart from ours, have been reported until now. Meanwhile we have observed this effect also at the 2p core levels of Fe (Roth et. al., 1993c, Hillebrecht et. al., 1994), at the Co 3p (Fanelsa et. al., 1994), the Ni 3p (Roth et. al., 1994) and at the Gd 4f level.

ANGULAR DEPENDENCE OF MLDAD

The spectra presented above on the magnetic linear dichroism were all taken under identical geometrical conditions. In circular dichroism, the strength of the effect depends on the angles between heilicity and magnetization, and magnetization and electron emission (Schneider et al., 1992; Venus et al., 1993). Also in linear magentic dichroism an angular dependence is to be expected. Since the interplay between exchange and spin-orbit interaction is the cause for the effect, one may consider the angular dependence of each of these effects as a guideline. Spin-orbit induced spin polarization is known to show an angular dependence according to $\sin\theta \times \cos\theta$, where θ is the angle between the electric field of the incoming light, and the direction of electron emission.

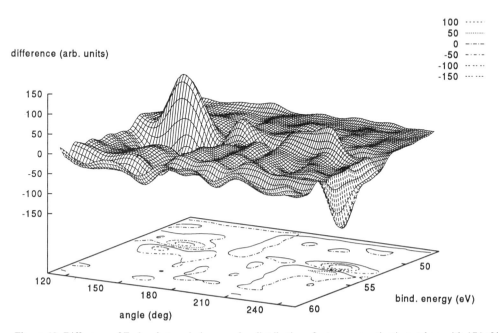

Figure 10. Difference of Fe 3p photoemission angular distributions for two magnetizations taken with 174 eV linearly polarized photons. Normal emission is at 180 degrees.

We measured the angular distribution of Fe 3p photoelectrons using a one dimensional display analyzer which measures the whole angular distribution in one plane simultaneuosly. This was performed for light incidence normal onto the sample, in the plane defined by the electric field of the radiation and the surface normal. The sample was magnetized normal to this plane. The angular distributions of photoelectrons were asymmetric with respect to normal emission. However, the distribution for one magnetization is the same as the mirror image of the other magnetization: $I(up,\theta)=I(down,-\theta)$. When the angular distributions for the two sample magnetizations are subtratcted from each other, one obtains a difference spectrum which is uneven with respect to reflection at normal emission: $I(up, \theta) - I(down, \theta) = -I(up, -\theta) + I(down, -\theta)$.

Fig. 10 shows a 3d-plot of the difference as function of binding energy and emission angle. Normal emission is at 180 degrees. One can see that at low binding energy (48 eV), i.e. below the Fe 3p photoemission peak, the angular distributions show the same intensities. Two strong structures appear at about 52 eV binding energy, i.e. in the Fe 3p peak. For emission corresponding to the right half of the figure, the structure is negative, while for the other side it is positive. For higher binding energies, one can recognize a sign change of the dichroism, in the form of a hill on the right side, and a valley on the left side of the figure. This 3d-plot shows the overall features of the angular distribution, while the finer structrures can only be seen in constant angle or energy sections. The sum rules state that the integrals over the difference either as function of binding energy or of emission angle should vanish. The measurement of the angular distribution confirms this expectation. Beyond that, it shows that there is quite rich structure in the angular distrubution, which is presumably due to photo-electron diffraction.

SUMMARY AND OUTLOOK

The results presented here show only a small fraction of what is known today about magnetic dichroism phenomena. When the first spin-resolved core level photoemission experiments on magnetic systems were performed, the intention was to learn something about the exchange interaction between the core hole and the magnetic valence electrons, and via that route to obtain information on the magnetism. Consequently, one was primarily interested in polarization phenomena caused by the exchange interaction, and usually spin-orbit inter- action was not considered to be important. Only in the course of studies of dichroism phenomena the influence of spin-orbit interaction became apparent, and it has been shown that spin-orbit interaction can lead to spin polarization as strong as that induced by the exchange interaction (Roth et al., 1994). Theoretical descriptions incorporating both exchange and spin-orbit interaction have been given, although until now primarily for atomic systems. When the results for real ferromagnets with delocalized d states are compared to atomic theories, one quite often finds some agreement with experiment in the gross features, although careful comparison ususally shows that the finer details cannot be described within such pictures. Experimentally, the future will lead to a comprehensive assessment of the effects which have been observed until now for a large set of materials and model systems, such that general trends can be distilled. The demonstration of the equivalence of linear and circular magnetic dichroim will open the field to a much larger group of researchers. The improved radiation sources which are going into operation at new synchrotron facilities in combination with improved spectroscopic facilities, will help to provide ever more "differential" experimental data, allowing ever more detailed comparison to theory. Since one is in electron spectroscopy always looking at a highly excited state, the outcome will be a better understanding of the spectroscopic effects. But even without a complete understanding of these spectroscopic techniques one can apply them to the study of magnetic phenomena. The examples given in this article and the whole course provide ample evidence for the usefulness of spectroscopic techniques.

Acknowledgements

The results presented in this short review were obtained in several runs of experiments at the BESSY and Hasylab synchrotron radiation sources in Berlin and Hamburg, respectively. We are grateful to the colleagues there who have set up and helped us operating the state-of the-art beamlines, which were essential for our research. The financial funding for the synchrotron experiments was provided by the Bundesministerium für Forschung und Technologie (BMFT) under grant no. 05 5PFDA B3, and for the laboratory experiments by the Deutsche Forschungsgemeinschaft (DFG) within SFB 166 / G7.

References

Ballentine, C.A., Fink, R.L., Araya-Pochet, J., and Erskine, J.L., 1989, Appl. Phys. A49, 459.
Baumgarten, L., Schneider, C. M., Petersen, H., Schäfers, F., and Kirschner, J., 1990, Phys. Rev. Lett. 65, 492.
Carbone, C., and Kisker, E., 1988, Sol. State Comm. 65, 1107.
Carbone, C., Kachel, T., Rochow, R., and Gudat, W., 1990, Z. Phys. B 79, 325.
Cherepkov, N. A., 1973, Zh. Eksp. Teor. Fiz. 65, 933.
Cherepkov, N. A., 1974, Sov. Phys. JETP 38, 463.
Dodt, Th., Tillmann, D., Rochow, R., and Kisker, E., 1988, Europhys. Lett. 6, 375.
Fanelsa, A., Rose, H.B., Hillebrecht, F.U., and Kisker, E., 1993, BESSY Annual Report, p. 391.
Fadley, C.S., and Shirley, D.A., 1970, Phys. Rev. A 2, 1109.
Fu, C.L., Freeman, A.J., and Oguchi, T., 1985, Phys. Rev. Lett 54, 2700.
Gradmann, U., and Waller, G., Surf. Sci. 116, 539 (1982).
Heinzmann, U., Schönhense, G., and Kessler, J., 1979, Phys. Rev. Lett. 42, 1603.
Hillebrecht, F.U., Fuggle, J.C., Bennett, P.A., Zolnierek, Z., and Freiburg, Ch., 1982, Phys. Rev. B 27, 2179.
Hillebrecht, F.U., Jungblut, R., and Kisker, E., 1990, Phys. Rev. Lett. 65, 2450.
Hillebrecht, F.U., Roth, Ch., Jungblut, R., Kisker, E., and Bringer, A., 1992, Europhys. Lett. 19, 711.
Hillebrecht, F.U., Roth, Ch., Rose, H.B., Park, W.G., and Kisker, E., 1994, Phys. Rev. B, in print.
Hüfner, S., and Wertheim, G.K., 1975, Phys. Lett. 51A, 299, 301.
Jungblut, R., Roth., Ch., Hillebrecht, F.U., and Kisker, E., 1991, J. Appl. Phys. 70, 5923.
Jungblut, R., Roth., Ch., Hillebrecht, F.U., and Kisker, E., 1992, Surf. Sci. 269/270, 615.
Kachel, T., Carbone, C., Gudat, W., 1993, Phys. Rev. B 47, 15391.
Kampen, D.G. van, Knieriem, M.L., and Klebanoff, L.E., 1991, Phys. Rev. B 43, 11668.
Kampen, D.G. van, Pouliot, R.J., and Klebanoff, L.E., 1993, Phys. Rev. B 48,
Kakehashi, Y., 1985, Phys. Rev. B 31, 7482.
Kakehashi, Y., and Fulde, P., 1985, Phys. Rev. B 32, 1595; ibid., 1607.
Kessler, J.,1985, "Polarized Electrons", Springer-Verlag Berlin.
Laan, G.v.d., Thole, B. T., Sawatzky, G. A., Goedkoop, J. B., Fuggle, J. C., Esteva, J.-M., Karnatak, R., Remeika, J. P., and Dabkowska, H. A., 1986, Phys. Rev. B 34, 6529.
Lee, C. M., 1974, Phys. Rev. A 10, 1598.
Newstead, D.A., Norris, C., Binns, C., and Stephenson, P.C., 1987, J. Phys. C 20, 6245.
Pappas, D., and Hopster, H., 1989, Rev. Sci. Instr. 60, 3068.
Rose, H.B., Roth, Ch., Hillebrecht, F.U., and Kisker, E., 1994, Sol. State Comm. 91, 129.
Rossi, G., Sirotti, F., Cherepkov, N.A., Combet-Farnoux, F., and Panaccione, G., 1994, Sol. State Comm. 90, 557 (1994).
Roth, Ch., Rose, H.B., Hillebrecht, F.U., and Kisker, E., 1993a, Sol. State Comm. 86, 647.
Roth, Ch., Hillebrecht, F.U., Rose, H.B., and Kisker, E., 1993b, Phys. Rev. Lett. 70, 3479.
Roth, Ch., Rose, H.B., Park, W.G., Hillebrecht, F.U., and Kisker, E., 1993c, HASYLAB Annual Report, 249.
Roth, Ch., 1994, Ph. D. Thesis, Heinrich-Heine-Universität Düsseldorf.
Roth, Ch., Kinoshita, T., Rose, H.B., Hillebrecht, F.U., and Kisker, E., 1994, J.Mag.Mag.Mat., in print.
Schmiedeskamp, G., Irmer, N., David, R., and Heinzmann, U., 1991, Appl. Phys. A53, 418 (1991).
Schneider, C.M., Bressler, P., Schuster, P., Kirschner, J., de Miguel, J.J., and Miranda, R., 1990, Phys. Rev. Lett. 64, 1059.
Schneider, Venus, D., and Kirschner, J., 1992, Phys. Rev. B 45, 5041.
Schönhense, G., 1980, Phys. Rev. Lett. 44, 640.

Sinkovic, B., and Fadley, C.S., 1985a, Phys. Rev. B 31, 4665.
Sinkovic, B., Hermsmeier, B., and Fadley, C.S., 1985b, Phys. Rev. Lett. 55, 1227.
Tamura, E., Feder, R., Krewer, J., Kirby, R.E., Kisker, E., Garwin, E.L., and King, F.K., 1985, Sol. State Comm. 55, 543.
Tamura, E., and Feder, R., 1991, Europhys. Lett. 16, 695.
Thole, B.T., van der Laan, G., 1991, Phys. Rev. Lett. 67, 3306; Phys. Rev. B 44, 12424.
Tillmann, D., Thiel, R., and Kisker, E., 1989, Z. Phys. B 77, 1.
Turtur, C., and Bayreuther, G., 1994, Phys. Rev. Lett. 72, 1557.
Unguris, J., Pierce, D.T., Celotta, R.J., 1986, Rev. Sci. Instr. 57, 131.
Venus, D., Baumgarten, L., Schneider, C.M., Boeglin, C., and Kirschner, J., 1993, J. Phys. Cond. Matter 5, 1239.
Victora, R.H., and Falicov, L.M., 1985, Phys. Rev. B. 31, 7335.
Qiu, Z.Q., Pearson, J., and Bader, S.D., 1993, Phys. Rev. Lett. 70, 1006.

SPIN DEPENDENT ELECTRON MEAN FREE PATH IN FERROMAGNETS

H. Hopster

Department of Physics and
Institute for Surface and Interface Science
University of California
Irvine, CA 92717

INTRODUCTION

The electron mean free path (MFP) plays a fundamental role in all applications of electron spectroscopic techniques in surface science. It determines the probing depth of the various methods. In the low-energy range used (typically below 1 keV) energetic electrons in solids are strongly scattered leading to a strong attenuation of the intensity with a 1/e attenuation length on the order of only a few atomic distances. In ferromagnetic materials, at least in principle, the mean free path can be expected to be spin dependent due to the spin dependence of the scattering processes. A spin dependent mean free path obviously will affect the measured polarizations in spin polarized spectroscopies, e.g., in photoemission. While there is general agreement that the mean free path of electrons in ferromagnets is indeed spin dependent at low energies the underlying mechanism is controversial. One point of concern is whether the main effects are due to a spin dependence of elastic or inelastic scattering. In this lecture I will summarize the evidence for a spin dependent electron mean free path in 3d transition metals. Some of the evidence is based on a direct measurement of spin dependent attenuation in ultrathin films whereas the evidence from other methods, like spin polarized secondary electron emission spectroscopy and spin polarized electron energy loss spectroscopy, is more indirect but agrees qualitatively with the results of the direct measurements. Spin polarized electron energy loss spectroscopy suggests that the main spin dependent scattering mechanism is due to Stoner excitations.

OVERLAYER METHOD

The attenuation of the intensity of a propagating electron "beam" in solids is described by an exponential decay exp(-x/L) where x is the distance traveled and L is the attenuation length. Although one has to distinguish between the attenuation length and the mean free path these two quantities are often equated. In addition, it is often implicitly assumed that the attenuation is due to inelastic scattering, thus equating the attenuation length with the inelastic mean free path (IMFP). A compilation of available data on many elements and compounds led to a phenomenological parametrization of the energy dependence of the IMFP by Seah and Dench[2]. This energy dependent function is known as the "universal curve" and appears in many textbooks on surface physics and electron spectroscopies and has been used extensively in the interpretation of electron spectra from solids. An important feature of this curve is that it has a minimum of about 0.5 nm in the energy range 50-100 eV and that the IMFP goes up to approximately 5 nm at very low kinetic energies, i.e. at the vacuum level. This would imply that electron spectroscopies using electrons at these very low energies probe rather deeply into the solid and therefore are not very surface sensitive.

Most of the measurements of the electron attenuation length have used the overlayer method. Here one measures the attenuation of a substrate signal (e.g., an Auger line or core level peak) as a uniform overlayer is deposited on the substrate. The decrease in intensity of the substrate signal is then fitted to an exponential decay from which the value of L in the overlayer material is determined. At very low energies some values have also been obtained by measuring directly the transmission through free-standing films. Obviously, the overlayer method can in principle be converted into a spin polarized version by choosing an overlayer with ferromagnetic order. If the intensity attenuation is spin dependent one will find a different attenuation for spin-up (majority) and spin-down (minority) electrons of the substrate peak as the electrons have to traverse the overlayer. Specifically, if the electrons from the substrate are originally unpolarized they will become polarized by the passage through the ferromagnetic film. Thus, one encounters a spin filter effect, equivalent to an optical polarization filter based on absorption. Spin resolved intensities can then be described by

$$I_+ = I_{0+} \cdot \exp(-d/L_+) \qquad I_- = I_{0-} \cdot \exp(-d/L_-)$$

where L_+, L_- are the mean free paths for up-spin and down-spin, respectively, and I_{0+}, I_{0-} the spin resolved intensities before transmission through the film. The attenuation of the total (spin integrated) intensity can then be described approximately by a single exponential

$$I = I_0 \cdot \exp(-d/L)$$

where $L = (L_+ + L_-)/2$ is the spin averaged attenuation length.

The first of these types of spin polarized overlayer experiments were performed by Pappas et. al.[3,4]. Since spin effects are expected to become less important with increasing energy these experiments were done at low kinetic energies. Instead of core levels or Auger lines the spin dependent attenuation of the Cu 3d valence band states were measured using photoemission spectroscopy at low photon energies. Ferromagnetic fcc Fe overlayers were grown epitaxially on a Cu(100) surface. In the ultrathin films used (below 5 atomic layers) the magnetization is perpendicular to the film plane. These types of experiments turn out to be experimentally quite demanding since there is a narrow range of an optimal overlayer thickness. For very small thicknesses one has a small substrate attenuation, i.e. large intensities, but the spin filter effect is small. Thus, the polarizations are small and therefore difficult to measure accurately. On the other hand, for larger thicknesses the spin filter effect leads to larger polarizations but the intensities are small, again making an accurate measurement difficult. The optimum thickness range is equal to the attenuation length. Thus, for experimental reasons the films used in Ref. 3 were restricted to a narrow range around 4 atomic layers. Figure 1 shows the experimental arrangement used. Electrons emitted normal to the surface are energy analyzed by a 90^0 electrostatic analyzer (quarter sphere) and the spin polarization is measured in a medium-energy (25 keV) retarding-field Mott detector. The spin detector consists of two orthogonal detector pairs (not shown in Fig. 1) so that the in-plane as well as the perpendicular polarization can be measured. Spin polarized photoemission spectra were taken on a bending magnet beam line at the Stanford Synchrotron Radiation Laboratory. The spin averaged attenuation lengths were determined separately by non-spin resolved photoemission using resonance lines (He I, He II, Ne I, Ar I) from a noble gas discharge lamp.

Figure 2 shows as an example the intensity spectrum (upper panel) and the measured spin polarization (lower panel) at 22 eV photon energy. The Cu 3d features are clearly visible in the spectrum in the binding energy range 2-3.5 eV. The curves labeled 'Cu' show for comparison the photoemission spectrum from the bare Cu(100) surface under otherwise identical conditions. The two spectra shown correspond to different energy resolutions. The spectrum shown as the solid line corresponds to the resolution used in the spin polarized spectrum with the Fe overlayer. It is obvious that the shape of the Cu features is not influenced by the Fe overlayer, thus validating the picture that the main effect is just an intensity attenuation. The Fe 3d states are clearly highly polarized as is

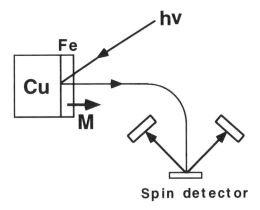

Figure 1: Schematic of the experiment

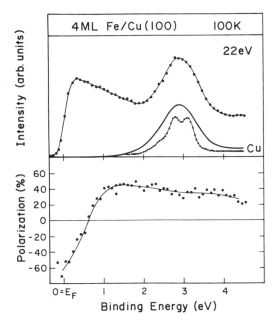

Figure 2: Intensity and polarization spectrum at 22 eV photon energy

obvious from the polarization at low binding energies. In the range of the Cu 3d peak we have two contributions. First, there are the attenuated Cu states and second, there is a background due to inelastically scattered electrons from the Fe states. The polarization spectrum already shows that the Cu 3d emission intensity must be polarized since there is hardly any decrease in the polarization in this energy range as would be expected if the Cu 3d electrons were unpolarized. From the intensity I and the polarization P one can calculate the spin resolved intensities I_+, I_- according to:

$$I_+ = I \cdot (1+P)/2 \qquad I_- = I \cdot (1-P)/2$$

These spin resolved intensity spectra are shown in Figure 3 for three different photon energies, 14 eV, 22 eV, and 44 eV. In all the spectra the Cu 3d states are clearly visible. The corresponding spectra from the bare Cu surface are indicated again as solid

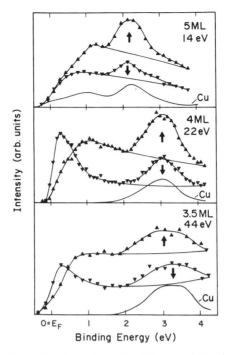

Figure 3: Spin resolved spectra at 14, 22, and 44 eV photon energy

lines. The inelastic background from the Fe overlayers is also indicated in each spectrum. From these spin resolved intensities the net polarization of the Cu 3d states can be obtained by integrating the Cu peak areas after the background has been subtracted. The non-zero net polarizations of the Cu states is very obvious in the 14 eV and 22 eV spectrum, whereas in the 44 eV spectrum the Cu 3d peak areas are nearly equal for spin-up and spin-down. This is of course within the expected trend, namely that spin effects become less important with increasing electron energy. From these net polarizations together with the total intensity attenuation of the Cu peaks one can directly determine the spin dependent attenuation length. Figure 4 shows the results of this analysis. The spin dependent attenuation length in atomic layers (1 ML=0.18 nm) is given for the three photon energies

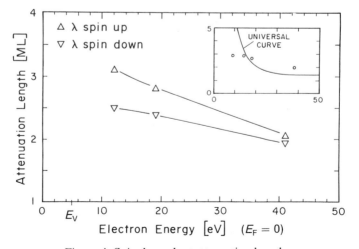

Figure 4: Spin dependent attenuation lengths

used. The line connecting the data points serves as a guide to the eye only. Clearly, at low energies the attenuation length is smaller for spin-down electrons than for spin-up electrons. The inset of Fig. 4 shows the spin averaged attenuation length determined using the four noble gas lines mentioned before. For comparison the universal curve is shown as a solid line. The large increase in L of the universal curve at very low energies is not found in the data.

It was suggested in Ref. 3 that the reason for the short MFP and for the spin dependence at low electron energies is inelastic scattering due to electron-hole pair excitations involving electron states around the Fermi level. However, it is not possible to infer the mechanism directly from the experimental data. Gokhale and Mills[5] performed spin polarized low-energy electron diffraction (SPLEED) calculations for this system. Specifically, they calculated the elastic transmission of electrons through thin layers of Fe as a function of energy and film thickness. They were able to reproduce the data at 12 eV and 19 eV based solely on spin dependent elastic scattering. The inelastic processes are described by an imaginary part of their scattering potential. The main point is that they do not need an explicitly spin dependent imaginary potential, thus contradicting the interpretation given in Ref. 3. Since their model is based on elastic diffraction effects their calculations predict distinct non-monotonic dependences on energy and film thickness. Thus, more energy dependent data than in Fig. 4 would be needed in order to test the model experimentally. Another test would be to measure the spin dependent attenuation at fixed electron energy but varying the overlayer thickness. These types of measurements have been performed by Getzlaff et al.[6]. They performed spin polarized photoemission on different systems, bcc Fe and hcp Co on W(110), for given photon energy (16,85 eV) but over a wide range of ferromagnetic layer thicknesses. The important result is that they found only a monotonic (exponential) attenuation of the substrate signal as the Fe or Co thickness increases and, in addition, that the spin polarization of the substrate signal increases monotonically with overlayer thickness. These findings are in agreement with the inelastic spin filter effect but do not support an elastic diffraction effect, as proposed by Gokhale and Mills. We should note, however, that calculations for the systems used in Ref. 6 have not been performed yet. The absolute values of the attenuation lengths found in Ref. 6 are somewhat higher than those from Fig. 4 but the spin asymmetries are quite similar. Thus, the two sets of data existing to date are in quite good agreement.

In a slight generalization of the experimental situation discussed so far, namely the polarization of originally unpolarized substrate electrons by transmission through a spin filter (ferromagnetic overlayer) we can allow for the situation in which the substrate electrons are already polarized. In this situation the spin filter effect will lead to an intensity variation depending on whether the substrate polarization and overlayer polarization are parallel or antiparallel. Exchange coupled ferromagnetic layers offer a realization of this situation. It recent years, many systems have been studied in which the magnetic coupling between two ferromagnetic layers through a spacer layer oscillates between ferromagnetic and antiferromagnetic with the thickness of the spacer layer. Unguris et. al.[7] studied the system Fe/Ag/Fe in which the Ag spacer is grown as a wedge, so that on one sample one has a continuously varying thickness of the spacer layer. Thus, as one scans across the wedge one encounters regions of ferromagnetic and antiferromagnetic alignment of the substrate and the overlayer. They use spin polarized secondary electron emission microscopy to measure the magnetization of the overlayer. They find indeed a variation in the secondary electron intensity as they scan across the wedge-shaped sample. The intensity depends on the relative alignment of substrate and overlayer only, not on the absolute direction of the overlayer or substrate magnetization individually. The intensity is larger in regions where there is ferromagnetic coupling compared to antiferromagnetic alignment. This behavior is qualitatively expected from the model of the spin filter effect, with spin-up electrons having a larger mean free path than spin-down electrons. However, the situation is too complex to extract quantitative information from these measurements since in the secondary electron spectrum the origin of the electrons is undetermined and one would have to take into account the exact mechanism of secondary electron excitations in all the layers involved, i.e. substrate, spacer layer, and overlayer.

SPIN POLARIZATION OF SECONDARY ELECTRONS

One of the earliest evidence of a spin dependent mean free path came from measurements of the spin polarization of low-energy secondary electrons, excited by UV photons or energetic electrons[8,9]. The familiar peak of true secondary electrons at very low kinetic energies is created by excitation of electrons from the valence bands. It was therefore expected that the polarization of these electrons would reflect the net magnetization of the valence bands, i.e. spin-up and spin-down electrons would be emitted according to the number electrons in the majority and minority spin bands, respectively. It turned out, however, that the polarization is strongly dependent on the energy and that the polarization at low energies (<10 eV) is much larger than the valence band polarization. It was immediately suggested that this effect is due to a spin filter effect during the transport of the excited electrons to the surface. Figure 5 shows a set of spectra of electron-excited secondary electrons from three different Fe-based metallic glasses[10]. One sees indeed that the polarization is largely enhanced at low energies, below 10 eV kinetic energy, and that it levels off to a nearly constant value above this range. The polarization value between 10 eV and 20 eV does correspond to the net (bulk) magnetization in each case.

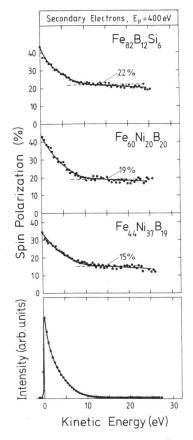

Figure 5: Intensity and polarization spectrum from three different ferromagnetic metallic glasses

The enhancement at low energies is a general feature found on all 3d transition metals. The secondary spectrum from single crystal surfaces show additional complications, however. Figure 6 shows a spectrum from a single crystalline Ni(110) surface[9]. Besides the enhancement structure at low energies there are several smaller structures, with the one around 16 eV kinetic energy being the most prominent one. These structures were explained by Tamura and Feder[11] as being due to the influence of spin dependent elastic

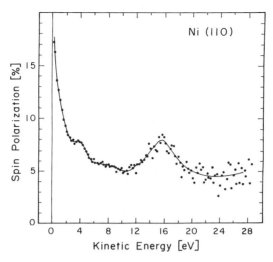

Figure 6: Polarization spectrum from Ni(100)

scattering on the emission probability of the secondary electrons (i.e. the same effects evoked by Gokhale and Mills to explain the spin dependent attenuation). The absence of similar structures on amorphous samples (as shown in Fig. 5) certainly support this picture. It is important to note that these LEED structures lead to polarizations of only a few percent, much smaller than the enhancement at low energies. Thus, together with the fact that the enhancement is "universal" on all 3d transition metal ferromagnets this leads us to believe that the enhancement is not due to elastic scattering effects but is instead due to a more general, material and structure independent cause, namely the spin dependence of inelastic scattering.

The effects of a spin dependent MFP on the polarization of secondary electrons can be estimated using a simple two-step model that takes into account the different escape probabilities of spin-up and spin-down electrons[3]. Assuming that secondary electrons are first excited with a polarization P_0, corresponding to the magnetization the spin filter effect attenuates the up and down electrons differently. The resulting polarization P is then given by $P=(P_0+A)/(1+P \cdot A)$, which approximately equals $P=P_0+A$, for A, and P_0 small. The quantity A in this equation is the spin asymmetry of the MFP: $A=(L_+-L_-)/(L_++L_-)$. One can only take estimates for the values of A at very low energies from the overlayer experiments since these data were taken at somewhat higher energies (above 10 eV). From the available data A is in the range of 20%. Thus, the polarization enhancement would be about this value. The measured maximum polarizations e.g. shown in Fig. 5 on the metglasses, are in very good agreement with this estimate. There are no data from overlayer experiments available for Ni. The enhancement of the secondary polarization over the magnetization in Ni is about 12% (17% measured versus 5.5% magnetization). Thus while the relative enhancement is very large in Ni (about a factor of 3) the absolute value is somewhat smaller than in Fe and Co. It would be interesting to test this by spin dependent overlayer experiments using Ni as a spin filter. One would expect slightly smaller spin asymmetries of the MFP than found on Fe and Co.

SPIN POLARIZED ELECTRON ENERGY LOSS SPECTROSCOPY

So far we have shown evidence that the mean free path is spin dependent and attributed the main effect to inelastic scattering but we have not yet addressed the detailed underlying scattering mechanisms. In spin polarized electron energy loss spectroscopy (SPEELS) one directly measures the spin dependences of the energy loss mechanisms. The experiment consists of scattering a monoenergetic beam of known polarization and measuring the intensity and polarization as a function of energy loss of the scattered electrons. In this way one can directly measure the four possible loss processes:

spin-up	---->	spin-up	non-flip up
spin-up	---->	spin down	flip up
spin down	---->	spin up	flip down
spin down	---->	spin down	non-flip down

SPEELS spectra have been taken on bulk Fe(110)[12] and Ni(110)[13] surfaces. Figure 7 displays a more recent example taken on thick bcc Fe(100) films grown epitaxially on Cr(100)[14].

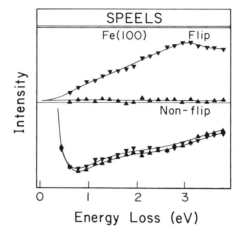

Figure 7: SPEELS spectra from Fe(100)

A general feature of all SPEELS spectra from 3d transition metals is that there is a large spin asymmetry of the flip scattering intensities. The flip scattering processes are due to exchange scattering. As seen in Fig. 7 the energy loss due to spin flip of incoming down electrons is the strongest loss process at small energy losses. The corresponding flip channel for incoming spin-up electrons is zero. The non-flip processes show no significant spin dependence in this case. The large spin dependence of the flip channels is expected from a ferromagnetic material. In theses processes the incoming electron falls into an empty state above the Fermi energy and an electron from an occupied state of opposite spin is emitted and detected as the scattered electron. Thus, the process appears as if one had essentially flipped a spin. These types of electron-hole-pair excitations, with a hole in the majority band and an electron in a minority band, are the so-called Stoner excitations. The energy loss involved to flip a spin is a measure of the exchange splitting. We note in passing that the occurrence of the flip maximum at about 2.9 eV can be taken as evidence of surface enhanced moments on this sample. Previous SPEELS data on bulk Fe(110)[12] found the maximum at 2.2 eV energy loss, which corresponds to the bulk exchange splitting of Fe.

Since there are more empty minority-spin states than majority-spin states the exchange process is more likely for spin-down electrons. The imbalance of the two flip channels leads to a different energy loss rate for spin-up and spin-down electrons, i.e. a spin asymmetry of the energy loss rate, with a higher rate for spin-down electrons. This, of course, would explain the spin dependent MFP, discussed before. However, a quantitative comparison between the SPEELS asymmetries and the spin dependent MFP is not directly possible since one would have to average the SPEELS scattering intensities over energy loss and scattering angle to arrive at the total energy loss rates. Also, the fact that the spin dependence of the scattering cross sections is mainly due to the difference of the flip processes makes the simple picture of a spin dependent mean free path more complicated. since now one has to take into account the change of spin of the scattered electrons. This means, e.g. that the simple spin filter model for the secondary electrons can be correct only qualitatively. A more realistic model would have to take into account that the scattered spin-down electrons appear as spin-up electrons after having undergone a spin-flip energy loss process. This adds another contribution to the spin polarization enhancement which is not included in the spin filter model.

In a more general context, it has also become clear that the electron mean free path in transition metals does not increase very much at very low energies, as predicted by the universal curve. The reason for this is that low-energy electron-hole pair excitations become a very important energy loss channel due to the high density of states around the Fermi level in transition elements. In a model, Siegmann[15] correlated the mean free path to the number of empty states. A comparision to available data shows that the electron scattering cross section is proportional to the number of empty d states.

CONCLUSIONS

There is ample evidence from various techniques that the electron mean free path in 3d transition metal ferromagnets is spin dependent at low energies. The most direct evidence comes from spin polarized transmission experiments through ferromagnetic films. The spin filter effects found could a priori be due to spin dependent elastic or inelastic scattering. The fact that one finds roughly the same magnitude of the spin asymmetries on very different overlayers (fcc Fe on Cu(100), bcc Fe and hcp Co on W(110)) is taken as an indication that inelastic scattering is the main cause. Also, the universal polarization enhancement of low-energy secondary electrons on 3d ferromagnetic surfaces is in favor of a more general mechanism than one based on elastic diffraction effects. Thus, we believe that the main cause is inelastic scattering due to electron-hole-pair excitations. The imbalance of spin-flip scattering due to Stoner excitations is mainly responsible for the shorter mean free path of spin-down electrons.

Financial support by the National Science Foundation is gratefully acknowledged.

REFERENCES

1. For a recent review and an historical account see e.g. C.J. Powell, Surf. Sci. 299/300, 34 (1994)
2. M.P. Seah and W.A. Dench, Surf. Interface Anal. 1, 2 (1979)
3. D.P. Pappas, K.-P. Kämper, B.P. Miller, H. Hopster, D.E. Fowler, C.R. Brundle, A.C. Luntz, and Z.-X. Shen, Phys. Rev. Lett. 66, 504 (1991)
4. Early spin polarized overlayer experiments suffered from the fact that the origin of the measured electrons could not be determined; see e.g., D.T. Pierce and H.C. Siegmann, Phys. Rev. B9, 4035 (1974)
5. M.P. Gokhale and D.L. Mills, Phys. Rev. Lett. 66, 2251 (1991)
6. M. Getzlaff, J. Bansmann, and G. Schönhense, Solid State Commun. 87, 467 (1993)
7. R.J. Celotta, J. Unguris, and D.T. Pierce, J. Appl. Phys. 75, 6452 (1994)
8. E. Kisker, W. Gudat, and K. Schröder, Solid State Commun. 44, 591 (1982)
9. H. Hopster, R. Raue, E. Kisker, G. Güntherodt, and M. Campagna, Phys. Rev. Lett. 50, 71 (1983)
10. H. Hopster, Phys. Rev. B 36, 2325 (1987)
11. E. Tamura and R. Feder, Phys. Rev. Lett. 57, 759 (1986)
12. J. Kirschner; Phys. Rev. Lett. 55, 973 (1985); D. Venus and J. Kirschner, Phys. Rev. B 37, 2199 (1988)
13. D. L. Abraham and H. Hopster, Phys. Rev. Lett. 62, 1157 (1989)
14. T.G. Walker, A.W. Pang, H. Hopster, and S.F. Alvarado, Phys. Rev. Lett. 69, 1121 (1992)
15. H.C. Siegmann, J. Phys.: Condens. Matter, 4, 8395 (1992)

Exchange interactions in magnetic surfaces and layer structures

J. Mathon

Department of Mathematics
City University
London EC1V 0HB, U.K.

The temperature dependence of the local magnetization in the spin-wave regime is discussed and it is shown that it can be used as a sensitive probe of local exchange interactions in magnetic surfaces and overlayers. The present status of the theory of oscillatory exchange interactions through transition and noble metal spacers is briefly reviewed. The quantum well theory of interlayer exchange interactions in magnetic multilayers is described and its application is illustrated for Co/Cu (001) trilayer. The relationship between oscillations in the exchange coupling and recently observed oscillations in the photoemission intensity in magnetic multilayers is explained within the framework of the quantum well theory.

Introduction

During last few years it has become possible to prepare magnetic layer structures based on transition, noble and simple metals with interfaces that are sharp on an atomic scale. This development means that one can engineer materials exhibiting new magnetic phenomena with great potential for novel devices [1]-[4]. To exploit these opportunities to the full good theoretical understanding of new multilayer systems is essential.

Perhaps the most exciting systems are sandwiches and superlattices with magnetic layers separated by nonmagnetic spacer layers. Typical examples are Fe/Cr, Co/Cr, Co/Ru and Co/Cu layer structures. It is found experimentally [4]-[10] that the magnetic moments of neighbouring magnetic layers are aligned ferromagnetically or antiferromagnetically depending on the thickness of the spacer layer. This implies an exchange coupling mediated by itinerant electrons in the spacer layer which oscillates as a function of its thickness. Since the strength of the interlayer exchange coupling can be measured by a number of experimental methods [4]-[10], magnetic multilayers provide a unique opportunity for studying exchange interactions in metals. Furthermore, the resistance of the structure is much higher in the antiferromagnetic configuration than in the ferromagnetic one. The relative change of the resistance can be very large, the current record is over 100% for Co/Cu structures. Change of the magnetic configuration from antiferromagnetic to ferromagnetic, and hence a change of the resistance, can be effected by an applied magnetic field and this mechanism can be exploited to read information from a magnetic disc.

Magnetic properties of transition metal ferromagnets are dominated by holes in the d band and both the exchange coupling and giant magnetoresistance effects can be understood in terms of different distributions of up- and down-spin d carriers (electrons or holes) in different parts of the layer structure. The total energy and hence the exchange coupling is determined directly by the exchange splitting of the d bands in the ferromagnet and the transport properties more indirectly through the Mott mechanism in which conduction electrons scatter into d band. For a simple account of the giant magnetoresistance effect see, for example, Ref.11.

The influence of layering on local exchange interactions was first studied for magnetic surfaces and overlayers. In the first part of this paper, I shall, therefore, describe how measurements of the temperature dependence of the local magnetisation can be combined with spin wave analysis to extract information about local exchange in magnetic surfaces and overlayers.

In the second part of the paper, the quantum well theory of oscillatory exchange coupling will be explained and illustrated for Co/Cu (001). The relationship between oscillations in the exchange coupling and oscillations in the photoemission intensity, that were recently observed [12, 13] in overlayers of noble metals on Co and Fe substrates, will be also discussed.

Temperature dependence of the local magnetization as a probe of local exchange interactions

Although spectacular progress has been made over the past ten years in calculating from first principles the ground state moment for magnetic surfaces and interfaces [14], it is not always easy to verify the theoretical predictions since it is very difficult to measure the absolute value of the local moment M(0) (see Ref.15). On the other hand, the relative change of the local moment, i.e. its temperature dependence $M(T)/M(0)$ can be determined very accurately by several methods such as photoemission, SPLEED, spin-polarized secondary electrons, Mossbauer spectroscopy, SMOKE, ECS (see eg., Refs.15-17). Since it is well known that $M(T)$ in bulk ferromagnets is a direct measure of exchange interactions, it is natural to ask whether we can exploit the high spatial resolution (on atomic scale) of modern experimental methods to determine from the measured local $M(T)$ the strength of local exchange interactions in magnetic surfaces and overlayers.

To extract information about local exchange from $M(T)$, we need a good theory of the temperature dependence of the magnetisation. One region where we have a good theory is the spin wave regime $0 < T < T_C/3$, where T_C is the Curie temperature. It has, therefore, been proposed [17, 18] to use the following strategy to access local exchange via spin waves.

First, a model exchange Hamiltonian H for a layer structure is assumed. Next the density of spin-wave states $N_i(E)$ in any atomic plane of the structure is determined from the local spin-wave Green function $G = (E - H)^{-1}$

$$N_i(E) = -\frac{1}{\pi} Tr\, Im\, G_{ii}(E, \vec{q}_{\|}) \tag{1}$$

where the trace is over the wave vector $\vec{q}_{\|}$ parallel to the structure surface and i labels atomic planes in the structure. Since spin waves are bosons, the temperature-induced reduction of the local magnetisation in any atomic plane i is given by

$$M_i(0) - M_i(T) = \int_0^{+\infty} 2\mu_B N_i(E)[exp(E/kT) - 1]^{-1} dE \tag{2}$$

where μ_B is the Bohr magneton.

Finally, the exchange parameters in the trial Hamiltonian H are deduced from the best fit of Eq.(2) to the observed $M(T)$.

To calculate the local spin-wave Green function we have developed a method which can be viewed as the theoretical equivalent of an MBE machine. One starts with a homogeneous semi-infinite ferromagnetic substrate with exchange J and spin S whose surface spin-wave Green function (density of states) is known [21, 22] from the classical spin wave theory. The next step is to "deposit" one-by-one all the atomic planes of the structure we wish to study. With every new adlayer the surface Green function is updated using the Dyson equation. At every stage of the deposition process we have, therefore, the exact surface Green function G^s expressed in terms of the known substrate Green function G^0. When the layer structure is complete we stop the deposition process and evaluate the local Green function in every atomic plane from the exact G^s (see Ref.20). We have demonstrated elsewhere [19, 20] that the method is so fast and accurate that the local spin-wave Green function for magnetic multilayers of $10^3 - 10^4$ at. planes can be calculated essentially exactly. It follows that one can determine from Eqs.(1) and (2) the local temperature dependence $M_i(T)$ for any magnetic multilayer with an arbitrary distribution of exchange integrals and with an arbitrary distribution of local spin.

The whole strategy outlined above is based on the hypothesis that a variation of local exchange on an atomic scale is reflected in the local $M(T)$. This can be easily tested for a ferromagnetic surface.

Consider first a perfect surface with no changes to the surface exchange and spin. Classical spin wave theory [21] predicts that the surface magnetisation $M_s(T)$ obeys a Bloch law

$$M_s(T)/M_s(0) = 1 - kCT^{3/2} \tag{3}$$

with k=2 (C is the bulk prefactor), i.e. the surface magnetisation should decrease twice as fast as in the bulk. Measurements of $M_s(T)$ for ferromagnetic metals using SPLEED [23], Mossbauer spectroscopy [24] and the polarization of secondary electrons [25, 26] confirm that M obeys a Bloch law but with a prefactor which is different for different surfaces and can be as large as k=5.7 (see Ref.27).

It is natural to assume that a faster decrease of $M_s(T)$ is due to a softenig of surface exchange. Such a softening can be clearly modelled by an overlayer with a suitable distribution of exchange deposited on a homogeneous ferromagnetic substrate.

The method of adlayers [19] gives the exact initial surface density of spin wave states (DOS) for an overlayer consisting of N atomic plabes with arbitrary (ferromagnetic) exchange interactions within and between the planes and with an arbitrary local spin S_i in each plane

$$N_s(E) = (S_N/S_{N-1})...(S_1/S)2N_B(E) = (S_N/S)2N_B(E) + O(E^{3/2}) \quad (4)$$

where $N_B(E) \propto E^{1/2}$ is the bulk DOS and S_N is the surface spin.

The most remarkable feature of the DOS in Eq.(4) is that it is quite independent of the exchange in the overlayer. Moreover, normalisation to $M_s(0)$ in Eq.(3) means that the surface spin is cancelled and we recover from Eqs.(2) and (4) the classical result k=2. The spin wave theory, therefore, predicts that the surface magnetisation should initially always decrease twice as fast as in the bulk irrespective of the surface exchange and spin. Not only this exact result contradicts the experiment but it also seems to indicate that no information about local exchange can be gained from the Bloch law for $M_s(T)$.

Fortunately, quite a different picture emerges when the surface DOS is computed over a wider energy range. This is illustrated in Fig.1a for an overlayer consisting of a single atomic plane which is coupled to the substrate by a weaker exchange $J_\perp/J = 1.0, 0.6, 0.3, 0.1$ (J is the bulk exchange). It can be seen that the classical result (4) breaks down at energies as low as $0.01kT$. The surface DOS at experimentally relevant energies is determined entirely by the surface to bulk exchange $J_\perp$ and it increases rapidly as the exchange $J_\perp$ weakens. We expect, therefore, that $M_s(T)$ should decrease more rapidly with increasing temperature. This is illustrated in Fig.1b, where $M_s(T)$ calculated from Eqs.(2) and (4) is plotted against the bulk magnetisation $M_B(T)$ for the same range of values of $J_\perp/J$ as in Fig.1a.

The results shown in Fig.1b allow us to make two firm theoretical predictions: (i) $M_s(T)$ for a ferromagnetic surface with a weaker surface to bulk exchange should always obey a good $T^{3/2}$ law; (ii) the prefactor k in this "second" $T^{3/2}$ law should increase with decreasing $J_\perp$.

These predictions of the spin-wave theory were tested by Mauri et al. [27] using secondary electrons. They first deposited on a clean permalloy surface a thin spacer layer of nonmagnetic Ta and then covered it with a very thin permalloy film. The Ta spacer layer thus formed a weak exchange link between the bulk and surface of permalloy whose strength could be controlled by varying the thickness of Ta. The experimental results of Mauri et al. [27] are reproduced in Fig.2., where the curves a, b, c correspond to a progressively weaker exchange link (thicker Ta interlayer; 0.5, 1.0 and 1.5 at. layers) and the curve 0 is for clean permalloy surface. The curves a-c demonstrate clearly that $M_s(T)$ decreases faster with decreasing $J_\perp$ following in all cases a good $T^{3/2}$ law with prefactors k=2.5, 4.1, and 5.7.

This pioneering work on magnetic overlayers shows that surface exchange interactions on the path perpendicular to the surface can be engineered to be weaker than in the bulk by the introduction of a suitable spacer layer. On the other hand, when a permalloy surface is covered with a submonolayer of Fe, it is found [28] that the surface magnetization decreases with increasing temperature much more slowly than

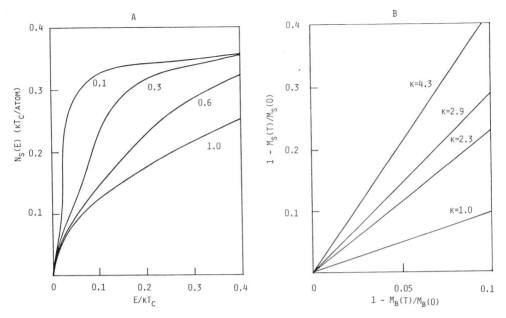

Figure 1: Surface density of spin wave states (a) and the plots of $M_s(T)$ against the bulk $M_B(T)$ (b) for an overlayer with a weaker surface to bulk exchange $J_\perp/J = 1.0, 0.6, 0.3, 0.1$.

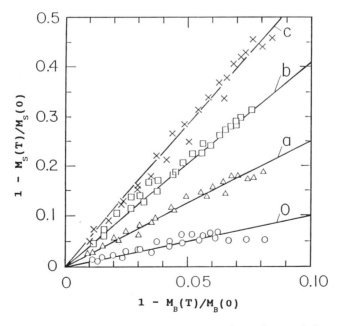

Figure 2: Experimental results for the temparature dependence of the surface magnetization in FeNi/Ta/FeNi overlayers.

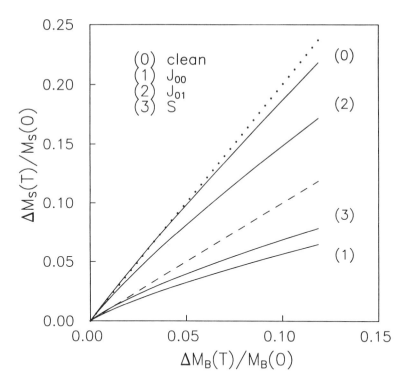

Figure 3: Temperature dependence of the surface magnetization for a surface with strengthened parallel exchange (curve 1), perpendicular exchange (curve 2) and for an enhanced surface spin (curve 3).

for pure permalloy surface. Using again spin wave analysis of $M_s(T)$, this was interpreted [28] as being due to an enhancement of the exchange in the surface layer $J_\parallel$. A value of the exchange $J_\parallel$ of approximately three times greater than the bulk J would be required to account for the observed $M_s(T)$. An alternative interpretation is that the faster temperature dependence of $M_s(T)$ is due to a higher value of the surface spin of the Fe atoms coating the surface. Figure 3 illustrates the effects on $M_s(T)$ of strengthening of $J_\parallel$ (curve 1 corresponds to $J_\parallel = 3J$) and of $J_\perp$ (curve 2 corresponds to $J_\perp = 4J$), and also the effect of an enhancement of the surface spin (curve 3 corresponds to $S^{surf} = 2S$). For comparison, the bulk $T^{3/2}$ law is indicated in Fig.3 by a dashed line, curve 0 is the computed result for a clean surface and the dotted line is the classical Rado result. It can be seen that strengthening of $J_\perp$ has only a small effect but an enhancement of the surface spin by a factor of two can itself account for the observed effect of Fe coating. Since the iron moment is almost exactly twice as large as that of permalloy, this may well be a more plausible interpretation of the experimental results of Ref.28.

Determination of the local exchange in magnetic overlayers by the method I have described, requires sophisticated experimental probes of the local magnetization with a resolution on an atomic scale and a careful spin wave anlysis of $M(T)$ to separate the surface and bulk contributions. Moreover, as the discussion of the permalloy surface coated with iron illustrates, one also needs some information about the surface spin in the ground state. These problems do not arise for magnetic sandwiches and superlattices because we do not have here the bulk background. Sandwiches and superlattices with magnetic layers separated by nonmagnetic spacer layers are, therefore, even more promising candidates for studying interlayer exchange.

Quantum well theory of oscillatory exchange interactions through a non-magnetic spacer layer

As already discussed in the Introduction, it is observed that the magnetic moments of neighbouring magnetic layers in layered structures consisting of alternating ferromagnetic and non-magnetic metals are aligned parallel or antiparallel depending on the thickness of the non-magnetic spacer layer. This implies an exchange coupling $J(N)$ between the magnetic layers which oscillates as a function of the spacer layer thickness N. Multilayers which exhibit such an oscillatory coupling are typically based on ferromagnetic transition metals and the non-magnetic spacer layers are usually transition, noble or simple metals. The oscillation periods in multilayers with noble and transition metal spacers can be as long as 18Å for Cr but short periods have also been observed [4]-[10] for Cu, Au and Cr spacers.

The strength of the coupling (amplitude) decreases asymptotically as $1/N$ and the maximum strength depends strongly on the spacer material. In fact, it is found [6] that for a given ferromagnet (Co or Fe), the maximum strength decreases exponentially when spacer materials on the right of the periodic table (Rh, Ru) are substituted by those on the left (e.g. Ti, Zr or Hf). The strength also decreases when one moves down in the periodic table from 3d to 5d metals.

A number of theories have been proposed to explain the oscillatory exchange coupling in magnetic multilayers. The most direct way would seem to be to compute numerically the total energies of the ferromagnetic and antiferromagnetic configurations of the magnetic layers and determine the interlayer coupling as the difference between them. Quite apart from the fact that such an approach does not bring any physical insight, it is numerically very difficult because the total energy difference is only a small fraction of the energy of each configuration. In fact, the exchange coupling per surface atom can be as weak as $10^{-6} - 10^{-7} E_F$, where E_F is the Fermi energy. Direct computations of the exchange coupling may, therefore, be only feasible for very thin spacer (and ferromagnetic) layers [29, 30].

Another approach that has been developed treats the effect of two atomic planes of localised spins modelling the ferromagnetic layers as a weak perturbation to the bulk spacer metal. It is just an adaptation of the conventional RKKY theory for dilute impurities to the layer geometry [31, 32]. Such an RKKY-type theory is useful in that it relates in a simple way the oscillation periods to the spacer layer bulk Fermi surface [33]. However, RKKY type theories cannot be used to determine the coupling strength. Moreover, the fundamental problem with all calculations based on perturbation theory is that they fail in the presence of bound states. This is not a serious problem with conventional RKKY since sufficiently weak three-dimensional impurity potentials do not create bound states. However, in the layer geometry one deals with quasi-one-dimensional potentials that are always binding and perturbation theory is, therefore, not applicable. In fact, it will be shown here that even for Co/Cu system, which has been regarded as a typical candidate for RKKY treatment, bound states are created in the Cu layer and RKKY theory thus fails.

The third, although historically first, approach is the quantum well theory of exchange coupling [34, 35]. It is a nonperturbative calculation of the total energy difference which exploits the fact that a non-magnetic spacer layer inserted between two thick ferromagnetic layers creates a quasi-one-dimensional potential well (barrier) for electrons of a given spin orientation travelling across the layer structure. Because the up- and down-spin electrons in the ferromagnet see different potentials (see, e.g. Ref. 11) the potential wells in the spacer layer are spin dependent and have, therefore, different effects on the total energy of the parallel and antiparallel configurations of the ferromagnetic layers. The key result of the quantum well theory is that quantum well states or resonances created in the spacer layer cross periodically the Fermi surface as the thickness of the spacer layer is varied. Such a periodic passage of quantum well states across E_F is the physical origin of oscillations in the exchange coupling. The mechanism involved is quite analogous to de Haas-van Alphen oscillations. Asymptotically exact analytic results for the exchange coupling can be thus derived [34, 35] using essentially the same formalism as in the theory of dHvA effect [36].

The quantum well formalism will be now described for a trilayer consisting of two thick ferromagnetic layers separated by N atomic planes of a non-magnetic spacer layer. The ferromagnetic layers are Fe, Co or Ni and the spacer can be any non-magnetic transition, noble or simple metal. Electrons in the trilayer are described by a tight-binding Hamiltonian

$$H = \sum_{i,j,\alpha,\beta,\sigma} t_{i\alpha j\beta} c^+_{i\alpha\sigma} c_{j\beta\sigma} + \sum_{i,\alpha} U_{i\alpha} n_{i\alpha\uparrow} n_{i\alpha\downarrow} \qquad (5)$$

where the first term includes the electron kinetic energy and a spin-independent local atomic potential, $c^+_{i\alpha\sigma}$ ($c_{j\beta\sigma}$) creates (annihilates) an electron of spin σ in an orbital α (β) on a site i (j), and $n_{i\alpha\sigma} = c^+_{i\alpha\sigma} c_{i\alpha\sigma}$. Finally, $U_{i\alpha}$ is the intra-atomic Coulomb integral which is non-zero only in the ferromagnet. The tight-binding parameters $t_{i\alpha j\beta}$ and local potentials in the ferromagnetic and non-magnetic layers are chosen to reproduce the respective bulk band structures of the constituent metals determined from first-principle spin-density functional calculations [37]. This leads to a fully realistic band structure of the trilayer and leaves no adjustable parameters in the theory.

The interlayer exchange coupling $J(N)$ is defined as the total energy difference per unit area between the ferromagnetic (FM) and antiferromagnetic (AF) configurations of the magnetic layers. It is given in terms of the thermodynamic potentials Ω^σ for electrons of spin σ by

$$J(N) = [(\Omega^\uparrow(N) + \Omega^\downarrow(N))_{FM} - (\Omega^\uparrow(N) + \Omega^\downarrow(N))_{AF}]/A \qquad (6)$$

where A is the cross-sectional area of the trilayer. The thermodynamic potential for a given magnetic configuration at temperature T is given by

$$\Omega^\sigma = -k_B T \int_{-\infty}^{+\infty} \ln\{1 + \exp[(\mu - E)/k_B T]\} \mathcal{N}^\sigma(E, N) dE \qquad (7)$$

where μ is the chemical potential and $\mathcal{N}^\sigma(E, N)$ is the total density of states for particles of spin σ in the trilayer having that configuration. Because of the translation invariance in the direction parallel to the layers, we can express the total density of states in terms of the spectral (partial) density $\mathcal{D}^\sigma(E, \vec{k}_\parallel, N)$

$$\mathcal{N}^\sigma(E, N) = \sum_{\vec{k}_\parallel} \mathcal{D}^\sigma(E, \vec{k}_\parallel, N) \tag{8}$$

Finally, the spectral density itself is given by

$$\mathcal{D}^\sigma(E, \vec{k}_\parallel, N) = -(1/\pi) \sum_i Tr \, Im G_{ii}(E, \vec{k}_\parallel, N) \tag{9}$$

where $G_{ii}(E, \vec{k}_\parallel, N)$ is the local one-electron Green function in the atomic plane i, the summation is over all atomic planes in the trilayer, the trace is over all atomic orbitals and the wave vector $\vec{k}_\parallel$ parallel to the layers is from the first two-dimensional Brillouin zone.

For any fixed $\vec{k}_\parallel$ the calculation of the spectral density thus reduces to an effective one-dimensional problem. Within the tight-binding scheme (5), the spacer differs from the ferromagnetic layers only by its atomic potentials and hopping integrals $t_{i\alpha j\beta}$, and its effect on magnetic carriers of a given spin is, therefore, equivalent to the effect of a quasi-one-dimensional potential well or barrier. A potential barrier for electrons amounts to a potential well for holes and vice versa. The descriptions in terms of electrons and holes are completely equivalent. We use the term magnetic carrier to describe either holes or electrons bearing in mind that in most cases they are holes experiencing a potential well in the spacer region. Because there is a finite exchange splitting between the majority and minority bands in the ferromagnetic layers, up- and down-spin carriers moving across the structure experience different potential wells in the FM and AF configurations of the trilayer. There are clearly two symmetric wells of different depths for up- and down-spin carriers in the FM configuration and two equivalent asymmetric wells for particles of either spin in the AF configuration. They are shown schematically in Fig.4. The depths of the potential wells for specific materials can be easily obtained from the offsets of the bulk bands of the two metals forming a trilayer. This is illustrated in Fig.5 for Co/Cu(001) trilayer in the case of carriers travelling in the [001] direction perpendicular to the layers. In this case, the wave vector is $\vec{k}_\parallel = 0$ and the relevant band structure is, therefore, in the $\Gamma - X$ direction. It can be seen that the bands of fcc Co and Cu are very similar, they are essentially merely displaced vertically relative to one another and it is this displacement (band offset) that determines the degree of confinement in the spacer layer potential well.

Insight gained from the quantum well description is useful but, at this stage, the numerical evaluation of Ω^σ for a fully realistic tight-binding band structure (5) remains as difficult as any direct total energy calculation, i.e. computationally not feasible. The key observation that allows us to carry out the $\vec{k}_\parallel$ and energy sums in Ω^σ analytically, and thus render the problem tractable, is that the spectral density $\mathcal{D}^\sigma(E, \vec{k}_\parallel, N)$ oscillates as a function of N. The physical origin of the oscillations is easy to understand. All states in the trilayer can be classified into propagating states and bound quantum well states localized in the spacer potential well. Consider first the bound states. The spectral density $D^\sigma(E, \vec{k}_\parallel, N)$ measures the probability that an electron of spin σ, energy E and parallel wave vector $\vec{k}_\parallel$ is found in the well region. It

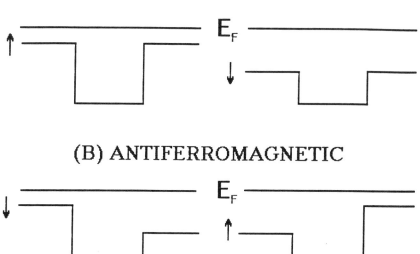

Figure 4: Potential wells for up- and down-spin magnetic carriers in the ferromagnetic and antiferromagnetic configurations of the trilayer.

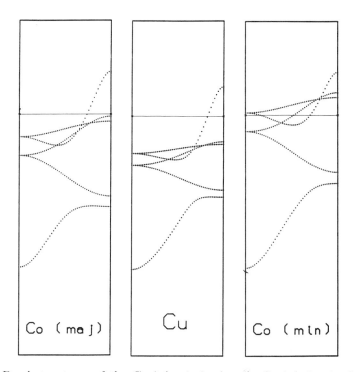

Figure 5: Band structures of the $Co \uparrow$ (majority band), $Co \downarrow$ (minority band) and Cu in the direction perpendicular to the layers of a Co/Cu(001) trilayer.

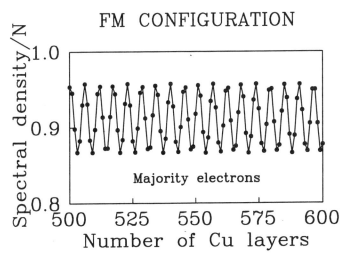

Figure 6: Calculated dependence of the normalized spectral density of up-spin carriers in Co/Cu(001) trilayer in its ferromagnetic configuration on the thickness of the Cu layer.

is convenient to define an electron wavelength $\lambda_\perp$ in the direction perpendicular to the layers by $\lambda_\perp = 2\pi/k_\perp$, where $k_\perp$ is the perpendicular wave vector. For given E and $\vec{k}_\parallel$, the wavelength $\lambda_\perp$ is fixed and the spectral density contains a delta function peak whenever an integral number of half-wavelengths fits in the well, i.e. $n(\lambda_\perp/2) = Nd$, where $n = 1, 2, ...$ and d is the distance between two neighbouring atomic planes in the spacer. It follows that $\mathcal{D}^\sigma(E, \vec{k}_\parallel, N)$ is a periodic array of delta functions with a period $\lambda_\perp/2$.

The propagating states do not differ qualitatively from the bound states since they can be regarded as resonances in the well region having a finite lifetime. Electron waves at energies above the edge of a potential well undergo multiple scattering from the well edges and all the partial waves interfere constructively only if $n(\lambda_\perp/2) = Nd$, which is the same condition as for the bound states. It follows that $D\mathcal{D}^\sigma(E, \vec{k}_\parallel, N)$ is now a periodic array of finite peaks but the period remains the same as in the case of true quantum well states, i.e. $\lambda_\perp/2$.

The essential mathematical device which allows us to treat the bound states and resonances on the same footing is the representation of the spectral density per atomic plane of the spacer $(1/N)D\mathcal{D}\sigma(E, \vec{k}_\parallel, N)$ by a Fourier series in N. It should be noted that the spectral density itself cannot be so represented because it oscillates about its average value which is proportional to the spacer thickness.

As a typical example, the periodic behaviour of the majority spin spectral density $(1/N)\mathcal{D}^\sigma(E, \vec{k}_\parallel, N)$ for a Co/Cu(001) trilayer in its FM configuration is shown in Fig.6. The spectral density was computed by the method of adlayers described in Sec.2 for a fully realistic tight-binding band structure of the Co/Cu trilayer with s, p and d bands and hopping to nearest and second-nearest neighbours. The value of $\vec{k}_\parallel = 0$ was chosen so that the oscillations of the calculated spectral density can be compared with the observed oscillations [12, 13] of the perpendicular ($\vec{k}_\parallel = 0$) photoemission intensity from an overlayer of Cu on Co(001). We recall that the spectral density is the principal factor that controls the angle-resolved photoemission intensity. Both the photoemission intensity [12, 13] and the calculated spectral density oscillate with the same period of about 5.7 atomic planes.

Having established that the spectral density per atomic plane of the spacer is a periodic function of N, we can use its Fourier representation to write the coupling given by Eqs. (6) and (7) in the form

$$J(N) = -k_B T N \, Re \sum_{p=1}^{\infty} \sum_{\vec{k}_\parallel} \int_{-\infty}^{+\infty} \ln\{1 + \exp[(\mu - E)/k_B T]\} \mid c_p \mid e^{2i(pN k_\perp d + \psi_p)} dE$$

(10)

where $\pi/k_\perp(E, \vec{k}_\parallel)$ is an oscillation period and $\mid c_p(E, \vec{k}_\parallel) \mid$ and $\psi_p(E, \vec{k}_\parallel)$ are the modulus and phase of the Fourier coefficient of the normalized spectral density difference $(1/N)\Delta \mathcal{D}(E, \vec{k}_\parallel, N) = (1/N)\{[\mathcal{D}^\uparrow + \mathcal{D}^\downarrow]_{FM} - [\mathcal{D}^\uparrow + \mathcal{D}^\downarrow]_{AF}\}$. The term $p = 0$ is absent since it corresponds to the bulk contributions to the thermodynamic potentials of the FM and AF configurations which cancel in Eq.(6).

It might seem from an inspection of Eq.(10) that $J(N)$ is a superposition of a large number of oscillations with periods $\pi/k_\perp(E, \vec{k}_\parallel)$, where each period depends on the energy and parallel wave vector. However, for large spacer thickness N, the imaginary exponential in Eq.(10) oscillates rapidly as a function of $\vec{k}_\parallel$ and the dominant contribution to the integral with respect to $\vec{k}_\parallel$ comes from the vicinity of the points $\vec{k}_\parallel^0$ at which the period reaches an extremum. We can, therefore, evaluate the $\vec{k}_\parallel$ integral in Eq.(10) analytically using the stationary phase approximation [34, 35]. The energy integral in Eq.(10) can also be evaluated analytically for large N. In fact, having made the stationary phase approximation in the $\vec{k}_\parallel$ integral, we are left with an imaginary exponential $e^{2ipNdk_\perp(E,\vec{k}_\parallel^0)}$ in the energy integral. For large N the exponential oscillates rapidly as a function of E. This results in cancellations and the only contribution to the energy integral thus comes from energies in the vicinity of the chemical potential μ at which the integral terminates abruptly. One, therefore, arrives [35] at the following asymptotic formula valid for large spacer thickness (in practice, $N > 5, 6$ is large):

$$J = Re \sum_{\vec{k}_\parallel^0} \sum_{p=1}^{\infty} \frac{\tau}{2p} \frac{c_p e^{2ipNdk_\perp}}{(2pNd\frac{\partial k_\perp}{\partial E} + \frac{\partial \psi_p}{\partial E})} \frac{k_B T d}{\sinh[\pi k_B T (2pNd\frac{\partial k_\perp}{\partial E} + \frac{\partial \psi_p}{\partial E})]} \left| \frac{\partial^2 k_\perp}{\partial k_x^2} \frac{\partial^2 k_\perp}{\partial k_y^2} \right|^{-1/2}$$

(11)

The sum over $\vec{k}_\parallel^0$ covers all the stationary points of $k_\perp$, $\tau = i$ when both second derivatives in Eq.(11) are positive, $\tau = -i$ when they are negative, and $\tau = 1$ when the derivatives have opposite signs. The perpendicular wave vector $k_\perp$, Fourier coefficients c_p and all the derivatives in Eq.(11) are evaluated at $E = \mu$ and at the stationary point $\vec{k}_\parallel = \vec{k}_\parallel^0$. It can be seen that the main effect of the stationary phase approximation is to select from all possible periods the extremal periods, and these are the only ones that are observed.

The principal results following from Eq.(11) are: (i) The oscillation periods of the exchange coupling $J(N)$ are determined by the extremal radii of the bulk spacer Fermi surface (FS) in the direction perpendicular to the layers. (ii) The amplitude of $J(N)$ decreases at zero temperature as $1/N^2$, which is as observed. (iii) The temperature dependence of the coupling is determined by the Fermi velocity of carriers at the extremal point. (iv) The overall coupling strength is controlled by two factors. The first factor is the curvature and carrier velocity at the spacer FS and the second one is the the difference between the degrees of confinement in the FM and AF configurations (magnetic contrast). The latter factor depends on band matching across the trilayer and is contained entirely in the Fourier components of the spectral density. Similarly, the phase of oscillations in the exchange coupling is determined not only by the type of spacer FS extremum (the factor τ) but also by band matching across the interfaces (the Fourier coefficient $c_p(E, \vec{k}_\parallel)$).

The great advantage of calculating $J(N)$ from Eq.(11) is that we have a nonperturbative total energy calculation but the prohibitive computational effort involved in evaluating the Brillouin zone and energy sums has been completely eliminated. In fact, we only need to know the one-particle states of the trilayer at $E = \mu$ and for the parallel wave vector $\vec{k}_\parallel^0$ at which the oscillation period of the spectral density reaches its extremum.

We now illustrate the application of the general formula (11) to a Co/Cu(001) trilayer. First of all, we need to determine the oscillation periods $\pi/k_\perp(\mu, \vec{k}_\parallel^0)$. For the (001) orientation of the layers, $k_\perp(\mu, \vec{k}_\parallel^0)$ are simply the extremal radii of the Cu FS in the direction perpendicular to the layers associated with the FS extrema at $\vec{k}_\parallel^b = (0,0)$ (FS belly) and $\vec{k}_\parallel^n a = (\pm 2.53, \pm 2.53)$ (necks), where a is the lattice constant of Cu. The corresponding periods are 5.7 (belly) and 2.6 atomic planes (necks).

The FS curvature and the derivative $\partial k_\perp / \partial E$ which are required in Eq.(11) can be readily obtained from the bulk band structure of Cu at the extremal points $\vec{k}_\parallel^0$ of its FS.

The only remaining problem is to calculate the Fourier components of the spectral density $(1/N)\Delta \mathcal{D}(E, \vec{k}_\parallel^0, N)$. We recall that it is this term which controls the overall coupling strength for each ferromagnet-spacer combination. The Fourier components are determined numerically from the spectral density which is calculated by the method of adlayers [20] adapted to the multi-orbital tight-binding band structure of bulk Cu and ferromagnetic fcc Co [37]. The up-spin spectral density in the FM configuration $(1/N)\mathcal{D}^\uparrow(N)$ for the belly extremum $\vec{k}_\parallel^0 = 0$, obtained by this method, has already been shown in Fig.6. For each extremum $\vec{k}_\parallel^0$, three spectral densities are required, i.e. $[\mathcal{D}^\uparrow(\mu, \vec{k}_\parallel^0, N)]_{FM}$, $[\mathcal{D}^\downarrow(\mu, \vec{k}_\parallel^0, N)]_{FM}$, and $[\mathcal{D}^{\uparrow,\downarrow}(\mu, \vec{k}_\parallel^0, N)]_{AF}$. These are evaluated for a large Cu thickness ($N \cong 500 - 600$) and typically a hundred values of N are used to Fourier analyze each component of the normalized spectral density.

All the steps described above have been fully implemented [38] for both the belly ($\vec{k}_\parallel^b = (0,0)$) and neck ($\vec{k}_\parallel^n a = (\pm 2.53, \pm 2.53)$) extrema of the Cu Fermi surface. The total calculated coupling $J(N)$ is shown in Fig.7. The left-hand scale in Fig.7 gives the coupling in mRy per atom in the (001) surface. The right-hand scale gives the conversion to the units (mJ/m^2) commonly used by experimentalists. To clarify the relative importance of the long-period (FS belly) and short-period (FS necks) oscillations, the belly contribution is shown separately in the inset in Fig.7. It can be seen that the magnitude of the total calculated coupling at the first antiferromagnetic peak is in good quantitative agreement with the experiment value [8] of about $0.4 mJ/m^2$. The calculated results also show that the coupling is completely dominated by the

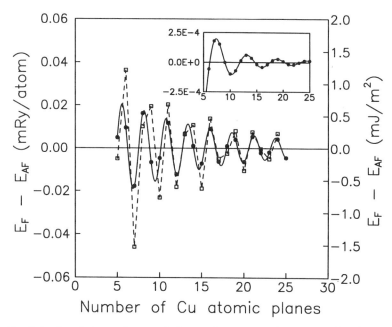

Figure 7: Calculated dependence of the exchange coupling in Co/Cu(001) trilayer on the thickness of the Cu layer. The contribution of the Cu Fermi surface belly is shown in the inset.

short-period neck component. To understand this rather surprising result, we have to examine all the components of the spectral density that contribute to the coupling. In the case of the belly extremum, we find that the peaks in all the components $\mathcal{D}^\sigma$ of the spectral density correspond to broad resonances, as illustrated in Fig.6. This implies a weak partial confinement in both spin channels in the FM and AF configurations and, therefore, a weak coupling.

We now turn to the short-period neck contribution which is much more interesting. Examination of all three components of the spectral density reveals that carriers of both spin orientations in the AF configuration, and also the malority spin carriers in the FM configuration, are only weakly confined (broad resonances). However, the minority spin carriers become completely confined in a deep quantum well in the FM configuration. Their spectral density is a set of delta functions and this is demonstrated in Fig.8, which shows our Fourier series fit to the normalized spectral density in the first period (continuous Cu thickness is measured in units of interatomic distance). Strong confinement in only one of the three channels leads necessarily to a strong coupling since the contribution of all the other channels, that might compensate one another in Eq.(6), is negligible.

The example of a Co/Cu(001) trilayer illustrates the importance of the compensation effect (or lack of it!) in Eq.(6), which clearly depends on the degree of confinement in the two spin channels in the FM and AF configurations. In fact, this compensation mechanism was invoked by Mathon et al. [39] to explain the trends in the coupling strength observed by Parkin [6].

We have described the general formulation of the quantum well theory of oscil-

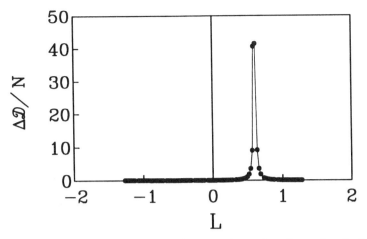

Figure 8: Spectral density of down-spin (minority) carriers from the Cu Fermi surface neck region which are fully confined in Co/Cu(001) trilayer in its ferromagnetic configuration.

latory exchange coupling valid for an arbitrary trilayer in which the magnetic layers are Fe, Co or Ni and the non-magnetic layers are transition, noble or simple metals. It has been shown that the underlying physical mechanism that determines the coupling are oscillations of the spectral density which occur when the thickness of the spacer is varied. Since the spectral density also controls the photoemission intensity, we have established a direct link between oscillations in the exchange coupling and in the photoemission intensity observed [12, 13] for overlayers of noble metals on Co and Fe.

Application of the quantum well theory to Co/Cu (001) shows that the long-period oscillation of the exchange coupling is determined by the belly extremum of the Cu FS and the oscillation period is in perfect agreement with the observed period of oscillations of the photoemission intensity [12, 13]. However, the total coupling is dominated by the short-period belly contribution whose large strength is due to complete confinement of minority-spin carriers in the ferromagnetic configuration.

Although the success of the calculation for a Co/Cu(001) trilayer confirms quantitative validity of the quantum well formulation, the theory is far from complete. It needs to be applied in future to other systems and other crystal orientations and should be also generalized to include the effect of an interfacial roughness. In particular, we expect quantum well calculations for different crystal orientations to be very interesting. Moreover, since such calculations provide, as a byproduct, oscillatory spectral densities that are directly accessible to photoemission and possibly other spectroscopic experiments, they should lead to fruitful interplay between theory and experiment.

Acknowledgements

Most of the theoretical ideas and calculations I have described are the product of collaboration with my colleagues Murielle Villeret, David Edwards and Bechara Muniz and without their contribution this paper could not have been written. We are also grateful to the Science and Engineering Research Council, United Kingdom, CNPq of Brazil and Nuffield Foundation for financial support.

References

[1] Shun-ichi Iwasaki, IEEE Trans. Magn. 20: 657 (1984).

[2] P. Grunberg, R. Schreiber, Y. Pang, M. B. Brodsky and. H Sower, Phys. Rev. Lett. 57: 2442 (1986).

[3] M.N. Baibich, J.M. Broto, A. Fert, F. Nguyen Van Dau, F. Petroff, P.Etiene, G. Creuset, A. Friederich and J. Chazelas, Phys. Rev. Lett. 21:272 (1988).

[4] B. Heinrich and J.F. Cochran, Adv. Phys. 42: 523 (1993).

[5] S.S.P Parkin, N. More and K.P. Roche, Phys. Rev. Lett. 64: 2304 (1990).

[6] S.S.P. Parkin, Phys. Rev. Lett. 67: 3598 (1991).

[7] P. Grunberg, S. Demokritov, A. Fuss, M. Vohl, and J.A. Wolf, J. Appl. Phys. 69: 4789 (1991).

[8] R. Coehoorn, M.T. Johnson, W. Folkerts, S.T. Purcell, N.W.E. McGee, A. De Vierman, and P.J.H. Bloemen, Magnetism and Structure in Systems of reduced dimension, Ed. by R.F.C. Farrow et al., Plenum Press, N.Y. (1993).

[9] P. Grunberg, S. Demokritov, A. Fuss, R. Schreiber, J.A. Wolf, and S.T. Purcell, J. Magn. Magn. Mater. 104-107: 1734 (1992).

[10] J. Unguris, R.J. Celotta, and D.T. Pierce, Phys. Rev. Lett. 67: 140 (1991).

[11] J. Mathon, Contemporary Phys. 32:143 (1991).

[12] J.F. Ortega and F.J. Himpsel, Phys. Rev. Lett. 69: 844 (1992).

[13] J.E. Ortega, F.J. Himpsel, G.J. Mankey, and R.F. Willis, Phys. Rev. B 47: 1540 (1993).

[14] A.J. Freeman, R. Wu, and C.L. Fu, J. Magn. Magn. Mater. 100: 497 (1991).

[15] J. Mathon, Rep. Prog. Phys. 51: 1 (1988).

[16] R. Feder (ed.), Polarized Electrons in Surface Physics, World Scientific, Singapore (1985).

[17] H.C. Siegmann, D. Mauri, D. Scholl and E. Kay, J. Phys. (Paris) 49: C8-9(1988).

[18] J. Mathon, in: Magnetic Properties of Low-Dimensional Systems II (ed. by L.M. Falicov and J.L. Moran-Lopez, Springer Verlag, Berlin (1990), p. 167.

[19] J. Mathon, Physica B 149: 31 (1988).

[20] J. Mathon, J. Phys.: Condens. Matter. 1: 2505 (1989).

[21] G.T. Rado, Bull. Am. Phys. Soc. 2: 127 (1957); D.L. Mills and A.A. Maradudin, J. Phys. Chem. Solids 28: 1855 (1967).

[22] J. Mathon and S.B. Ahmad, Phys. Rev. B 36: 660 (1988).

[23] D.T. Pierce, R.J. Celotta, J. Unguris and H.C. Siegmann, Phys. Rev. B 261: 2566 (1982).

[24] J.C. Walker, R. Droste, G. Stern and J. Tyson, J. Appl. Phys. 55: 2500 (1984).

[25] D. Mauri, D. Scholl, H.C. Siegmann and E. Kay, Phys. Rev. Lett. 61: 758 (1988).

[26] M. Landolt, in: Polarized Electrons in Surface Physics (ed. by R. Feder), World Scientific, Singapore (1985).

[27] D. Mauri, D. Scholl, H.C. Siegmann and E. Kay, Phys. Rev. Lett. 62: 1900 (1989).

[28] D. Scholl, M. Donath, D. Mauri, E. Kay, J. Mathon, R.B. Muniz, and H.C. Siegmann, Phys. Rev. B 43: 13309 (1991).

[29] F. Herman, J. Sticht, and M. van Schilfgaarde, J. Appl. Phys. 69: 4783 (1991); Int. J. Mod. Phys. B 7: 456 (1993).

[30] P. Lang, L. Nordstrom, R. Zeller, and P.H. Dederichs, Phys. Rev. Lett. 71: 1927 (1993).

[31] Y. Yafet, J. Appl. Phys. 61: 4058 (1987).

[32] W. Baltensberger and J.S. Helman, Appl. Phys. Lett. 57: 2954 (1990).

[33] P. Bruno and C. Chappert, Phys. Rev. Lett. 67: 1602 (1991).

[34] D.M. Edwards, J. Mathon, R.B. Muniz, and M.S. Phan, Phys. Rev. Lett. 67: 493 (1991).

[35] D.M. Edwards, J. Mathon, R.B. Muniz, and M.S. Phan, J. Phys.: Condens. Matter 3: 4941 (1991).

[36] A.A. Abrikosov, Introduction to the Theory of Normal Metals, Academic Press, New York (1972).

[37] D.A. Papaconstantopoulos, Handbook of the Band Structure of Elemental Solids (Plenum Press, New York, 1986); V.L. Moruzzi, J.F. Janak, and A.R. Williams, Calculated Electronic Properties of Metals (Pergamon Press, Oxford 1978).

[38] J. Mathon, M.A. Villeret, R.B. Muniz, and D.M. Edwards (to be published).

[39] J. Mathon, M.A. Villeret, and D.M. Edwards, J. Magn. Magn. Mater. 127: L261 (1993).

MAGNETIC CIRCULAR DICHROISM IN PHOTOEMISSION FROM RARE-EARTH MATERIALS: BASIC CONCEPTS AND APPLICATIONS

G. Kaindl, E. Navas, E. Arenholz, L. Baumgarten*, and K. Starke

Institut für Experimentalphysik, Freie Universität Berlin
Arnimallee 14, D-14195 Berlin-Dahlem, Germany

INTRODUCTION

Magneto-optical Kerr effect and Faraday effect provide the basis of established methods for studying the magnetic properties of matter by polarized light in the visible spectral range. It was only quite recently that an analogous effect in the x-ray region, magnetic circular dichroism in x-ray absorption, was first observed by Gisela Schütz et al. for the near-edge fine structure at the K edge of ferromagnetic iron.[1] Later on, magnetic circular x-ray dichroism (MCXD) was also observed at the $L_{II,III}$ thresholds of rare-earths[2] and 3d transition metals,[3] opening up the possibility for element-specific analyses of magnetic moments in compound magnets and multilayers. Today MCXD is mainly used as a tool at the $L_{II,III}$ x-ray absorption thresholds of 3d transition metals, where relatively large MCD asymmetries in the white lines upon reversal of either sample magnetization or circular polarization (photon spin) of the absorbed light are observed. MCXD can be understood in the simplest way in a one-electron picture by taking the spin polarization of the excited electron due to the inner-shell spin-orbit coupling (Fano effect[4]) into account as well as the spin-split density of final states at and above the Fermi level.[5] More rigorous theoretical treatments have been given,[6,7,8] which allow to recognize the three important ingredients for magnetic circular dichroism: (i) Exchange interaction as the driving force for long-range spin order; (ii) use of circularly polarized light with preferential propagation along the magnetic quantization axis; (iii) spin-orbit interaction providing the mechanism for an effective coupling between the angular momentum of the circularly polarized photon and the magnetically ordered electron spins.

The observation of large MCXD effects, which provide a preferentially bulk-sensitive probe, has stimulated the search for similar effects in photoemission (PE), motivated by the well-known merits of PE: High surface sensitivity and tuneable sampling depth by varying the kinetic energy of the photoelectrons. The first successful study was made for the 2p core level of Fe,[9] with the $2p_{1/2}$ and $2p_{3/2}$ electrons being emitted into the continuum rather than excited into states just above the Fermi level. In this case one cannot expect a spin-filter effect analogous to MCXD, since continuum states far above threshold should have negligible spin dependence. The observed MCD effect could be explained within a fully relativistic single-particle model[10] as well as in a manybody treatment.[8]

---

* Present address: Institut für Festkörperforschung, Forschungszentrum Jülich, D-52425 Jülich, Germany

The relatively small MCD effect observed for the $2p_{1/2}$ and $2p_{3/2}$ core-level PE lines is due to exchange interaction between the magnetically ordered 3d electrons and the 2p core electrons ($\Delta E_{ex} \cong 0.5$ eV),[10] causing an exchange splitting of the core level into $|M_J\rangle$ sublevels. Their relative intensities, given by the dipole-transition probabilities to continuum states, depend on the selection rules $\Delta M_J = +1$ or $\Delta M_J = -1$, i.e. on the relative orientation between photon spin and sample magnetization. The observed MCD effect (originally an asymmetry of $\cong 2$ % was found) is small compared to MCXD effects, mainly because the exchange splitting is smaller than the intrinsic width of the 2p core level; this means that individual $|M_J\rangle$ components cannot be resolved. Larger MCD effects in transition metals might be expected in PE from magnetic 3d band states, since the exchange interaction of these states is larger. However, the MCD effects were found to be of similar size as for core-level PE (up to $\cong 8$ % asymmetry).[11] This is due to an inefficient transfer of the photon-momentum orientation to the (magnetically ordered) spin of the 3d electrons, caused by weak spin-orbit coupling and the well-known quenching of the orbital angular momentum in 3d transition metals.

The situation is entirely different for the localized 4f states in rare earths (RE). Being closely bound inside the filled 5s and 5p shells, the 4f states maintain their atomic character also in the solid state and are normally not participating in chemical bonding. Crystal fields, which quench the orbital moment in 3d ferromagnets, have thus little effect on the RE 4f orbitals; i.e. the large orbital momenta exist also in the solid state. At each crystal-lattice site, the 4f spin-orbit coupling (of typically 0.1 to 1.5 eV) aligns orbital and spin momentum according to Hund's rules. As a further direct consequence of their localized nature, the 4f electrons are subject to a stronger Coulomb correlation than 3d electrons in transition metals. This is beautifully reflected in the 4f PE spectra, shown for the example of Tb metal in Figure 1.[12] The eight 4f electrons of Tb form the $^7F_6$ (L=3, S=3) ground state, i.e. seven are spin-up and one is spin-down. The energy distribution curve (EDC) of a photoemitted 4f

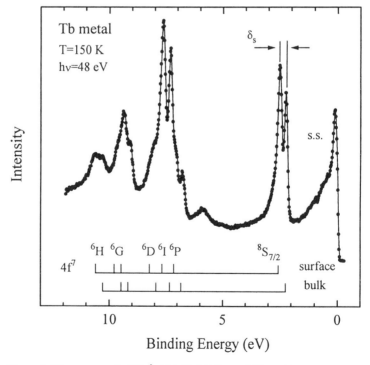

Figure 1. PE spectrum of a 130-Å-thick Tb(0001) metal film at T = 130 K grown on W(110) (from Navas et al.[12]). The solid curve through the data points serves as a guide to the eyes. The vertical-bar diagrams indicate the strongest components of the $4f^7$ final-state multiplets from bulk and surface.

electron shows many lines reflecting different correlation energies of the remaining seven 4f electrons: The "high-spin" state $^8S_{7/2}$ (all electrons are spin-up) is separated by $\Delta E \cong 5$ eV from the ^6X group of "low-spin" states (one electron is spin-down). ΔE can be regarded as a rough measure of the intra-atomic exchange interaction in the PE final state; the low-spin lines reflect different correlation energies for various angular distributions of the remaining electrons around each atom.

The high-resolution PE spectrum of Figure 1, taken from a well-ordered Tb(0001) metal surface at T = 150 K, demonstrates also the phenomenon of the surface core-level shift, δ_s, i.e. two $4f^7$ final-state multiplets are observed, one from the bulk and one from the topmost surface layer, with the latter shifted by $\delta_s = 0.26 \pm 0.03$ eV to higher binding energies.[12] This allows a separate spectroscopy of the topmost surface layer. Note also the surface state (s.s.) directly below the Fermi level in addition to the triangular-shaped bulk conduction band states, typical for close-packed RE metal surfaces.[13]

Let us now consider the case of long-range ferromagnetic order, where the total 4f moments at all lattice sites are oriented parallel. The indirect exchange interaction primarily couples neighboring 4f spin moments; however, also the orbital angular momenta assume long-range order by spin-orbit coupling at each lattice site. In PE experiments with circularly polarized light, the selection rule $\Delta M = \pm 1$ causes the excitation probability to depend on the relative orientation of the photon momentum and the 4f orbital momentum. Thus large MCD effects are expected in 4f-PE from RE like Tb, where PE multiplet lines with different final-state angular momenta can be resolved (see Figure 1).

EXPERIMENTAL DETAILS

The PE experiments were performed with circularly polarized soft x-rays from two different monochromators at the Berliner Elektronenspeicherring für Synchrotronstrahlung (BESSY): The plane-grating SX700/III located at a bending magnet and the U2-FSGM behind the crossed undulator.[14] The SX700/III is equipped with two premirrors selecting synchrotron radiation from above or from below the storage-ring plane.[15] It provides light with a high degree of circular polarization at photon energies exceeding $\cong 100$ eV, which are well suited for bulk-sensitive PE studies (e.g. $S_3 \cong 0.9$ at $h\nu = 256$ eV for an off-plane angle of $\psi = 0.9$ mrad;[16] S_3 is the Stokes parameter measuring the degree of circular polarization[17]). For surface sensitive measurements, we used the crossed undulator, which supplies a high photon flux at lower energies around $h\nu = 50$ eV, with $S_3 \cong 0.5$.[14] At present, there is no alternative to the use of synchrotron radiation for circularly polarized soft x-rays.

The preparation of clean and well-ordered single-crystalline RE metals is a delicate matter. Due to the high chemical reactivity of RE metals, contaminants (mainly O, H, and C)[18] in bulk crystals can be hardly depleted by the usual UHV-sputter/anneal cycles. Much cleaner samples are obtained by preparing epitaxial films in UHV, e.g. by vapor deposition of the RE onto single-crystalline substrates. The base pressure in our experimental chamber was $<3 \cdot 10^{-11}$ mbar; during evaporation, it rose briefly to $<2 \cdot 10^{-10}$ mbar. We used W(110) as a substrate for several reasons: (i) It is easy to prepare; (ii) it does not alloy with RE metals; (iii) heavy RE metals grow on W(110) by forming hexagonally close-packed (0001) surfaces. The lattice mismatch of $\cong 4$ % for Gd,[19] e.g., leads to some strain in monolayer-thin films, which is released near the surface of thick films.

In order to suppress the formation of islands during deposition, the substrate was kept at temperatures below 300 K resulting in flat films (80 Å to 150 Å thick) with little lateral order. Well-ordered films of Gd, Tb, and Dy, as confirmed by LEED, were obtained by subsequent annealing for 5 minutes at $T_{an} = 700$ K to 900 K, with T_{an} depending on film thickness and RE element.[20] The intensities of the well-known d-like surface states[13] in the valence-band PE spectra just below the Fermi level were used as a sensitive measure of film quality. The film thickness was monitored by a quartz microbalance and calibrated via the relative intensities of the 4f-PE lines from W and from the RE. Chemical cleanliness was checked via 1s-PE intensities of O and C as well as by the O-2p PE signal; in the latter case, O_2 exposures as low as 1/100 of a Langmuir could be easily monitored.

The experimental geometry is shown schematically in the inset of Figure 2. The circularly polarized light was incident at an angle of 15° with respect to the film plane, and the photoelectrons were collected around the surface normal by a hemispherical electron-energy analyzer with a moderate angular resolution of ± 10°. The PE spectra were taken with the films remanently magnetized in-plane. Remanent magnetization was compulsory, since an application of external fields is not compatible with low-energy PE. The in-plane magnetization was checked in-situ by magneto-optical Kerr effect (MOKE), with a MOKE setup that has been described elsewhere.[21]

Note that these PE experiments are quite simple as far as energy and angular resolution are concerned, while they are quite demanding in two other respects: (i) While many RE materials have low Curie temperatures (e.g. Ho with $T_c \cong 20$ K), the substrate is cleaned by flashing it to $\cong 2000$ K. This requires an optimized compromise regarding the thermal contact between substrate and cooling stage. In this respect, Gd, which has the highest spin moment ($7\mu_B$) and the highest Curie temperature ($T_c \cong 290$ K) of all RE metals, is least demanding. (ii) All magnetic RE metals, except for Gd and Eu, have a non-spherical 4f charge distribution, which - by interaction with the low-symmetry hexagonal crystal field - gives rise to the well-known large single-ion anisotropy energies. They lead to high coercive fields that do not readily allow to change the sample magnetization by small external fields. For Gd (half-filled 4f shell), the orbital angular momentum and thus the single-ion anisotropy vanishes. The coercive fields are therefore small enough to allow the sample to be magnetized by field pulses (duration < 1 s) of only a few 100 A/cm applied through a closeby selenoid. This easy magnetization procedure in case of Gd allows to measure a pair of MCD spectra (both magnetization directions for a fixed photon spin) in a quasi-simultaneous way by reversing the magnetization pulse at each electron-kinetic energy point. By this differential measurement of the MCD signal, long-term instabilities of the apparatus (e.g. of the storage-ring current) cancel out.

MAGNETIC CIRCULAR DICHROISM IN 4f PHOTOEMISSION

Gadolinium Metal

The first MCD-in-4f-PE experiments were reported in 1993 by Starke et al. for ferromagnetically ordered Gd metal.[22] Figure 2 (a) shows 4f-PE spectra from a magnetized Gd-metal film at $T \cong 50$ K obtained with circularly polarized light from the SX700/III monochromator. For nearly parallel orientation of photon spin and sample magnetization, the $4f^6$-^7F_J final-state multiplet assumes a peaked shape, which changes into a rounded shape upon reversal of magnetization. The intensity asymmetry $(I^{\uparrow\uparrow}-I^{\downarrow\uparrow})/(I^{\uparrow\uparrow}+I^{\downarrow\uparrow})$, calculated from the raw experimental data, amounts up to 17 % (Figure 2 (b)). It is considerably larger than the MCD asymmetries obtained from 3d transition metals. An identical MCD effect was observed for a given magnetization when the photon helicity was reversed (spectra not shown here).[22]

Due to the localized nature of the 4f orbitals, we expect that the observed MCD effect can be described in an atomic model. In the following we derive, using the LS-coupling scheme, the intensities of the seven ^7F_J final-state PE-multiplet components for the two cases of parallel and antiparallel orientation of photon spin and sample magnetization. We shall see that the MCD effect is a simple consequence of the dipole-selection rule $\Delta M = +1$ or $\Delta M = -1$, depending on whether photon spin and magnetization are parallel or antiparallel, respectively.

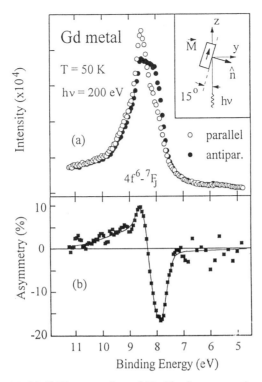

Figure 2. (a) Gd 4f PE spectra (hν = 200 eV) of a remanently magnetized Gd(0001)/W(110) film at T = 50 K. Open (filled) dots represent parallel (antiparallel) orientation of photon spin and sample magnetization. (b) The intensity asymmetry, calculated from the raw data in (a), amounts up to 17 %. The inset shows the experimental geometry.

The Gd ground state $^8S_{7/2}$-$|J,M\rangle$ is characterized by the total angular momentum quantum number $J = 7/2$ and the magnetic quantum number M. The ground state is connected via the dipole-selection rules $\Delta J = 0, \pm 1$ and $\Delta M = \pm 1$ with the complete final state $|J',M'\rangle$, obtained by coupling all angular momenta of the ^7F_J PE final state (L = 3, S = 3) and of the detected photoelectron. In order to calculate the MCD effect, i.e. the influence of the ΔM selection on the intensities of the final-state PE-multiplet components, we first use the complete final state and then decouple the angular momentum of the photoelectron. For the photoelectron, we only consider 4f→εg transitions, since contributions from 4f→εd are expected to be small for PE final states far above the continuum threshold.[23]

In a first step, we consider the four complete final states $|J',M'\rangle$ that can be reached from the fully magnetized ground state $|J = 7/2, M = -7/2\rangle$ (see Figure 3): For $\Delta M = -1$, the only possible transition is $\Delta J = +1$, whereas all three ΔJ transitions are allowed for $\Delta M = +1$. By help of the Wigner-Eckart theorem, we separate the ΔM-dependence of the transition probability $|\langle J'M'|P^{\Delta M}|JM\rangle|^2$ (with $P^{\Delta M}$ = dipole operator) and hereby obtain statistical weights for all four transitions given in Figure 3. While $|J' = 9/2\rangle$ is the only allowed state for $\Delta M = -1$, this state is hardly reached by the $\Delta M = +1$ transitions, with a negligible weight of 1/36. Most of the $\Delta M = +1$ transitions (3/4) reach the $|J' = 5/2\rangle$ state, reflecting a dominance of transitions with $\Delta J = -\Delta M$. This dominance is essential for the occurence of MCD effects, as shown in the following.

		|J', M'⟩	ΔJ	relative weight
	ΔM= +1	|9/2,−5/2⟩	+1	1/36
		|7/2,−5/2⟩	0	2/9
		|5/2,−5/2⟩	−1	3/4
|7/2,−7/2⟩				
		|9/2,−9/2⟩	+1	1
	ΔM= −1	−		
		−		

Figure 3. Relative dipole transition probabilities for Gd from the magnetized ground state |7/2,−7/2⟩ to the 4 allowed total final states |J'M'⟩. The relative weights (right column) are calculated in the LS coupling scheme. They reflect the dominance of transitions with ΔJ = −ΔM.

In a second step we decouple the photoelectron momenta ($l = 4$, $s = 1/2$) from the final state |J',M'⟩, and in this way obtain the intensity distribution over the individual ^7F_J multiplet components of the PE final state. It is drastically different for the different ΔJ: In case of ΔJ = −1, the J = 6 final-state component carries about 50 % of the intensity, whereas it assumes only about 5 % in case of ΔJ = +1.

The results of this atomic-multiplet calculation are shown graphically in Figure 4 (a) for ΔM = +1 (parallel orientation) and (b) for ΔM = −1 (antiparallel orientation). The values agree well with the results of recent intermediate-coupling calculations by van der Laan & Thole using the Cowan computer code.[23,24] Also included in Figure 4 are the experimental spectra from Figure 2, however, normalized to complete circular polarization of the photon beam ($S_3 = 1$), in order to facilitate a comparison. Note the good qualitative agreement between the shapes of the normalized experimental spectra and the theoretical multiplets.

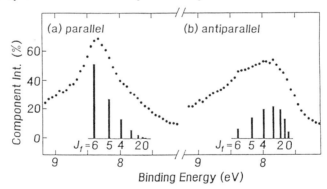

Figure 4. Vertical bars: Calculated relative intensities of the $4f^6$-^7F_J final-state multiplet components in the PE spectrum of Gd for (a) parallel (ΔM = +1) and (b) antiparallel (ΔM = −1) orientation of photon spin and sample magnetization. Filled dots: Experimental spectra from Figure 2, normalized to complete circular polarization.

Terbium Metal

Terbium, with its many well-resolved 4f-PE multiplet lines (see Figure 1) represents an ideal case for the observation of MCD in 4f-PE. However, Tb metal orders ferromagnetically only below $T_c = 220$ K. Since the MCD effect is expected to vanish above T_c and to reach a maximum at saturation magnetization, $M(T)/M(0) = 1$, the sample temperature has to be significantly lower than 220 K in order to observe a sizeable effect. Because of the relatively large coercive fields in Tb metal, remanent magnetization could be only achieved by cooling the sample in the presence of an external magnetic field of 400 A/cm from a temperature close to T_c down to the temperature of measurement. This small magnetic field was also found to be sufficient for reversing the film magnetization at temperatures close to T_c, where magneto-crystalline anisotropies and coercivity are small.[25]

Figure 5 (a) shows a pair of 4f-PE spectra of Tb metal at $T = 110$ K, with parallel and antiparallel orientation of photon spin and sample magnetization. The intensity difference ($I^{\uparrow\uparrow}-I^{\downarrow\uparrow}$), also called "MCD spectrum", is shown in the lower panel (Figure 5 (b), upper curve). All 4f-multiplet components reveal non-vanishing MCD; it is most pronounced for components with the lowest and the highest orbital angular momenta, respectively, i.e. for the isolated $^8S_{7/2}$ component (with L=0) and - with opposite sign - for the strong low-spin component, ^6I (with L = 6). By comparison with the results of the intermediate-coupling calculations[23] reproduced in Figure 5 (b) (lower curve), we see that experimental and theoretical MCD curves agree well even in small details.

The isolated high-spin $^8S_{7/2}$ component at a binding energy of $\cong 2.4$ eV deserves further attention. It is separated by more than 4 eV from all other Tb multiplet components and displays the largest MCD effect of all PE components of Tb (corresponding to an asymmetry of $A_{exp} \cong 30$ %). For this isolated component, a quantitative comparison with

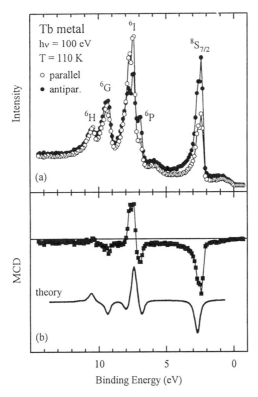

Figure 5. (a) Tb 4f PE spectra (hv = 100 eV) of a remanently magnetized Tb(0001)/W(110) film (150 Å thick; T = 110 K). Open (filled) dots are for nearly parallel (antiparallel) orientation of photon spin and sample magnetization. (b) Filled squares: Intensity difference (MCD) of the experimental spectra in (a); the solid curve at the bottom of (b) reproduces the theoretical MCD spectrum of van der Laan & Thole.[23]

theory is much simpler than for the unresolved ^7F_J-multiplet components of Gd. By use of atomic multiplet theory, as described in the previous section, we arrive at an MCD-intensity ratio for the Tb-$^8S_{7/2}$ component of 28 to 1; this corresponds to a theoretical asymmetry of A_{th} = 93 %. The value refers to the ideal case of fully aligned photon spin and sample magnetization and of completely circularly polarized light. Note that this LS-coupling-scheme result agrees well with the one from intermediate-coupling calculations, 27.7 to 1.[23,24]

The much smaller MCD effect observed experimentally is mainly due to the incomplete circular polarization ($S_3 \cong 0.5$ in the present case) as well as to the relatively high sample temperature of T = 110 K, corresponding to a reduced temperature of $t = T/T_c \cong 0.5$. For t = 0 (ferromagnetic ground state), only the lowest-lying magnetic M level would be occupied, i.e. $\langle M \rangle = -J$. Yet, at elevated temperatures, also higher M levels get populated, thus reducing the MCD effect; the latter is expected to vanish for $\langle M \rangle \to 0$, i.e. for equal population of all M levels.

We use a spin-wave picture for a qualitative understanding of the influence of temperature on the MCD effect. The 4f magnetic moments precess around the sample magnetization giving rise to a perpendicular magnetization component, which will grow with increasing temperature. Since the perpendicular component rotates slowly on the time scale of the PE process, it is seen by the polarized photon as a stationary magnetization component. In the experimental grazing-incidence geometry used here (see inset of Figure 2), this magnetization component is nearly perpendicular to the photon spin, thus giving rise to $\Delta M = +1$ and $\Delta M = -1$ transitions with equal probability; hereby the net MCD effect gets smaller. We note that with this spin-wave model, the magnitude of the observed MCD effect can be quantitatively described by assuming a circular polarization of $S_3 = 0.55$, which is a realistic value (see Table 1).

Dysprosium Metal

Another case of a ferromagnetic RE metal, for which MCD in 4f PE has been observed, is dysprosium metal, with a $4f^9$ ground-state configuration.[26] It orders in the bulk with a helical spin structure below T_N = 176 K, and ferromagnetically below T_c = 88.3 K. As in the case of Tb metal, the $\cong$ 150-Å thick Dy(0001) films grown on a W(110) substrate could be remanently magnetized by cooling them from a temperature close to T_c down to the ferromagnetic phase in the presence of an external magnetic field of 400 A/cm; this allowed to reverse the magnetization of the films.

The PE spectra of Dy metal at 55 K, in the binding-energy region of the $4f^8$ final-state multiplet, are displayed in Figure 6 for the two different situations, for almost parallel (open dots) and antiparallel (filled dots) orientation of photon spin and sample magnetization. Despite the fact that the reduced temperature t was only $\cong 0.62$ in the present case, a large MCD effect was observed amounting to an MCD asymmetry of $\cong$ 25 % (uncorrected for background) for the shallowest ^7F multiplet component. Note that also other components of the $4f^8$ PE multiplet of Dy exhibit large MCD effects of both signs. The MCD spectrum, calculated from the raw data, is again plotted in (b) and compared with the results of the intermediate-coupling calculations (lower solid curve).[23] The agreement between experiment and theory is again very good. Minor deviations stem from the surface-shifted $4f^8$ multiplet and the inelastic background in the PE spectra, which were not considered in the theoretical work. Dy is thus another favorable case for the study of MCD in 4f PE.

MCD IN 4d CORE-LEVEL PE FROM Gd METAL

As described in the introduction, MCD in PE had been first observed by Baumgarten et al. in 2p core-level PE from ferromagnetic Fe metal.[9] In this case, the magnetic exchange splitting of the $2p_{1/2}$ and $2p_{3/2}$ core holes is unresolved, causing only shifts in the PE lines

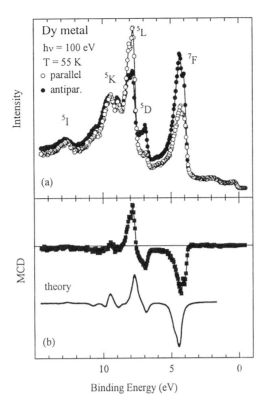

Figure 6. (a) Dy 4f PE spectra (hν = 100 eV) of a remanently magnetized Dy(0001)/W(110) film (150 Å thick; T = 55 K). Open (filled) dots are for nearly parallel (antiparallel) orientation of photon spin and sample magnetization. (b) Filled squares: Intensity difference (MCD) of the experimental spectra in (a); the solid curve at the bottom of (b) represents the theoretical MCD spectrum of van der Laan & Thole.[23]

for the two different orientations of photon spin and sample magnetization, and hence smaller MCD asymmetries than in cases of the well-resolved $4f^{n-1}$ PE multiplets of the RE. Similar situations exist in deep core-level PE from RE materials, as e.g. in PE from the 4d and 5p core levels.

We have therefore studied, in an exploratory way, the MCD effect in PE from the 4d core levels of Gd.[26] The results are shown in Figure 7, with the PE spectra obtained with 438-eV photons from the SX700/III beamline at BESSY given in (a) for the two orientations of photon spin and sample magnetization.

The MCD asymmetry, obtained from the raw data, amounts to ≃ 6 %; it is plotted across the 4d PE spectrum in Figure 7 (b). Coming from low binding energies, the asymmetry exhibits a plus-minus behavior at the $4d_{5/2}$ subspectrum, and a weaker minus-plus behavior at the $4d_{3/2}$ subspectrum. In this way, it resembles the original observation of Baumgarten et al. for the MCD in 2p PE from ferromagnetic Fe metal.[9] Note that a splitting of particularly the $4d_{5/2}$ subspectrum for antiparallel alignment into apparently equally-spaced components is visible in the raw data.

The PE spectra of Figure 7 were accordingly least-squares fitted by a superposition of six equally-spaced components for $4d_{5/2}$ and four components for $4d_{3/2}$, with freely varying relative intensities. This accounts for exchange splitting due to (2J+1) orientations of the 4d-hole angular momentum J with respect to the magnetization direction. Note that this

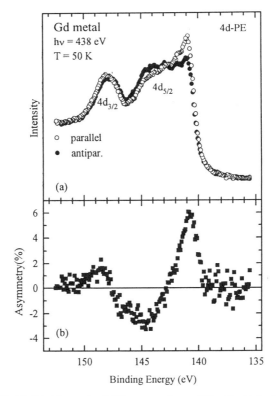

Figure 7. (a) Gd 4d core-level PE spectra (hν = 438 eV) obtained from a remanently magnetized Gd(0001)/W(110) film (80 Å thick; T = 50 K). Open (filled) dots are for nearly parallel (antiparallel) orientation of photon spin and sample magnetization. The MCD asymmetry derived from the raw data is plotted in (b).

model has only approximate character, since - due to exchange splitting - J can no longer be considered to be a good quantum number.[10] The results of this analysis are shown in Figure 8 as solid curves through the data points; the dash-dotted curve represents an integral background. As a result, we obtain exchange splittings of 0.98 eV (0.63 eV) between neighboring $4d_{5/2}$ ($4d_{3/2}$) sublevels, and a spin-orbit splitting of Δ_{so} (4d) $\cong$ 4.8 eV.

MCD IN RESONANT 4d→4f PHOTOEMISSION

Resonant PE at the 4d→4f giant resonances[27] and the 3d→4f absorption thresholds[28] of RE elements is a very effective tool to substantially increase the 4f PE cross section and in this way enhance the element specificity of 4f PE.[29] At photon energies corresponding to the 4d→4f giant resonance, the direct PE channel interferes with an indirect one due to Super-Coster-Kronig decay of the $4d^94f^{n+1}$ intermediate state, giving rise to a Fano resonance of the 4f PE cross section.

Figure 9 (top panel) shows 4d→4f constant-final-state (CFS) spectra of remanently magnetized Gd(0001)/W(110) at T = 150 K recorded via 40-eV secondary electrons for the two orientations of photon spin and sample magnetization;[30] it is known that the secondary-electron yield is approximately proportional to the photoabsorption cross section. The dashed vertical line divides the spectrum into a relatively weak pre-edge region with narrow lines and the giant-resonance region with a broad peak. The fine structure of this spectrum

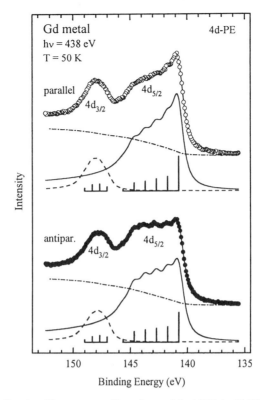

Figure 8. Results of least-squares fit analyses of the MCD-in-4d-PE spectra of Gd metal from Figure 7 (a) by superpositions of 6 (4) equally spaced components for the $4d_{5/2}$ ($4d_{3/2}$) subspectra. For details see text.

has been explained on the basis of exchange interaction of the $4d^9 4f^{n+1}$ configuration, where the higher levels lie above threshold and hence autoionize rapidly.[31]

In the lower panel of Figure 9, the MCD asymmetry, as obtained from the raw data, is plotted showing pronounced structures in the pre-edge region and a strong positive peak at the leading edge of the giant resonance. As discussed above, the dipole-selection rule leads to a predominant excitation of states with $\Delta J = -\Delta M$ (see Figure 3). Starting from the fully magnetized ground state $|J = 7/2, M = -7/2\rangle$ of Gd, for parallel orientation ($\Delta M = +1$) intermediate states with predominantly $J' = 5/2$ (and $7/2$) are excited (positive MCD asymmetry), while for antiparallel orientation ($\Delta M = -1$) states with $J' = 9/2$ (negative MCD asymmetry). This allows a comparison with the results of an atomic multiplet calculation by Sugar plotted in form of a bar diagram in the center of Figure 9.[27] This qualitatively explains the observed MCD asymmetry, where three peaks with negative asymmetry are observed in the pre-edge region, corresponding to three theoretical resonances with $J' = 9/2$; in addition, the positive asymmetry in the leading edge of the giant resonance corresponds to the theoretical resonances with $J' = 5/2, 7/2$ in this energy region. Within this interpretation, MCD allows the experimental determination of the J'-character of the $4d^9 4f^{n+1}$ intermediate state. A more recent many-body calculation of the fine structure of the $4d^9 4f^8$ absorption spectrum by Imada & Jo[32] agrees better with the experimental data, but lacks an explicit J' labelling of the multiplet states.

We have also studied MCD in resonant 4d→4f PE of remanently magnetized Gd metal for various photon energies given by vertical arrows in the upper panel of Figure 9. The PE spectra, displayed in Figure 10 for parallel and antiparallel orientation of photon spin and

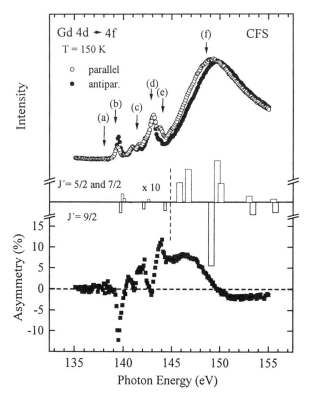

Figure 9. MCD in the Gd 4d→4f constant-final-state (CFS) spectrum of remanently magnetized Gd(0001)/W(110), ≅ 80 Å thick, T = 150 K, measured by recording 40-eV secondary electrons for parallel (open dots) and antiparallel (filled dots) orientation of photon spin and sample magnetization. The asymmetry, calculated from the raw data, is plotted in the lower panel (filled squares). The vertical bar diagram in the center of the figure reproduces the theoretical results of Sugar for the relative positions and energies as well as the J' character of the individual multiplet components; the lengths of the bars represent the theoretical excitation probabilities.[27]

sample magnetization, exhibit dramatic changes when the photon energy is tuned to the various $4d^9 4f^{n+1}$ resonances. It is clear that due to selective excitation of J' = 9/2 and J' = 5/2,7/2 intermediate states, the MCD of individual PE components can be varied substantially.[30]

APPLICATIONS

We have shown that MCD in PE from RE materials can probe magnitude and orientation of the sample magnetization. No time consuming spin analysis of the photoelectrons, as in the established technique of spin-resolved PE, is required. In this way, the new technique can be applied as a magnetometer for RE surfaces and thin magnetic films, where one is able to distinguish between individual magnetic sublattices by separating the element-specific PE multiplets. On the other hand, the linear dependence of the MCD effect on the degree of circular polarization suggests its use as an x-ray polarimeter. In a discussion of applications, we shall begin with the latter aspect.

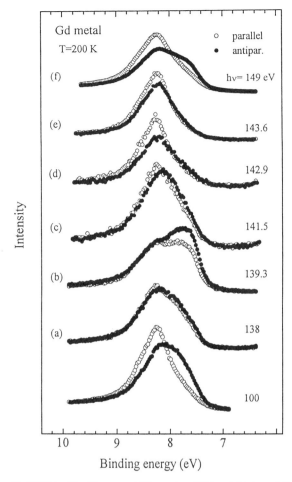

Figure 10. MCD in 4d→4f resonant PE spectra of Gd metal taken at the given photon energies; the latter are also indicated by the correspondingly designated vertical arrows in the CFS spectrum of Figure 9 (a). The bottom spectrum was taken off resonance.

X-ray Polarimeter

In order to achieve a small reduced temperature $t = T/T_c$, one can either lower the sample temperature or increase T_c, e.g. by alloying Tb with Fe. Intermetallic compounds of the form $TbFe_x$, especially cubic $TbFe_2$ ($T_c \cong 1040$ K), have the additional advantage of much smaller magneto-crystalline anisotropies as compared to Tb metal. We therefore prepared a thin film of $TbFe_x$ by first depositing $\cong 10$ monolayers (ML) Fe on W(110), followed by $\cong 2$ ML Tb and subsequent annealing. Using LEED, a cubic structure was observed for low electron energies, indicating the formation of $TbFe_2$ near the film surface. Coercivities were found to be so small that the magnetization at 100 K could be switched by small field pulses as in case of Gd metal.[25]

The 4f-PE spectra of $TbFe_x$, taken at hν = 152 eV in 4d→4f resonance, are presented in Figure 11 (a). The $^8S_{7/2}$ and ^6I components reveal the largest MCD effect so far observed in PE as well as in x-ray absorption; the intensity asymmetry (see Figure 11 (b)), as

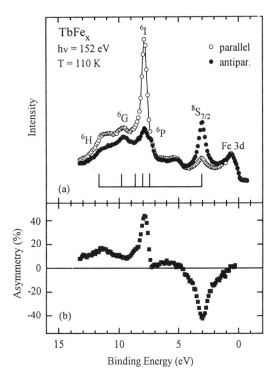

Figure 11. (a) Tb 4d→4f resonant PE spectra (hν = 152 eV, T = 110 K) of a remanently magnetized TbFe$_x$ film grown on W(110). Open (filled) dots are for nearly parallel (antiparallel) orientation of photon spin and sample magnetization. (b) Asymmetry calculated from the raw experimental spectra.[25]

calculated from the raw experimental spectra without background subtraction, exceeds 40 %. Hardly any MCD effect is noticeable in the emission from Fe-3d states close to the Fermi level. Intermixing of Tb and Fe is reflected in the binding energy of the $^8S_{7/2}$ line (at ≅ 3.1 eV), in comparison to that of Tb metal (at ≅ 2.4 eV).

The isolated $^8S_{7/2}$ component in the PE spectrum of TbFe$_x$ is well suited for measuring - on the basis of MCD - the degree of circular polarization of x-rays. Unlike Tb metal, films of the intermetallic compound TbFe$_x$ require only ordinary ultra-high vacua (in the 10^{-10}-mbar range), and can thus be used for long periods of time. For a first demonstration, we used two different settings of the SX700/III monochromator, accepting synchrotron radiation at angles of ψ = 0.6 mrad and ψ = 0.9 mrad with respect to the storage-ring plane.

As expected, larger MCD asymmetries were observed for the ψ = 0.9-mrad setting. From least-squares fits of the peak intensities, the experimental peak asymmetries, A_{exp}, were obtained, resulting in (51 ± 5) % for 0.6 mrad and (62 ± 6) % for 0.9 mrad. To extract the degree of circular polarization of the soft x-rays (described by the Stokes parameter S_3), one has to consider a correction factor, C, to account for the non-vanishing angle between light propagation direction and sample magnetization. C is obtained by considering the additional excitation probabilities for ΔM = 0, ±1, which are caused by magnetization components perpendicular to the photon spin; C = 0.96 in the present case.[25] From the equation A_{exp}= C S_3 A_{th} we obtain S_3=(57 ± 6) % for the ψ = 0.6-mrad case and S_3 = (69 ± 7) % for 0.9 mrad. Table 1 shows that these values fit well into the results of optical polarization measurements, since S_3 is known to increase with increasing photon energy at the SX700/III beamline.

Table 1. Comparison of experimental results for the degree of circular polarization (in %) of soft x-rays from the SX700/III monochromator at BESSY.

hν (eV)	0.6 mrad	0.9 mrad	References
70	55	–	Petersen et al.[15]
152	57±6	69±7	present work
265	75	90	Di Fonzo et al.[16]

Surface Magnetization

The magnetization of the topmost atomic layer of RE metal surfaces has attracted considerable attention since the observation of surface-enhanced magnetic order and magnetic-surface reconstruction in 1985, when Weller et al. had found long-range magnetic order of the surface layer up to some 10 K above the bulk Curie temperature T_c^b, as well as some indication for an antiparallel orientation with respect to the bulk.[33] In order to detect the relative orientation of surface and bulk, they used the surface core-level shift δ_s. In case of Gd metal, the surface and bulk components of the broad ^7F_J multiplet (see Figure 2) could not be resolved. However, by spin-resolved PE, opposite spin polarizations were observed on the high-binding-energy (surface) side and on the low-binding-energy (bulk) side of the unresolved ^7F_J peak. In a recent repetition of this experiment, with somewhat better statistics and at lower temperatures, the spin polarizations, however, were found to be identical on both sides of the ^7F_J peak, giving thus strong evidence for parallel orientation of surface and bulk.[34]

The very different shapes of the Gd-^7F_J MCD-PE spectra for parallel and antiparallel orientation of magnetization and photon spin (see Figure 2) suggest immediately that this effect should allow to distinguish between parallel and antiparallel orientation of surface and bulk magnetization. In order to enhance the surface contribution, analogous spectra as in Figure 2 were taken at lower photon energies. Figure 12 shows a pair of MCD-PE spectra of Gd(0001) at hν = 47 eV for (a) parallel and (b) antiparallel orientation of sample magnetization (bulk) and photon spin. Both spectra exhibit clearly visible shoulders on the high-binding-energy side, which are a consequence of the intense surface contributions (the surface-to-bulk intensity ratio is $\cong 1.1$); such shoulders had not previously been resolved with linearly polarized light.[33,34,35] In the upper spectrum, which contains predominantly the pointed ΔM=+1 multiplet (compare with Figure 4 (a)), the shoulder is much sharper than in the bottom spectrum, where both the bulk and surface contributions are rounded. This clearly indicates that the magnetizations of surface and bulk are oriented essentially parallel to each other.[22]

For a quantitative description, the two spectra in Figure 12 were simultaneously least-squares fitted. We used a common parameter set and Doniach-Sunjic line shapes[36] assuming the calculated ^7F_J multiplet intensities. The resulting spectral shapes were convoluted by a Gaussian to account for finite experimental resolution. For the assumption of parallel alignment of the magnetizations of the (0001) surface layer (grey-shaded components) and of the Gd bulk (solid subspectra), the results are shown in Figure 12 as solid curves through the data points. Assuming antiparallel alignment of surface and bulk or a paramagnetic surface layer, substantial misfits were obtained (not shown here). The MCD data thus clearly rule out antiparallel orientation or lack of long-range magnetic order of the (0001) surface layer. They provide further strong evidence for an essentially parallel magnetic orientation of surface and bulk for the studied Gd(0001) film. This result is corroborated by recent spin-resolved 4f-PE experiments on Gd films[35] as well as by spin analysis of secondary electrons;[37] the latter indicate a non-vanishing normal magnetization component on the Gd(0001) surface.

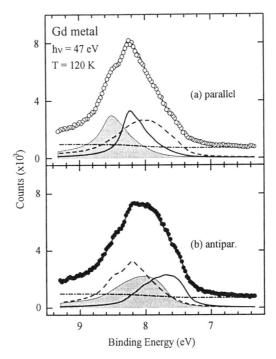

Figure 12. Surface-sensitive MCD-in-4f-PE spectra from Gd(0001)/W(110) for (a) nearly parallel and (b) antiparallel orientation of photon spin and sample magnetization. The spectra show clear shoulders on the high-binding-energy side due to an intense contribution from the (0001) surface layer. Solid curves through the data points represent least-squares-fit results assuming parallel orientation of surface layer (grey-shaded subspectra) and bulk (solid subspectra). Dashed subspectra summarize bulk and surface transitions with $\Delta M = 0$ and opposite ΔM ($\Delta M = -1$ in (a), $\Delta M = +1$ in (b)) due to incomplete circular polarization, a non-vanishing angle between photon spin and sample magnetization, as well as finite temperature.

An enhancement of the surface Curie temperature had previously also been reported for Tb(0001).[38] On this surface, the isolated $^8S_{7/2}$ 4f-PE component (compare Figures 1 and 5) allows a clear spectroscopic separation of the topmost surface layer from the underlying bulk. Figure 13 displays surface-sensitive MCD-PE spectra in the region of the $^8S_{7/2}$ component of Tb(0001). The PE component is split into a bulk signal at $\cong$ 2.3-eV binding energy (BE) and a clearly resolved surface signal, shifted by $\delta_s = (0.26\pm0.03)$ eV to higher BE.[12] The well-resolved surface component of the Tb $^8S_{7/2}$ component allows a separate observation of MCD for the surface layer and for the bulk, i.e. a clear spectroscopic distinction between the magnetizations of the topmost surface layer and the bulk.

The MCD effect is significantly larger for the topmost surface layer than for the bulk, indicating an enhanced in-plane magnetization within the (0001) surface layer. Yet, a quantitative analysis must include the possibilities of (i) a depolarization of the circularly polarized light upon transmission into the bulk and (ii) a changing photoelectron-angular distribution upon reversal of light helicity.[39] The latter would induce an additional intensity variation due to the finite electron-detection angle; both mechanisms are presently investigated in our laboratory.

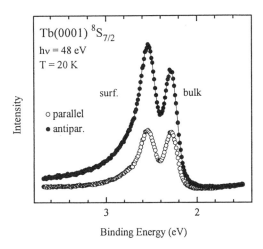

Figure 13. Surface-sensitive MCD-in-4f-PE spectra of Tb(0001). The surface core-level shift of the isolated Tb-$^8S_{7/2}$ PE component allows a separate observation of the MCD effect for the topmost surface layer and for the bulk.[25]

Element-Specific Magnetization

Many of the advanced magnetic systems presently in use contain RE elements, e.g. Gd/Tb, Nd/Tb, and Nd/Dy in storage media for magneto-optical recording,[40] or Ce, Pr, and Sm in permanent magnets. In this last section we want to demonstrate that MCD in 4f-PE can be used to measure the magnetization of RE elements in an element-specific way.

As an example, we take the binary system 1-ML-Eu/Gd(0001).[41] Eu is the element preceding Gd in the periodic table, but it has the same $4f^7$ configuration as Gd (due to the stability of the half-filled shell); Eu is therefore divalent in the metal. One can expect the formation of a thermodynamically stable monolayer of Eu on the Gd(0001) surface, without interdiffusion at higher temperatures.[42] The reason for this is given by the fact that the reduced coordination at the surface is energetically less favorable for the stronger-bound trivalent Gd atoms as compared to divalent Eu atoms.

The geometrical structure of 1-ML Eu on Gd(0001) was studied by LEED at low temperatures (from 20 K to 300 K). In addition to the low-background hexagonal LEED pattern from clean Gd(0001), an array of sharp hexagonally-ordered extra spots was found for Eu thicknesses around one monolayer. This pattern indicates a 6×6 phase: The divalent Eu atoms form a commensurate two-dimensional hexagonal lattice with a nearest-neighbor distance of 4.3 Å, which is $\simeq 20$ % larger than the distance between the trivalent Gd atoms (3.58 Å). Thus in 1-ML-Eu/Gd(0001), the Eu-Eu distance is elongated by $\simeq 9$ % as compared to the elemental Eu lattice. On the basis of LEED patterns taken for different sample temperatures and electron energies, we propose a structure model analogous to the 5×5 reconstruction of the divalent (0001) surface on bulk-trivalent Sm metal.[43] The present results show that 1-ML-Eu/Gd(0001) has indeed a sharp interface of well-ordered Eu and Gd atoms, as expected from thermochemical arguments.

4f-PE spectra of 1-ML-Eu/Gd(0001) at 35 K are presented in Figure 14 (a). Both Gd and Eu give rise to analogous $4f^6$-^7F_J PE multiplets, with binding energies differing by $\simeq 6$ eV; this allows an unambiguous separation of the PE signals. The binding energy of the Eu signal was found to assume a minimum value of 2.6 eV (highest-BE component J=6) upon completion of one monolayer; this was used to calibrate the quartz-microbalance. A similar lowering of the 4f binding energy with increasing coverage had previously been found for the submonolayer regime of Yb/Mo(110).[44] MCD spectra taken at 15 K are

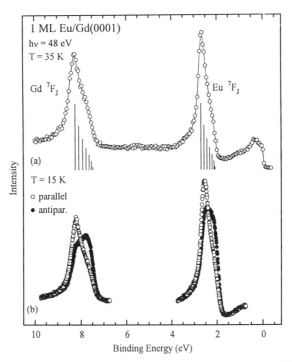

Figure 14. 4f PE spectra from 1-ML-Eu/Gd(0001): (a) The $4f^{6}$-^7F_J PE multiplets of Eu and Gd are clearly separated (vertical bars indicate the unresolved multiplet components). (b) The very similar MCD effects of the two ^7F_J peaks, observed for a sample temperature of 15 K, readily reveal long-range ferromagnetic order of the Eu adlayer as well as approximately parallel orientation of the Eu and Gd magnetizations at the interface.

presented in (b). The Gd spectra clearly show ferromagnetic order. They are similar to the ones from the uncovered Gd(0001) surface (see Figure 2), yet without a surface component on the high-BE side, which has been fully quenched upon adsorption of a complete Eu monolayer. It is evident from inspection by eye of the MCD spectra that the Eu monolayer is ferromagnetically ordered, with parallel orientation of the 4f moments of Eu and of the adjacent Gd layer. Note that Eu metal orders antiferromagnetically below 90 K in the elemental bcc lattice.

Furthermore, the Eu-MCD signal is found to decay rapidly with increasing temperature (followed up to 100 K), whereas it stays nearly constant for Gd, thus indicating very different exchange-coupling constants $J_{\parallel}$, within the Eu layer, and $J_{\perp}$, between the Eu and the adjacent Gd layer.[45] It is obvious that the sharp interface between magnetically-ordered Eu and Gd can serve as a model system for studying the relative exchange-coupling strengths within divalent and trivalent RE metals (having identical 4f moments) as well as the coupling between di- and trivalent magnetic layers.

SUMMARY AND PERSPECTIVES

The use of circularly polarized light in PE from ferromagnetically ordered RE materials leads to very large magnetic circular dichroism (MCD) effects: The intensities of the 4f and 4d PE components depend strongly on the magnitude and orientation of sample magnetization with respect to the photon spin of the exciting circularly polarized soft x-rays.

In some relevant cases, like the Tb-$^8S_{7/2}$ PE component, a sizeable PE signal is only observed for one magnetization direction with respect to the photon spin. MCD in PE from RE materials thus allows to measure the magnetization of a given sample, similar as spin-resolved PE, yet without the need for time-consuming electron-spin analysis.

Due to the localized nature of the 4f electrons, MCD in 4f-PE can be described in an atomic picture. By use of a simple multiplet theory, we can understand the effects as a consequence of the dipole-selection rules. In a few relevant cases (Gd, Tb-$^8S_{7/2}$) one can use the illustrative LS-coupling scheme in conjunction with the dipole selection rules to calculate the MCD effect. The results agree reasonably well with those obtained by more accurate intermediate-coupling calculations.[23,24]

We suggest to use the MCD-PE asymmetry as an x-ray polarimeter over a wide photon-energy range, from $\cong$ 30 eV up to the hard x-ray region, and we give a first demonstration of its feasibility. At the (0001) surface of RE metals, MCD in 4f-PE can separate the magnetization within the topmost atomic surface layer from the bulk magnetization, making use of the surface core-level shift. For Gd(0001), it reveals the existence of a large in-plane surface magnetization oriented parallel to the bulk.

By applying MCD in 4f-PE to materials containing different RE elements, individual magnetic sublattices can be distinguished by separating the element-specific PE multiplets. This was first demonstrated using the hetero-magnetic interface 1-ML-Eu/Gd(0001) as an example: The Eu-4f moments are found to be ferromagnetically ordered at low temperatures and oriented parallel to the neigboring Gd moments.

MCD in 4f-PE from magnetically ordered RE materials opens new perspectives in the analysis of surface and thin-film magnetism: (i) It allows to measure the magnitudes and the relative orientation of bulk and surface magnetizations without the need for electron-spin analysis, i.e. with the speed of conventional PE experiment. (ii) The sheer magnitude of the MCD effect is very attractive for domain-imaging applications, offering high magnetic contrast, comparable with or even higher than the one obtained presently by MCD in x-ray absorption.[46] As an advantage, the surface sensitivity is higher and can be varied continuously. This potential has already motivated first experiments for magnetic imaging.[47] (iii) Exploiting the recently formulated sum-rules,[23] MCD in 4f-PE has the potential to yield the orbital 4f moment, carried by each individual RE element in e.g. transition-metal/RE intermetallic compounds. (iv) 4f-PE experiments, in which excitation by circularly polarized light is combined with photoelectron-spin analysis, will directly measure the expectation value of the inner product $(\vec{L}\cdot\vec{S})$; in such experiments, the orbital angular momentum $\vec{L}$ is defined through the circularly polarized light (MCD) and the spin $\vec{S}$ through the axis of the electron-spin detector. An experimental access to the strength of 4f spin-orbit coupling is important for an improved understanding of the role of single-ion anisotropies in the search for technologically relevant thin magnetic films with perpendicular spontaneous magnetization.

Acknowledgements

We thank the staff of BESSY, in particular G. Reichart and M. Willmann, for their indispensable experimental help. We are also grateful to G. van der Laan for communicating theoretical MCD multiplet intensities prior to publication. This work was supported by the Bundesminister für Forschung und Technologie, project No. 05-5KEAXI-3/TP01, and the SfB-290/TPA6 of the Deutsche Forschungsgemeinschaft.

REFERENCES

1. G. Schütz, W. Wagner, W. Wilhelm, P. Kienle, R. Zeller, R. Frahm, and G. Materlik, Phys. Rev. Lett. 58: 737 (1987).
2. G. Schütz, M. Knülle, R. Wienke, W. Wilhelm, W. Wagner, P. Kienle, and R. Frahm, Z. Physik B 73: 67 (1988).
3. C.T. Chen, F. Sette, Y. Ma, and S. Modesti, Phys. Rev. B 42: 7262 (1990).
4. U. Fano, Phys. Rev. 178: 131 (1969).
5. G. Schütz, Phys. Blätter 46: 475 (1990).
6. H. Ebert, P. Strange, and B.L. Gyorffy, Z. Phys. B 73: 77 (1988).
7. P. Carra and M. Altarelli, Phys. Rev. Lett. 64: 1286 (1990).
8. B.T. Thole, P. Carra, F. Sette, and G. van der Laan, Phys. Rev. Lett. 68: 1943 (1992).
9. L. Baumgarten, C.M. Schneider, H. Petersen, F. Schäfers, and J. Kirschner, Phys. Rev. Lett. 65: 492 (1990).
10. H. Ebert, L. Baumgarten, C.M. Schneider, and J. Kirschner, Phys. Rev. B 44: 4406 (1991).
11. C.M. Schneider, M.S. Hammond, P. Schuster, A. Cebollada, R. Miranda, and J. Kirschner, Phys. Rev. B 44: 12066 (1991).
12. E. Navas, K. Starke, C. Laubschat, E. Weschke, and G. Kaindl, Phys. Rev. B 48 - Rap. Commun.: 14753 (1993).
13. A.V. Fedorov, A. Höhr, E. Weschke, K. Starke, V.K. Adamchuk, and G. Kaindl, Phys. Rev B 49 - Rap. Commun.: 5117 (1994); and references therein.
14. J. Bahrdt, A. Gaupp, W. Gudat, M. Mast, K. Molter, W.B. Peatman, M. Scheer, Th. Schroeter, and Ch. Wang, Rev. Sci. Instrum. 63: 339 (1992).
15. H. Petersen, M. Willmann, F. Schäfers, and W. Gudat, Nucl. Instrum. Methods A 333: 594 (1993).
16. S. Di Fonzo, W. Jark, F. Schaefers, H. Petersen, A. Gaupp, and J.H. Underwood, Appl. Optics 33, 2624 (1994).
17. M. Born and E. Wolf. "Principles of Optics", Pergamon Press, London (1959).
18. B.J. Beaudry and K.A. Gschneidner, in: "Handbook of Physics and Chemistry of Rare Earths", Vol. 1, K.A. Gschneidner and L.R. Eyring eds., North-Holland, Amsterdam (1978).
19. J. Kolaczkiewicz and E. Bauer, Surf. Sci. 175: 487 (1986).
20. M. Farle, K. Baberschke, U. Stetter, A. Aspelmeier, and F. Gerhardter, Phys. Rev. B 47: 11571 (1993).
21. K. Starke, K. Ertl, and V. Dose, Phys. Rev. B 46: 9709 (1992).
22. K. Starke, E. Navas, L. Baumgarten, and G. Kaindl, Phys. Rev. B 48 - Rap. Commun.: 1329 (1993).
23. G. van der Laan and B.T. Thole, Phys. Rev. B 48: 210 (1993).
24. G. van der Laan, private communication.
25. K. Starke, L. Baumgarten, E. Arenholz, E. Navas, and G. Kaindl, Phys. Rev. B - Rap. Commun.: in print (1994).
26. E. Navas, E. Arenholz, K. Starke, and G. Kaindl, to be published.
27. J. Sugar, Phys. Rev. B 5: 1785 (1972).
28. C. Laubschat, E. Weschke, G. Kalkowski, and G. Kaindl, Physica Scripta 41: 124 (1990).
29. F. Gerken, J. Barth, and C. Kunz, Phys. Rev. Lett. 47: 993 (1981).
30. L. Baumgarten, E. Arenholz, E. Navas, K. Starke, and G. Kaindl, to be published.
31. J.L. Dehmer, A.F. Starace, U. Fano, J. Sugar, and J.W. Cooper, Phys. Rev. Lett. 26: 1521 (1971).
32. S. Imada and T. Jo, J. Phys. Soc. Jap. 59: 3358 (1990).
33. D. Weller, S.F. Alvarado, W. Gudat, K. Schröder, and M. Campagna, Phys. Rev. Lett. 54: 1555 (1985).

34. G.A. Mulhollan, K. Garrison, and J.L. Erskine, Phys. Rev. Lett. 69: 3240 (1992).
35. E. Vescovo, C. Carbone, and O. Rader, Phys. Rev. B 48 - Rap. Commun.: 7731 (1993).
36. S. Doniach and M. Sunjic, J. Phys. C 3: 285 (1970).
37. H. Tang, D. Weller, T.G. Walker, J.C. Scott, C. Chappert, H. Hopster, A.W. Pang, D.S. Dessau, and D.P. Pappas, Phys. Rev. Lett. 71: 444 (1993).
38. C. Rau, Appl. Phys. A 49: 579 (1989).
39. Y.U. Idzerda and D.E. Ramaker, Mat. Res. Soc. Symp. Proc. 313: 659 (1993).
40. J. Daval and B. Bechevet, J. Magn. Magn. Mater. 129: 98 (1994).
41. E. Arenholz, K. Starke, E. Navas, and G. Kaindl, to be published.
42. A.R. Miedema, P.F. de Chatel, and F.R. de Boer, Physica B (Amsterdam) 100: 1 (1980).
43. A. Stenborg, J.N. Andersen, O. Björneholm, A. Nilsson, and N. Mårtensson, Phys. Rev. Lett. 63: 187 (1989).
44. A. Stenborg, O. Björneholm, A. Nilsson, N. Mårtensson, J.N. Andersen, and C. Wigren, Surf. Sci. 211/212: 470 (1989).
45. J. Mathon and S.B. Ahmad, Phys. Rev. B 37 - Rap. Commun.: 660 (1988).
46. J. Stöhr, Y. Wu, M.G. Samant, B.D. Hermsmeier, G. Harp, S. Koranda, D. Dunham, and B.P. Tonner, Science 259: 658 (1993).
47. T. Kachel, W. Gudat, and K. Holldack, Appl. Phys. Lett. 64: 655 (1994).

DICHROIC PHOTOEMISSION FOR PEDESTRIANS

Gerrit van der Laan

Daresbury Laboratory
Warrington WA4 4AD
United Kingdom

1. INTRODUCTION

Until very recently it was commonly believed that core level photoemission would not depend on the linear or circular polarization of the incident photon, except in the near threshold region. However, in localized materials such as transition metal, rare earth, and actinide compounds there is a strong interaction between the core hole and the valence electrons which can result in spin and orbital polarization of the core level and consequently in dichroic photoemission.[1] This phenomenon differs from magnetic x-ray dichroism (MXD),[2] where the core electron is excited into the unoccupied part of the $3d$ level. The dichroism is then primarily caused by the Pauli exclusion principle, i.e. only electrons with specific spin and orbital magnetic quantum numbers can be excited into the empty valence states. In x-ray photoemission the electron is excited into continuum states which have no magnetic structure far above the Fermi level and the dichroism is induced by core-valence interaction.

Baumgarten *et al.*[3] observed magnetic circular dichroism (MCD) in x-ray photoemission from the $2p$ core level of ferromagnetic iron. Since then various studies have confirmed the presence of MCD [4-9] as well as shown the existence of linear dichroism in photoemission.[10] Contrary to its counterpart in x-ray absorption, the MCD in angle dependent photoemission does not disappear when the magnetization direction is perpendicular to the light polarization.[4] Moreover, linear magnetic dichroism in the angular dependence (LMDAD) has been observed in a chiral geometry by taking the difference in the photoemission upon reversal of the magnetization direction perpendicular to the plane of measurement.[11,12] Circular dichroism has also been observed in resonant photoemission.[13]

Theoretical studies have been performed using a single particle model [14] as well as using a many body approach.[15,16] Applications have been found in the analysis of the core level satellite and multiplet structure [17] and in surface magnetism, e.g. the strong MCD in the Ni 2p of ferromagnetic Ni(110) gives evidence for an enhanced surface orbital magnetic moment.[6]

Here we will we introduce the basic theory of circular and linear dichroism in angular dependent and spin resolved core level photoemission using the established methods of angular momentum coupling.[18-20] A more extensive treatment of the theory can be found in Ref. 21-23. The outline is as follows. In Sec. 2 we show how by rewriting the angle integrated transition probability for polarized light new linear combinations can be made which correspond to the isotropic spectrum, the circular dichroism and the linear dichroism, respectively. Sum rules for these spectra are derived in Sec. 3. For closed shells the dichroism signals are zero but for open shells they are proportional to the ground state magnetic moments. In Sec. 4 we give a more formal derivation, which will allow us to generalize the previous results. Spin detection is treated in Sec. 5 and angle dependence in Sec. 6. The geometrical dependence can be used to measure odd magnetic moments with linearly polarized light and even magnetic moments with circularly polarized light as is shown in Sec. 7. Expressions for the angular dependent emission from a p shell are given in Sec. 8 and the spectral shape is discussed in Sec. 9. Emission from a d shell is treated in Sec. 10. Some conclusions are given in Sec. 11.

2. FUNDAMENTAL SPECTRA

Consider a ground state $|g\rangle$ of the configuration l^n. Dipole transitions are allowed to a final state which contains any state f of the configuration l^{n-1} and a continuum electron with orbital momentum $c = l \pm 1$ and orbital component γ. Omitting the reduced matrix elements, the angle integrated transition probability for photoemission with q polarized light is obtained by summing over the m levels of the ground state as

$$I_q = \sum_m \langle g | l_m | f \rangle \langle f | l_m^\dagger | g \rangle \sum_\gamma \begin{pmatrix} l & 1 & c \\ -m & q & \gamma \end{pmatrix}^2 , \qquad (1)$$

where $l_m^\dagger$ and l_m create and destroy an electron from the shell l with azimuthal quantum number m. We assume cylindrical symmetry so that there are no cross terms mm'. The components of the polarization vector of the light are $q = -1, 0, +1$ which denote right-circularly, Z-linearly, and left-circularly polarized radiation, respectively. The sum over γ gives the dependence of the intensity on the light polarization and orbitals in angle integrated photoemission. The angle dependent case will be treated in Sec. 6. The coefficient in the round brackets is a 3-j symbol that gives the coupling of the three angular momentum vectors lm, $1q$ and $c\gamma$. The rotation invariance of the cylinder symmetry requires that $\gamma = m-q$. Because the spin only enters when we measure the photoelectron spin we implicitly assume a summation over up and down spin.

We can consider

$$I(m) \equiv \langle g | l_m | f \rangle \langle f | l_m^\dagger | g \rangle , \qquad (2)$$

as the fundamental properties of the system which contain the information about the one-electron properties connected with the shell l of the atom in the state $|g\rangle$ and also the one-electron excitations to the final states $|f\rangle$.

Thus we can write Eq. (1) as

$$I_q = \sum_m I(m)\, t_{mq}, \tag{3}$$

where t_{mq} gives the transition probability for photoemission from a level m with light of polarization q:

$$t_{mq} \equiv \sum_\gamma \begin{pmatrix} l & 1 & c \\ -m & q & \gamma \end{pmatrix}^2 = \begin{pmatrix} l & 1 & c \\ -m & q & m-q \end{pmatrix}^2. \tag{4}$$

The squared 3-j symbol can be written in powers of m as

$$t_{mq} = \tfrac{1}{3}A_{0lc} + A_{1lc}\frac{qm}{2l} + A_{2lc}\left(\tfrac{3}{2}q^2 - 1\right)\frac{m^2 - \tfrac{1}{3}l(l+1)}{l(2l-1)}. \tag{5}$$

where the coefficients are given as

$$A_{x,l,l-1} = \frac{1}{2l+1} \quad \forall\, x, \tag{6}$$

$$A_{0,l,l+1} = \frac{1}{2l+1}, \tag{7}$$

$$A_{1,l,l+1} = -\frac{l+1}{l(2l+1)}, \tag{8}$$

$$A_{2,l,l+1} = \frac{(l+1)(2l+3)}{l(2l-1)(2l+1)}. \tag{9}$$

Using Eq. (3) and (5) we can make new linear combinations J^x for the transition probabilities

$$J^0 \equiv I_1 + I_0 + I_{-1} = A_{0lc}\sum_m I(m) \equiv A_{0lc}\, I^0, \tag{10}$$

$$J^1 \equiv I_1 - I_{-1} = \frac{A_{1lc}}{l}\sum_m m\, I(m) \equiv A_{1lc}\, I^1, \tag{11}$$

$$J^2 \equiv I_1 - 2I_0 + I_{-1} = \frac{A_{2lc}}{\tfrac{1}{3}l(2l-1)}\sum_m [m^2 - \tfrac{1}{3}l(l+1)]\, I(m) \equiv A_{2lc}\, I^2, \tag{12}$$

which are the isotropic spectrum, the circular dichroism and the linear dichroism, respectively.

3. INTEGRATED INTENSITIES

We can now derive the sum rules for the I^x spectra. The one-electron properties of the ground state are obtained by summing over all $|f\rangle$ in Eq. (2). Using the closure relation this yields the occupation numbers n_m of the m levels

$$\rho(m) \equiv \sum_f I(m) = \sum_f \langle g | l_m | f \rangle \langle f | l_m^\dagger | g \rangle = \langle g | l_m l_m^\dagger | g \rangle$$

$$= \langle g | n_m | g \rangle = \langle n_m \rangle \ . \tag{13}$$

Thus, the $\rho(m)$'s tell us which electron states in $|g\rangle$ are occupied or unoccupied. Using Eq. (3) and (13) the integrated photoemission for q polarized light is given as

$$\rho_q = \sum_{fm} \rho(m) \ t_{mq} = \sum_m \langle n_m \rangle \ t_{mq} \ . \tag{14}$$

We can make linear combinations $\rho^x = \sum_f I^x$ as

$$\rho^0 \equiv \frac{\rho_1 + \rho_0 + \rho_{-1}}{A_{0lc}} = \sum_m \langle n_m \rangle = \langle n_l \rangle \ , \tag{15}$$

which means that the integrated isotropic signal is equal to the number of electrons in the l shell,

$$\rho^1 \equiv \frac{\rho_1 - \rho_{-1}}{A_{1lc}} = l^{-1} \sum_m \langle n_m \rangle \, m = l^{-1} \langle L_z \rangle \ , \tag{16}$$

thus the integrated circular dichroism is equal to the orbital magnetic moment, and

$$\rho^2 \equiv \frac{\rho_1 - 2\rho_0 + \rho_{-1}}{A_{2lc}} = \frac{1}{\frac{1}{3}l(2l-1)} \sum_m \langle n_m \rangle \, [m^2 - \tfrac{1}{3}l(l+1)] = \frac{1}{\frac{1}{3}l(2l-1)} \langle Q_{zz} \rangle \ , \tag{17}$$

thus the integrated linear dichroism is proportional to the quadrupole moment

$$Q_{zz} = \sum_i l_z^2(g) - \tfrac{1}{3}l(l+1) \ . \tag{18}$$

It follows that for closed shells $\rho^{x \neq 0} = 0$.

4. MAGNETIC MOMENTS RECOUPLING

In Sec. 3 we found useful relations for angle integrated photoemission just by rewriting the expression for the squared 3-j symbol [Eq. (5)]. We will now give a more formal derivation, which will allow us to generalize the previous results.

In the expression for the photoemission transition probability [Eq. (1)] the angular momentum vectors lm of the electron and $1q$ of the light are coupled to a resultant $c\gamma$ and the conjugated vectors are coupled accordingly. The total result is a scalar, which gives the angle-integrated intensity. There are two ways in which four angular momentum vectors $a\alpha$, $b\beta$, $c\gamma$ and $d\delta$ can be coupled to a zero resultant, and the relation between these two ways is given by

$$\sum_{\varepsilon} (-)^{e-\varepsilon} \begin{pmatrix} e & a & c \\ \varepsilon & \alpha & \gamma \end{pmatrix} \begin{pmatrix} e & b & d \\ -\varepsilon & \beta & \delta \end{pmatrix}$$

$$= \sum_{f} [f] (-)^{b+c+e+f} \begin{Bmatrix} a & b & f \\ d & c & e \end{Bmatrix} \sum_{\phi} (-)^{f-\phi} \begin{pmatrix} f & a & b \\ \phi & \alpha & \beta \end{pmatrix} \begin{pmatrix} f & c & d \\ -\phi & \gamma & \delta \end{pmatrix}, \quad (19)$$

where $[f]$ is shorthand for $2f+1$. The coefficient with the curly brackets is a 6-j symbol, which can be written as sum over 3-j symbols:

$$\begin{Bmatrix} a & b & f \\ d & c & e \end{Bmatrix} = \sum (-)^{a+f+c-\alpha-\phi-\gamma} \begin{pmatrix} a & e & c \\ \alpha & \varepsilon & -\gamma \end{pmatrix} \begin{pmatrix} c & d & f \\ \gamma & \delta & -\phi \end{pmatrix} \begin{pmatrix} f & b & a \\ \phi & \beta & -\alpha \end{pmatrix} \begin{pmatrix} b & d & e \\ \beta & \delta & \varepsilon \end{pmatrix}, \quad (20)$$

where the sum is to be taken over all components α, β, etc.

The recoupling in Eq. (19) changes the resultant $e\varepsilon$ to $f\phi$, which can be considered as an auxiliary quantum number with component ϕ. If the momentum vectors are conjugated, so that $a\alpha = -(b\beta)^*$ and $c\gamma = -(d\delta)^*$, the component ϕ is zero.

Applying Eq. (19) to Eq.(1) and using the symmetry properties of the 3-j symbols

$$\begin{pmatrix} l & 1 & c \\ m & q & \gamma \end{pmatrix}^* = \begin{pmatrix} l & 1 & c \\ -m & -q & -\gamma \end{pmatrix} = (-)^{l+1+c} \begin{pmatrix} l & 1 & c \\ m & q & \gamma \end{pmatrix}, \quad (21)$$

results in

$$I_q = \sum_{mx} \langle g | l_m | f \rangle \langle f | l_m^\dagger | g \rangle (-)^{l-m} \begin{pmatrix} l & x & l \\ -m & 0 & m \end{pmatrix} \begin{Bmatrix} l & x & l \\ 1 & c & 1 \end{Bmatrix} (-)^{1-q} [x] \begin{pmatrix} 1 & x & 1 \\ -q & 0 & q \end{pmatrix}, \quad (22)$$

where we introduced the auxiliary quantum number x which has component 0, since $lm = -(lm)^*$. The triad $(1x1)$ gives the triangular condition $x = 0, 1, 2$.

We can separate the geometry from the part describing the physical properties of the atom. The latter are defined by the spectra

$$I^x \equiv n_{lx}^{-1} \sum_{m} \langle g | l_m | f \rangle \langle f | l_m^\dagger | g \rangle (-)^{l-m} \begin{pmatrix} l & x & l \\ -m & 0 & m \end{pmatrix}, \quad (23)$$

where we used a numerical factor n_{lx} to remove the square roots

$$n_{lx} \equiv \begin{pmatrix} l & x & l \\ -l & 0 & l \end{pmatrix} = \frac{(2l)!}{\sqrt{(2l-x)!\,(2l+1+x)!}}. \quad (24)$$

The 3-j symbol in Eq. (23) imposes $0 \le x \le 2l$. Substitution of Eq. (23) into (22) gives

$$I_q = \sum_{mx} I^x \, n_{lx} \begin{Bmatrix} l & x & l \\ 1 & c & 1 \end{Bmatrix} (-)^{1-q} [x] \begin{pmatrix} 1 & x & 1 \\ -q & 0 & q \end{pmatrix}, \tag{25}$$

The linear combinations of intensities [cf. Eq. (10-12)] are defined as

$$J^x = \sum_q n_{1x}^{-1} (-)^{1-q} \begin{pmatrix} 1 & x & 1 \\ -q & 0 & q \end{pmatrix} I_q. \tag{26}$$

Substitution of Eq. (25) into (26) gives

$$J^x = \sum_q n_{1x}^{-1} (-)^{1-q} \begin{pmatrix} 1 & x & 1 \\ -q & 0 & q \end{pmatrix} \sum_{x'} I^{x'} n_{lx'} \begin{Bmatrix} l & x' & l \\ 1 & c & 1 \end{Bmatrix} (-)^{1-q'} [x'] \begin{pmatrix} 1 & x' & 1 \\ -q' & 0 & q' \end{pmatrix}. \tag{27}$$

Using the orthogonality relations for 3-j symbols

$$\sum_{\alpha\beta} [c] \begin{pmatrix} a & b & c \\ \alpha & \beta & \gamma \end{pmatrix} \begin{pmatrix} a & b & c' \\ \alpha & \beta & \gamma' \end{pmatrix} = \delta(cc') \, \delta(\gamma\gamma'), \tag{28}$$

and

$$\sum_{c\gamma} [c] \begin{pmatrix} a & b & c \\ \alpha & \beta & \gamma \end{pmatrix} \begin{pmatrix} a & b & c \\ \alpha' & \beta' & \gamma \end{pmatrix} = \delta(\alpha\alpha') \, \delta(\beta\beta'), \tag{29}$$

reduces Eq. (27) to

$$J^x = A_{xlc} I^x, \tag{30}$$

[cf. Eq. (10-12)] where

$$A_{xlc} \equiv \begin{Bmatrix} l & x & l \\ 1 & c & 1 \end{Bmatrix} n_{lx} \, n_{1x}^{-1}. \tag{31}$$

[cf. Eq. (6-9)] The integrals of the I^x in Eq. (23) are

$$\rho^x \equiv n_{lx}^{-1} \sum_m \langle n_m \rangle (-)^{l-m} \begin{pmatrix} l & x & l \\ -m & 0 & m \end{pmatrix}, \tag{32}$$

which generalizes the expressions for ρ^0, ρ^1 and ρ^2 in Eq. (15-17).

5. SPIN RESOLVED PHOTOEMISSION

When the spin of the photoelectron is detected along the z axis we can define the fundamental spectra I^{xy} with integrals

$$\rho^{xy} \equiv n_{lx}^{-1} \, n_{(1/2)y}^{-1} \sum_{\sigma m} \langle n_{m\sigma} \rangle (-)^{l-m} \begin{pmatrix} l & x & l \\ -m & 0 & m \end{pmatrix} (-)^{(1/2)-\sigma} \begin{pmatrix} 1/2 & y & 1/2 \\ -\sigma & 0 & \sigma \end{pmatrix}, \tag{33}$$

where $n_{m\sigma}$ is the occupation number of the $m\sigma$ sublevel. For $y=0$ the signals for spin up and down are added, so this is the signal when we do not measure the spin polarization, thus $\rho^{x0} = \rho^x$. For $y=1$ the signals for spin up and down are subtracted resulting in the spin difference spectra with sum rules:

$$\rho^{01} = \sum_m \langle n_{m\sigma} \rangle = \langle S_z \rangle , \tag{34}$$

$$\rho^{11} = l^{-1} \sum_m \langle n_{m\sigma} \rangle m = l^{-1} \langle \sum l_z(i) s_z(i) \rangle , \tag{35}$$

where l_z and s_z are defined as one-electron operators acting only on l shell functions, contrary to L and S which are sums of these over all electrons, and measure only properties of l shell electrons.

$$\rho^{21} = \frac{1}{\frac{1}{3}l(2l-1)} \sum_m \langle n_{m\sigma} \rangle [m^2 - \tfrac{1}{3}l(l+1)] = \frac{1}{\frac{1}{3}l(2l-1)} \langle \sum q_{zz}(i) s_z(i) \rangle , \tag{36}$$

where q_{zz} is defined as the one-electron operator $l_z^2 - \tfrac{1}{3}l(l+1)$.

6. ANGLE DEPENDENT PHOTOEMISSION

It is straightforward to generalize the results from Sec. 4 to the angular dependent case. The dipole transition probability using polarized radiation from a ground state $\langle g|$ to a final state $|f\rangle$ under emission of an electron in the direction ε from an open or closed shell with angular momentum l is

$$|D_{qq'}(\varepsilon)|^2 = \sum_m \langle g | l_m^\dagger | f \rangle \langle f | l_m | g \rangle \sum_{cc' \gamma\gamma'} \begin{pmatrix} l & 1 & c \\ -m & q & \gamma \end{pmatrix} \begin{pmatrix} l & 1 & c' \\ -m & q' & \gamma' \end{pmatrix} Y_\gamma^c(\varepsilon) Y_{\gamma'}^{c'}(\varepsilon)^*$$

$$\times \langle l \| r C^{(1)} \| c \rangle \langle c' \| r C^{(1)} \| l \rangle e^{i(\delta_c - \delta_{c'})} , \tag{37}$$

where the $\langle l \| r C^{(1)} \| c \rangle$ is the reduced dipole matrix element and δ_c is the phase shift of the continuum function with orbital momentum c, where $c=l\pm 1$. The $Y_\gamma^c(\varepsilon)$ are spherical harmonics. We allow for interference between the two final state channels c and c'. We introduce both q and q' to allow in Eq. (44) for an arbitrary direction of the light, which need not be along the z-axis.

Compared to the angle integrated case, the four angular momentum vectors are now not coupled to a scalar but to a vector representing the direction of the outgoing photoelectron. Applying similar rules as given in Eq. (20) and introducing the sums over the auxiliary quantum numbers a, b and x of the photon, the photoelectron and the atom, respectively, Eq. (37) can be recoupled to

$$|D_{qq'}(\varepsilon)|^2 = \sum_{xm} \langle g | l_m^\dagger | f \rangle \langle f | l_m | g \rangle (-)^{l-m} \begin{pmatrix} l & x & l \\ -m & 0 & m \end{pmatrix}$$

159

$$\times \sum_{ab} [abx] A_{abx}^{cc'l} \begin{pmatrix} 1 & a & 1 \\ -q' & \alpha & q \end{pmatrix} \begin{pmatrix} a & x & b \\ -\alpha & 0 & \alpha \end{pmatrix} \sum_{\gamma\gamma'} \begin{pmatrix} c' & b & c \\ -\gamma' & \alpha & \gamma \end{pmatrix} Y_\gamma^c(\epsilon) Y_{\gamma'}^{c'}(\epsilon)^*$$

$$\times \langle l \| r C^{(1)} \| c \rangle \langle c' \| r C^{(1)} \| l \rangle e^{i(\delta_c - \delta_{c'})}, \qquad (38)$$

where the numerical factor A is proportional to a 9-j symbol:

$$A_{abx}^{cc'l} \propto \begin{Bmatrix} l & x & l \\ c & b & c' \\ 1 & a & 1 \end{Bmatrix}. \qquad (39)$$

The 9-j symbol can be written as a sum over 6-j or 3-j symbols:

$$\begin{Bmatrix} l & x & l \\ c & b & c' \\ 1 & a & 1 \end{Bmatrix} = \sum_j (-)^{2j} [j] \begin{Bmatrix} l & c & 1 \\ a & 1 & j \end{Bmatrix} \begin{Bmatrix} x & b & a \\ c & j & c' \end{Bmatrix} \begin{Bmatrix} l & c' & 1 \\ j & l & x \end{Bmatrix}$$

$$= \sum \begin{pmatrix} l & x & l \\ -m & 0 & m \end{pmatrix} \begin{pmatrix} c & b & c' \\ -m & \alpha & \gamma' \end{pmatrix} \begin{pmatrix} 1 & a & 1 \\ -q & \alpha & q' \end{pmatrix} \begin{pmatrix} l & c & 1 \\ -m & \gamma & q \end{pmatrix} \begin{pmatrix} x & b & a \\ 0 & \alpha & \alpha \end{pmatrix} \begin{pmatrix} l & c' & 1 \\ -m & \gamma' & q' \end{pmatrix}, \qquad (40)$$

where the sum is to be taken over all components.

In Eq. (38) we again can separate the physical properties I^x defined in Eq. (23) from the part describing the geometry of the experiment.

As an exercise one may restore from Eq. (38) the angle integrated expression in Eq. (22) by taking $b=0$ and $a=x$ which gives the following simplifications:

$$\begin{Bmatrix} l & x & l \\ c & 0 & c \\ 1 & x & 1 \end{Bmatrix} = (-)^{l+1+x+c} [xc]^{-1/2} \begin{Bmatrix} l & x & l \\ 1 & c & 1 \end{Bmatrix}, \qquad (41)$$

$$\sum_\gamma (-)^{c-\gamma} \begin{pmatrix} c & c & 0 \\ \gamma & -\gamma & 0 \end{pmatrix} = [c]^{1/2}, \qquad (42)$$

$$(-)^x \begin{pmatrix} x & x & 0 \\ 0 & 0 & 0 \end{pmatrix} = [x]^{1/2}. \qquad (43)$$

7. GEOMETRICAL DEPENDENCE

Now that the importance of the fundamental spectra has become clear, we have to know how to measure them. First we discuss the polarization of the light. For this we will define special linear combinations of the intensities measured with different kinds of light:

$$J^a(\mathbf{P}, \epsilon) = \sum_{qq'} r_{qq'}^{1a}(\mathbf{P}) |D_{qq'}(\epsilon)|^2. \qquad (44)$$

The r factors can be defined in such a way that $J^a(\mathbf{P}, \epsilon)$ has a simple interpretation when we rotate the coordinate system to bring the Z-axis along the vector $\mathbf{P}$. Then because for the normalized spherical harmonic we have $C_\alpha^a(\mathbf{Z}) = \delta_{\alpha 0}$, forcing $q=q'$, we have for the coefficients

$$r_{qq'}^{1a}(\mathbf{P}) \equiv \sum_\alpha n_{1a}^{-1} C_\alpha^a(\mathbf{P}) \begin{pmatrix} 1 & a & 1 \\ -q & \alpha & q' \end{pmatrix} = \delta_{qq'} n_{1a}^{-1} (-)^{1-q} \begin{pmatrix} 1 & a & 1 \\ -q & 0 & q' \end{pmatrix} . \qquad (45)$$

From the 3-*j* symbol

$$\begin{pmatrix} 1 & a & 1 \\ -q & 0 & q \end{pmatrix} , \qquad (46)$$

we see that for $a=0$ we add $q=-1,0,+1$ and so measure the isotropic spectrum. For $a=1$ we subtract the spectra for $q=1$ and -1, i.e. left and right circularly polarized light propagating along $\mathbf{P}$ and so measure circular dichroism. For $a=2$ we measure the linear dichroism, subtracting the intensity for light polarized perpendicular ($q=1$ and -1) and parallel ($q=0$) to $\mathbf{P}$. Introducing the angle dependent functions

$$U^{abx}(\mathbf{P},\varepsilon) \propto \sum_\alpha \begin{pmatrix} a & b & x \\ -\alpha & \alpha & 0 \end{pmatrix} Y_\alpha^a(\mathbf{P}) Y_{-\alpha}^b(\varepsilon) , \qquad (47)$$

we can write Eq. (44), the total emission intensity in direction ε for the light specified by a and $\mathbf{P}$ as

$$J^a(\mathbf{P},\varepsilon) = \frac{1}{4\pi} \sum_x I^x \sum_b U^{abx}(\mathbf{P},\varepsilon) \sum_{cc'} A_{abx}^{cc'l} R^{lc} R^{lc'} e^{i(\delta_c - \delta_{c'})} , \qquad (48)$$

where the R are the radial matrix elements. This is the master equation. We see explicitly that in each geometry we measure a linear combination of the same set of spectra I^x. Each I^x produces a limited set of angular distributions ("waves") U with contributions from each channel as a numerical factor A times the radial matrix elements and phase shifts. The three physical properties in the photoemission process are the light, the photoelectron and the emitting shell which have polarizations along $\mathbf{P}$, ε and $\mathbf{Z}$-axis, respectively with moments a, b and x. [c.f. Eq. (47)] For a chosen value of a each I^x sends out a set of different waves U^{abx} with $b = |x-a| \ldots x+a$ and b even. We see that for $x=0$ there are two waves with $b=a=0$ or 2. For every even $x \neq 0$ there is one wave $b=x$ for $a=0$, one wave for $a=1$, and three waves for $a=2$. For every odd x there are two waves for $a=1$ and three waves for $a=2$ (except for $x=1$). All these waves can be distinguished and so each can be measured in more than one way with coefficients which contain the R and δ and so these can be solved.

We can now see what special effects occur in the case of interference. Without interference ($c=c'$) the 9-*j* symbol in Eq. (40) has two identical columns ($lc1$). For odd permutations the value of the 9-*j* symbol is multiplied by $(-1)^S$, where S is the sum of all nine arguments. Then, since b is even due to parity, the value of $a+x$ has to be even. Thus with $a=0$ and $a=2$, i.e. the isotropic spectrum and the linear dichroism, we measure even poles of x. With $a=1$, i.e. with circular dichroism, we measure odd poles of x. We will call a wave odd or even when $a+b+x$ is odd or even. Odd waves can only occur in interference ($c \neq c'$). They are zero in co-planar geometries, i.e. when **P**, **ε** and **M** are in one plane, and also when they are all perpendicular. So in these geometries we only measure even waves and we need non-coplanar (chiral) geometries to see the odd waves. The measurement of these waves gives circular and magnetic linear dichroism in the angular dependence, which is not present in non-interference measurements. In the circular dichroism in the angular dependence (CDAD) we use $a=1$ to measure even moments x. Odd moments, if the sample is magnetic, can be removed by reversing the magnetization and adding the signals or taking the mirror image of ε in the plane of **P** and **M** and subtracting the signals. In the magnetic linear dichroism in the angular dependence (MLDAD) odd moments are studied using $a=0$ and 2. Even moments are again separated by reversing the magnetization or mirroring ε.

8. ANGULAR EMISSION FROM AN $l=1$ SHELL

It is instructive to write out the angle dependence of the emission $J^a(\mathbf{P},\varepsilon)$ for a p level

$$4\pi J^0 = I^0 \, U^{000}\left(\tfrac{1}{3}R^{00} + \tfrac{2}{3}R^{22}\right) + I^2 \, U^{022}\left[-\tfrac{1}{3}(R^{02} + R^{20}) - \tfrac{1}{3}R^{22}\right] , \quad (49)$$

$$4\pi J^1 = I^1 \left\{ U^{101}\left(\tfrac{1}{3}R^{00} - \tfrac{1}{3}R^{22}\right) + U^{121}\left[-\tfrac{1}{3}(R^{02} + R^{02}) + \tfrac{2}{3}R^{22}\right]\right\}$$
$$+ I^2 \, U^{122} \tfrac{5}{12}i \left(R^{02} - R^{20}\right) , \quad (50)$$

$$4\pi J^2 = I^0 \, U^{220}\left[-\tfrac{2}{3}(R^{02} + R^{20}) - \tfrac{2}{3}R^{22}\right]$$
$$+ I^1 \, U^{221} \tfrac{5}{4}i \left(R^{02} - R^{20}\right)$$
$$+ I^2 \left\{ U^{202}\left(\tfrac{1}{3}R^{00} + \tfrac{1}{15}R^{22}\right) + U^{222}\left[\tfrac{1}{3}(R^{02} + R^{20}) - \tfrac{2}{21}R^{22}\right] + U^{242}\left(\tfrac{36}{35}R^{22}\right)\right\} , \quad (51)$$

where $R^{20} = R^0 R^2 \, e^{i\delta}$ and $R^{02} = R^0 R^2 \, e^{-i\delta}$ with phase difference $\delta \equiv \delta_0 - \delta_2$. The interference terms R^{02} and R^{20} are of special interest. When $a+b+x$ is even the A factor is symmetric in c and c' so that

$$R^{20} + R^{02} = R^0 R^2 \left(e^{i\delta} + e^{-i\delta}\right) = 2R^0 R^2 \cos\delta . \quad (52)$$

When $a+b+x$ is odd, A is imaginary and antisymmetric in c and c' so that

$$R^{20} - R^{02} = R^0 R^2 \left(e^{i\delta} - e^{-i\delta}\right) = 2iR^0 R^2 \sin\delta . \quad (53)$$

Writing the bipolar spherical harmonics U^{abx} as dot and cross products of the light polarization **P**, the emission direction ε and the magnetization **M** we obtain

$$4\pi J^0 = \tfrac{1}{3}I^0 \left(R^{00} + 2R^{22}\right) - \tfrac{1}{6}I^2\left[3(\varepsilon\cdot\mathbf{M})^2 - 1\right]\left(R^{22} + 2R^0 R^2 \cos\delta\right) , \quad (54)$$

$$4\pi J^1 = \tfrac{1}{3}I^1 \left\{(\mathbf{P}\cdot\mathbf{M}) R^{00} + \left[3(\mathbf{P}\cdot\varepsilon)(\varepsilon\cdot\mathbf{M}) - 2(\mathbf{P}\cdot\mathbf{M})\right]R^{22}\right.$$

$$-\left[3\,(\mathbf{P}\cdot\boldsymbol{\varepsilon})\,(\boldsymbol{\varepsilon}\cdot\mathbf{M}) - (\mathbf{P}\cdot\mathbf{M})\right]R^0R^2\cos\delta\bigg\}$$

$$+ I^2\,(\boldsymbol{\varepsilon}\cdot\mathbf{M})\,\mathbf{P}\cdot(\boldsymbol{\varepsilon}\times\mathbf{M})\,R^0R^2\sin\delta \quad , \tag{55}$$

$$4\pi J^2 = -\tfrac{1}{3}I^0\left[3\,(\mathbf{P}\cdot\boldsymbol{\varepsilon}) - 1\right]\left(R^{22} + 2R^0R^2\cos\delta\right)$$

$$-3\,I^1\,\mathbf{P}\cdot(\boldsymbol{\varepsilon}\times\mathbf{M})\,(\mathbf{P}\cdot\boldsymbol{\varepsilon})\,R^0R^2\sin\delta$$

$$+ I^2\left\{\tfrac{1}{6}\left[3\,(\boldsymbol{\varepsilon}\cdot\mathbf{M})^2 - 1\right]R^{00}\right.$$

$$+ \tfrac{1}{2}\left[9\,(\mathbf{P}\cdot\boldsymbol{\varepsilon})^2(\boldsymbol{\varepsilon}\cdot\mathbf{M})^2 - 6\,(\mathbf{P}\cdot\boldsymbol{\varepsilon})(\boldsymbol{\varepsilon}\cdot\mathbf{M})(\mathbf{P}\cdot\mathbf{M}) - (\mathbf{P}\cdot\boldsymbol{\varepsilon})^2 - (\boldsymbol{\varepsilon}\cdot\mathbf{M})^2 + (\mathbf{P}\cdot\mathbf{M})^2\right]R^{22}$$

$$\left. + \tfrac{1}{3}\left[9\,(\mathbf{P}\cdot\boldsymbol{\varepsilon})(\boldsymbol{\varepsilon}\cdot\mathbf{M})(\mathbf{P}\cdot\mathbf{M}) - 3\,(\mathbf{P}\cdot\boldsymbol{\varepsilon})^2 - 3\,(\boldsymbol{\varepsilon}\cdot\mathbf{M})^2 - 3\,(\mathbf{P}\cdot\mathbf{M})^2 + 2\right]R^0R^2\cos\delta\right\} \quad , \tag{56}$$

The angle integrated intensities are obtained from Eq. (54-56) as

$$J_t^0 = \tfrac{1}{3}I^0\left(R^{00} + 2R^{22}\right) , \tag{57}$$

$$J_t^1 = \tfrac{1}{3}I^1\,\mathbf{P}\cdot\mathbf{M}\left(R^{00} - R^{22}\right) , \tag{58}$$

$$J_t^2 = \tfrac{1}{6}I^2\left[3(\mathbf{P}\cdot\mathbf{M})^2 - 1\right]\left(R^{00} + \tfrac{1}{5}R^{22}\right) . \tag{59}$$

Thus for every J_t^a the two final state channels have different coefficients and the interference term has disappeared. I^1 and I^2 are measured in circular and linear dichroism, respectively.

However, in angular dependent photoemission I^1 and I^2 can be measured using either linear or circular dichroism. This results in four different types of geometries which are summarized in Table 1.

Table 1. The four types of geometries with their symmetry properties and angle dependent factors U^{abx} for p emission. The acronyms make clear under which conditions the spectra can be measured. In circular dichroism (CD) we measure the difference between the two circular polarizations of the light ($a=1$). In linear dichroism (LD) we measure the difference between two orthogonal linear polarizations ($a=2$). The spectra with prefix M can also be obtained by reversing the magnetic moment (x=odd). Alternatively, MCD and LD can be referred to as MCDAD and LDAD, respectively, but here we reserve the postfix AD for the case that the measurement requires interference between the two channels, and therefore a chiral geometry ($a+b+x$ =odd).

acronym	a	x	$a+b+x$	U^{abx}
MCD	1	1	even	$U^{101} + U^{121}$
CDAD	1	2	odd	U^{122}
MLDAD	2	1	odd	U^{221}
LD	2	2	even	$U^{202} + U^{222} + U^{242}$

In the first geometry, which was reported by Baumgarten et al.,[3] the I^1 spectrum is measured in circular dichroism ($a=1$) and since an odd magnetic moment is required it is called magnetic circular dichroism (MCD). From Eq. (55) we obtain

$$J_{MCD} \equiv J^1(\text{odd}) = \tfrac{1}{12\pi} I^1 \left\{ (\mathbf{P}\cdot\mathbf{M}) R^{00} + \left[3(\mathbf{P}\cdot\boldsymbol{\varepsilon})(\boldsymbol{\varepsilon}\cdot\mathbf{M}) - 2(\mathbf{P}\cdot\mathbf{M}) \right] R^{22} \right.$$

$$\left. - \left[3(\mathbf{P}\cdot\boldsymbol{\varepsilon})(\boldsymbol{\varepsilon}\cdot\mathbf{M}) - \mathbf{P}\cdot\mathbf{M} \right] R^0 R^2 \cos\delta \right\}$$

$$= \tfrac{1}{3} U^{101}(R^{00} - R^{22}) + \tfrac{2}{3} U^{121}(R^{22} - R^0 R^2 \cos\delta) \ . \tag{60}$$

The last form of the expression shows more clearly what is going on. $U^{101} = \mathbf{P}\cdot\mathbf{M}$ gives the angle integrated emission ($b = 0$) [c.f. Eq. (58)]. For a given $\theta_1 = \angle(\mathbf{M},\mathbf{P})$ this function is a sphere. $U^{121} = \tfrac{3}{2}(\mathbf{P}\cdot\boldsymbol{\varepsilon})(\boldsymbol{\varepsilon}\cdot\mathbf{M}) - \tfrac{1}{3}(\mathbf{P}\cdot\mathbf{M})$ gives the angular dependent emission ($b \neq 0$), which is zero integrated over all emission angles because the U^{abx}'s are orthogonal. The function U^{121} is plotted in Fig. 1 for different values of θ_1. The shape changes from a $d(z^2)$ function for $\mathbf{M}//\mathbf{P}//z$ to a $d(xz)$ function for $\mathbf{M}//z$ and $\mathbf{P}//x$. Note that these functions have nodes and that there are positive and negatives lobes. The extrema in the emission are at $\theta = \theta_1/2 + n\times 90°$, $\phi = 0$. Also observe the way the doughnut changes into a lobe when θ_1 goes from 0 to 90°. As experimentally observed [4] for $\mathbf{M}\perp\mathbf{P}$ the angular dependent emission U^{121} does not vanish, although the angle integrated emission U^{101} is zero.

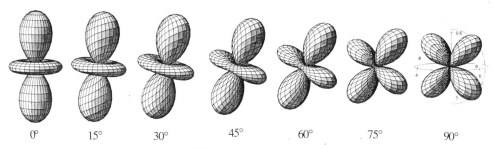

Fig. 1. The bipolar spherical harmonic U^{121} for different values of $\theta_1 = \angle(\mathbf{M},\mathbf{P})$ as indicated in degrees ($\phi_1 = 0$). $\mathbf{M}$ is always directed along the z axis ($\theta = 0$).

The I^1 spectrum can also be measured in linear dichroism ($a=2$).[11] From Eq. (56) we obtain

$$J_{MLDAD} \equiv J^2(\text{odd}) = -\tfrac{3}{4\pi} I^1 \ \mathbf{P}\cdot(\boldsymbol{\varepsilon}\times\mathbf{M})(\mathbf{P}\cdot\boldsymbol{\varepsilon}) R^0 R^2 \sin\delta \ , \tag{61}$$

which is called the magnetic linear dichroism in the angular dependence (MLDAD) and requires a chiral geometry.

The I^2 spectrum measured in linear dichroism (LD) is more lengthy and we give here only the expression for s polarized light as in the measurement by Roth et al.,[10] where $\varepsilon \perp \mathbf{P}$, $\varepsilon \perp \mathbf{M}$ and $\mathbf{P}//\mathbf{M}$. From Eq. (56) we obtain

$$J_{LD}(s \text{ pol}) \equiv J^2(\text{even, } s\text{-pol}) = \tfrac{1}{24\pi} I^2 (2R^{00} + 3R^{22} - 2R^0 R^2 \cos\delta) \; . \tag{62}$$

The I^2 spectrum measured in circular dichroism is obtained from Eq. (55) as

$$J_{CDAD} \equiv J^1(\text{even}) = \tfrac{1}{4\pi} I^2 (\varepsilon \cdot \mathbf{M}) \mathbf{P} \cdot (\varepsilon \times \mathbf{M}) R^0 R^2 \sin\delta \; , \tag{63}$$

which is called the circular dichroism in the angular dependence (CDAD) and requires a chiral geometry.

9. THE SHAPE OF THE $l=1$ SPECTRA

It is illustrative to give the fundamental spectra in the limit of jj coupling. The core hole spin-orbit interaction splits the photoelectron spectrum into a $j = l+s$ and a $j = l-s$ level with an intensity ratio of $l+1:l$, where the $j = l+s$ has the lowest binding energy. In the presence of a small exchange interaction both j levels split into $2j+1$ sublevels with equal energy spacing. The $j = l+s$ level splits into the $m_j = l+s, l+s-1/2, ..., -l-s$, whereas the $j = l-s$ level splits into the $m_j = l-s, l-s-1/2, ..., -l+s$ sublevels.

First we discuss the energy sequence of the sublevels. Taking the direction of the spin magnetic moment S_z of the valence electrons as reference we define spin up (down) photoelectrons as those with spin momentum parallel (antiparallel) to this direction. Note that S_z does not need to be parallel to the macroscopic magnetization direction. For a positive exchange interaction, the core hole with spin up has a higher binding energy than with spin down. In the $j = l+s$ level the spin up character increases with the value m_j and, therefore, the binding energy increases with m_j. In the $j = l-s$ level the spin up character and the binding energy decrease with m_j. Thus this determines the order of the m_j levels. For negative exchange interaction the core hole with spin down has higher binding energy and compared to the case with positive exchange interaction the sequence of the m_j levels within each j level is reversed.

The wavefunctions of the jm_j sublevels for a p core level are given in Table 2. The transition probabilities for light with polarization q ($= -\Delta m$) to the εs continuum $|0^\pm\rangle$ are given in Table 3. From this we can make the linear combinations in given Table 4 with spectra shown in Fig. 2. For example for the sublevel $j = 1/2$, $m_j = -1/2$ with $|\psi\rangle = |(\tfrac{1}{3})^{1/2} 0^- - (\tfrac{2}{3})^{1/2}(-1)^+\rangle$ the spin signal $\sigma^+ - \sigma^-$ is 1/3 and the circular dichroism, which is the difference between $\Delta m = -1$ and $\Delta m = +1$, is -2/3.

The isotropic spectrum, I^{00}, has the same transition probability for each sublevel so that it is independent of m. The spin spectrum, I^{01}, gives the polarization of the spin magnetic moment of the l shell. Both j levels show an intensity distribution with a positive and a negative peak. The sign of these distributions is the same since the core hole spin is coupled in the same way to the exchange field, i.e. spin antiparallel at low binding energy and spin parallel at high binding energy. The integrated intensity over each j level is zero, since the small exchange field does not mix the j levels. The first moment of the $j = 3/2$ edge is ten times larger than that of the $j = 1/2$ edge. This is because the $m_j = \pm 3/2$ in the $j = 3/2$ edge are pure spin states, whereas the $m_j = \pm 1/2$ are mixed spin states.

Table 2. The jm_j states of a p core level given as a linear combination over $m_l m_s$ states.

j	m_j	$\sum c_{m_l m_s} m_l m_s$
$\frac{3}{2}$	$-\frac{3}{2}$	$(-1)^-$
	$-\frac{1}{2}$	$\left(\frac{2}{3}\right)^{1/2} 0^- + \left(\frac{1}{3}\right)^{1/2} (-1)^+$
	$\frac{1}{2}$	$\left(\frac{1}{3}\right)^{1/2} 1^- + \left(\frac{2}{3}\right)^{1/2} 0^+$
	$\frac{3}{2}$	1^+
$\frac{1}{2}$	$\frac{1}{2}$	$\left(\frac{2}{3}\right)^{1/2} 1^- - \left(\frac{1}{3}\right)^{1/2} 0^+$
	$-\frac{1}{2}$	$\left(\frac{1}{3}\right)^{1/2} 0^- - \left(\frac{2}{3}\right)^{1/2} (-1)^+$

Table 3. The dipole transition probabilities for light with polarization q from the jm_j states of a p core level to an εs continuum with spin σ.

$q\sigma$	$j = \frac{3}{2}$				$j = \frac{1}{2}$	
	$-\frac{3}{2}$	$-\frac{1}{2}$	$\frac{1}{2}$	$\frac{3}{2}$	$\frac{1}{2}$	$-\frac{1}{2}$
$+1\uparrow$	-	-	-	1	-	-
$0\uparrow$	-	-	$\frac{2}{3}$	-	$\frac{1}{3}$	-
$-1\uparrow$	-	$\frac{1}{3}$	-	-	-	$\frac{2}{3}$
$+1\downarrow$	-	-	$\frac{1}{3}$	-	$\frac{2}{3}$	-
$0\downarrow$	-	$\frac{2}{3}$	-	-	-	$\frac{1}{3}$
$-1\downarrow$	1	-	-	-	-	-

Table 4. Dipole transition probabilities for the fundamental spectra of a one-electron p state.

I^{xy}	$j = \frac{3}{2}$				$j = \frac{1}{2}$	
	$-\frac{3}{2}$	$-\frac{1}{2}$	$\frac{1}{2}$	$\frac{3}{2}$	$\frac{1}{2}$	$-\frac{1}{2}$
I^{00}	1	1	1	1	1	1
I^{01}	-1	$-\frac{1}{3}$	$\frac{1}{3}$	1	$-\frac{1}{3}$	$\frac{1}{3}$
I^{10}	-1	$-\frac{1}{3}$	$\frac{1}{3}$	1	$\frac{2}{3}$	$-\frac{2}{3}$
I^{11}	1	$-\frac{1}{3}$	$-\frac{1}{3}$	1	$-\frac{2}{3}$	$-\frac{2}{3}$
I^{20}	1	-1	-1	1	0	0
I^{21}	-1	$\frac{5}{3}$	$-\frac{5}{3}$	1	$-\frac{4}{3}$	$\frac{4}{3}$

It is interesting to compare the I^{01} and the I^{10} spectrum. Since the photon does not act on the spin, the I^{10} depends only on the orbital part of the wavefunction. The orbit is not directly coupled to the exchange field, but only aligned to the core spin by the spin-orbit interaction. Because we sum over the two spin directions the I^{10} signals are twice as large as the I^{01} signals but they have to be multiplied by the geometrical spin-orbit factor $l \cdot s$, which is 1/2 for the $j = 3/2$ edge and -1 for the $j = 1/2$ edge. Thus the $j = 3/2$ signal is the same as in the I^{01} spectrum but the $j = 1/2$ signal is -2 times stronger.

The spin-orbit spectrum, I^{11}, gives the correlation between s_z and l_z of the core level which is positive for the $j=3/2$ and negative for the $j=1/2$ level. This signal does not require exchange interaction. The I^{20} spectrum gives the alignment of the quadrupole moment $l_z^2 - \frac{1}{3}l(l+1)$. The $j=1/2$ level cannot contain a quadrupole moment so that the signal is zero. The I^{21} spectrum shows the correlation between $l_z^2 - \frac{1}{3}l(l+1)$ and s_z.

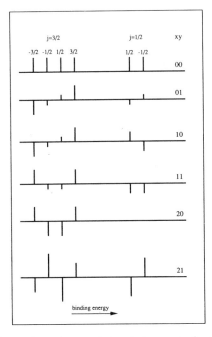

Fig. 2. Fundamental one-electron photoemission spectra for a p core level.

10. EMISSION FROM AN $l=2$ SHELL

We have seen that for emission from an $l=1$ shell the MCD and MLDAD are proportional to the I^1 spectrum and that the LD and CDAD are proportional to the I^2 spectrum. Within a defined geometry (Table 1) the emission angle changes only the magnitude but not the shape of these spectra. This is no longer true for emission from an $l>1$ shell, which has $2l+1$ spectra ($0 \leq x \leq 2l$) without spin detection ($y = 0$). I^0 is the isotropic spectrum, leaving l odd spectra and l even spectra, which can be observed in MCD/MLDAD and LD/CDAD, respectively. Writing out the dichroism for d emission gives

Table 5. The jm_j states of a d core level given as a linear combination over $m_l m_s$ states.

j	m_j	$\sum c_{m_l m_s} m_l m_s$
$\frac{5}{2}$	$-\frac{5}{2}$	$(-2)^-$
	$-\frac{3}{2}$	$\left(\frac{4}{5}\right)^{1/2}(-1)^- + \left(\frac{1}{5}\right)^{1/2}(-2)^+$
	$-\frac{1}{2}$	$\left(\frac{3}{5}\right)^{1/2} 0^- + \left(\frac{2}{5}\right)^{1/2}(-1)^+$
	$\frac{1}{2}$	$\left(\frac{2}{5}\right)^{1/2} 1^- + \left(\frac{3}{5}\right)^{1/2} 0^+$
	$\frac{3}{2}$	$\left(\frac{1}{5}\right)^{1/2} 2^- + \left(\frac{4}{5}\right)^{1/2} 1^+$
	$\frac{5}{2}$	2^+
$\frac{3}{2}$	$\frac{3}{2}$	$\left(\frac{4}{5}\right)^{1/2} 2^- - \left(\frac{1}{5}\right)^{1/2} 1^+$
	$\frac{1}{2}$	$\left(\frac{3}{5}\right)^{1/2} 1^- - \left(\frac{2}{5}\right)^{1/2} 0^+$
	$-\frac{1}{2}$	$\left(\frac{2}{5}\right)^{1/2} 0^- - \left(\frac{3}{5}\right)^{1/2}(-1)^+$
	$-\frac{3}{2}$	$\left(\frac{1}{5}\right)^{1/2}(-1)^- - \left(\frac{4}{5}\right)^{1/2}(-2)^+$

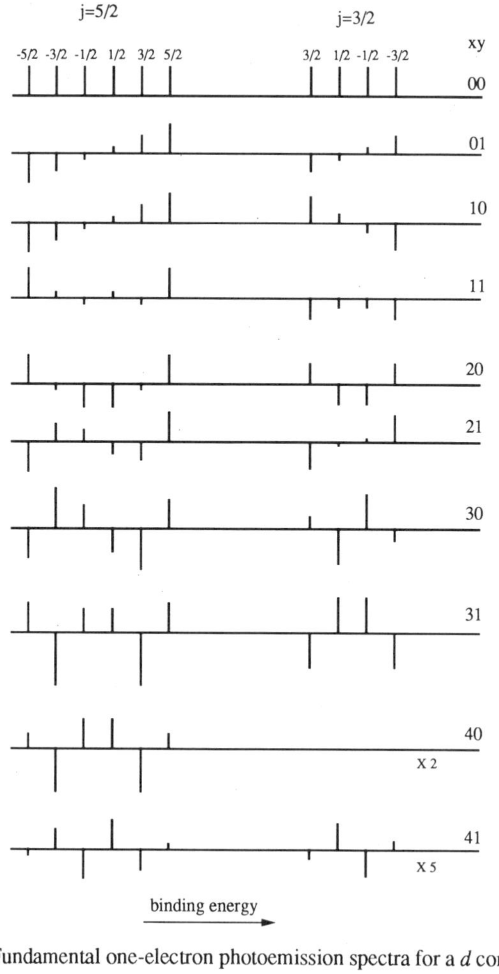

Fig. 3. Fundamental one-electron photoemission spectra for a d core level.

Table 6. The quadrupole transition probabilities for light with polarization q from the jm_j states of a d core level to an εs continuum with spin σ.

			$j=\frac{5}{2}$					$j=\frac{3}{2}$		
$q\sigma$	$-\frac{5}{2}$	$-\frac{3}{2}$	$-\frac{1}{2}$	$\frac{1}{2}$	$\frac{3}{2}$	$\frac{5}{2}$	$\frac{3}{2}$	$\frac{1}{2}$	$-\frac{1}{2}$	$-\frac{3}{2}$
$+2\uparrow$	−	−	−	−	−	1	−	−	−	−
$+1\uparrow$	−	−	−	−	$\frac{4}{5}$	−	$\frac{1}{5}$	−	−	−
$0\uparrow$	−	−	−	$\frac{3}{5}$	−	−	−	$\frac{2}{5}$	−	−
$-1\uparrow$	−	−	$\frac{2}{5}$	−	−	−	−	−	$\frac{3}{5}$	−
$-2\uparrow$	−	$\frac{1}{5}$	−	−	−	−	−	−	−	$\frac{4}{5}$
$+2\downarrow$	−	−	−	−	$\frac{1}{5}$	−	$\frac{4}{5}$	−	−	−
$+1\downarrow$	−	−	−	$\frac{2}{5}$	−	−	−	$\frac{3}{5}$	−	−
$0\downarrow$	−	−	$\frac{3}{5}$	−	−	−	−	−	$\frac{2}{5}$	−
$-1\downarrow$	−	$\frac{4}{5}$	−	−	−	−	−	−	−	$\frac{1}{5}$
$-2\downarrow$	1	−	−	−	−	−	−	−	−	−

Table 7. Quadrupole transition probabilities for the fundamental spectra of a one-electron d state.

			$j=\frac{5}{2}$					$j=\frac{3}{2}$		
I^{xy}	$-\frac{5}{2}$	$-\frac{3}{2}$	$-\frac{1}{2}$	$\frac{1}{2}$	$\frac{3}{2}$	$\frac{5}{2}$	$\frac{3}{2}$	$\frac{1}{2}$	$-\frac{1}{2}$	$-\frac{3}{2}$
I^{00}	1	1	1	1	1	1	1	1	1	1
I^{01}	-1	$-\frac{3}{5}$	$-\frac{1}{5}$	$\frac{1}{5}$	$\frac{3}{5}$	1	$-\frac{3}{5}$	$-\frac{1}{5}$	$\frac{1}{5}$	$\frac{3}{5}$
I^{10}	-1	$-\frac{3}{5}$	$-\frac{1}{5}$	$\frac{1}{5}$	$\frac{3}{5}$	1	$\frac{9}{10}$	$\frac{3}{10}$	$-\frac{3}{10}$	$-\frac{9}{10}$
I^{11}	1	$\frac{1}{5}$	$-\frac{1}{5}$	$\frac{1}{5}$	$-\frac{1}{5}$	1	$-\frac{7}{10}$	$-\frac{3}{10}$	$-\frac{3}{10}$	$-\frac{7}{10}$
I^{20}	1	$-\frac{1}{5}$	$-\frac{4}{5}$	$-\frac{4}{5}$	$-\frac{1}{5}$	1	$\frac{7}{10}$	$-\frac{7}{10}$	$-\frac{7}{10}$	$\frac{7}{10}$
I^{21}	-1	$\frac{3}{5}$	$\frac{2}{5}$	$-\frac{2}{5}$	$-\frac{3}{5}$	1	$-\frac{9}{10}$	$-\frac{1}{10}$	$\frac{1}{10}$	$\frac{9}{10}$
I^{30}	-1	$\frac{7}{5}$	$\frac{4}{5}$	$-\frac{4}{5}$	$-\frac{7}{5}$	1	$\frac{2}{5}$	$-\frac{6}{5}$	$\frac{6}{5}$	$-\frac{2}{5}$
I^{31}	1	$-\frac{9}{5}$	$\frac{4}{5}$	$\frac{4}{5}$	$-\frac{9}{5}$	1	$-\frac{6}{5}$	$\frac{6}{5}$	$\frac{6}{5}$	$-\frac{6}{5}$
I^{40}	1	-3	2	2	-3	1	0	0	0	0
I^{41}	-1	$\frac{17}{5}$	$-\frac{26}{5}$	$\frac{26}{5}$	$-\frac{17}{5}$	1	$-\frac{8}{5}$	$\frac{24}{5}$	$-\frac{24}{5}$	$\frac{8}{5}$

$$4\pi J_{\text{MCD}}$$

$$= I^1[(\tfrac{2}{5}U^{101}-\tfrac{4}{25}U^{121})R^{11}+(-\tfrac{2}{5}U^{101}+\tfrac{16}{25}U^{121})R^{33}+(\tfrac{4}{5}U^{101}-\tfrac{12}{25}U^{121})R^1R^3\cos\delta\,]$$

$$+ I^3[-\tfrac{6}{25}U^{123}R^{11}+\tfrac{18}{175}U^{123}R^{33}+(\tfrac{24}{175}U^{123}+\tfrac{12}{35}U^{143})R^1R^3\cos\delta \quad , \tag{64}$$

$$4\pi J_{\text{MLDAD}} = [I^1(-3U^{221}) + I^3(\tfrac{1}{2}U^{223}+\tfrac{9}{10}U^{243})]R^1R^3\sin\delta\,] \quad , \tag{65}$$

$$4\pi J_{\text{CDAD}} = [I^2(-\tfrac{5}{7}U^{122}) + I^4(\tfrac{27}{70}U^{144})]R^1R^3\sin\delta \quad . \tag{66}$$

Thus in each geometry we measure a sum over two different spectra with relative intensities that depend on the emission angle.

The wave functions of the jm_j sublevels for a d core level in the jj couplings limit are given in Table 5. Because the fundamental spectra do not contain any geometrical information, they can be calculated using arbitrary order of multipole transition. For an l shell it is convenient to use $2l$-pole radiation. The spectra $I_{q\sigma}$ obtained using quadrupole transition probabilities ($q=-2, ..., 2$) to the $|0^\pm\rangle$ continuum are given in Table 6. Then using the transformation

$$r_q^{Qa}(\mathbf{P})\,r_\sigma^{(1/2)y}(\mathbf{P_S}) = n_{Qa}^{-1}\,n_{(1/2)y}^{-1}\,(-1)^{Q-q}\,(-1)^{(1/2)-\sigma}\begin{pmatrix}Q & a & Q \\ -q & 0 & q\end{pmatrix}\begin{pmatrix}1/2 & y & 1/2 \\ -\sigma & 0 & \sigma\end{pmatrix}, \tag{67}$$

we can make the linear combinations I^{xy} shown in Table 7. Fig. 3 displays the fundamental spectra in graphical form. The I^{x0} spectra give the polarization of the x-th magnetic moment of the l shell, which is induced by the valence magnetic moment and the electrostatic core-valence interaction. The I^{x1} spectra give the correlation between the x-th magnetic moment and s_z of the l shell. In the I^{10} and I^{30} spectra both j levels have a first and third spectral moment, respectively, so that these spectra can be distinguished.

11. CONCLUSIONS

The different ways to orient the polarizations of the magnetization, electric vector of the light and the spin of the photoelectron allow measurements of different kinds of correlations between the corresponding atomic properties: the valence spin, core hole orbital momentum and core hole spin, respectively. We can define fundamental spectra as those linear combinations of the polarized spectra that are directly connected to physical properties. Magnetic dichroism in core level photoemission, which gives the alignment between the valence spin magnetic moment and the core hole orbital moment, requires both spin-orbit and electrostatic interactions. For the emission from an incompletely filled localized shell, such as the 4f in the rare earths, the integrated intensities of the magnetic circular dichroism and the spin spectrum are proportional to the ground state orbital and spin magnetic moment, respectively. In angle integrated photoemission the light polarization and the induced moment of the atom have the same parity, so that the isotropic and linearly polarized light measure only even moments, whereas circular polarization measures only odd (magnetic) moments. However, when the light polarization vector, the Z-axis of the system (e.g. the molecular axis or the magnetic axis) and the emission direction of the photoelectron are not coplanar, the experimental geometry can be chiral, i.e. the geometry is not a mirror-image of itself. Then the interference term between the $l+1$ and $l-1$ channel no longer cancels but depends on the radial matrix elements and the phase difference of these channels and we can measure even moments with odd polarized light and odd moments with even polarized light. Thus a spectrum measured in magnetic circular dichroism can also be obtained using linear polarized radiation in a chiral geometry.

REFERENCES

1. G. van der Laan, Phys. Rev. Lett. 66, 2527 (1991); J. Phys. Condens. Matter 3, 1015 (1991).
2. B.T. Thole, G. van der Laan and G.A. Sawatzky, Phys. Rev. Lett. 55, 2086 (1985).
3. L. Baumgarten, C.M. Schneider, H. Petersen, F. Schäfers, and J. Kirschner, Phys. Rev. B 44, 4406 (1991).
4. C.M. Schneider, D. Venus, and J. Kirschner, Phys. Rev. B 45, 5041 (1992).
5. G.D. Waddill, J.G. Tobin, and D.P. Pappas, Phys. Rev. B 46, 552 (1992).
6. G. van der Laan, M.A. Hoyland, M. Surman, C.F.J. Flipse, and B.T. Thole, Phys. Rev. Lett. 69, 3827 (1992).
7. D. Venus, L. Baumgarten, C.M. Schneider, C. Boeglin, and J. Kirschner, J. Phys. Condens. Matter 5, 1239 (1993).
8. C. Boeglin, E. Beaurepaire, V. Schorsch, B. Carrière, K. Hricovini, and G. Krill, Phys. Rev. B 48, 13123 (1993).
9. K. Starke, E. Navas, L. Baumgarten, and G. Kaindl, Phys. Rev. B 48, 1329 (1993).
10. Ch. Roth, F.U. Hillebrecht, H.B. Rose, and E. Kisker, Solid State Commun. 86, 647 (1993).
11. Ch. Roth, F.U. Hillebrecht, H.B. Rose, and E. Kisker, Phys. Rev. Lett. 70, 3479 (1993).
12. F.U. Hillebrecht and W.-D. Herberg, Z. Phys. B 93, 299 (1994).
13. L.H. Tjeng, C.T. Chen, P. Rudolf, G. Meigs, G. van der Laan, and B. T. thole, Phys. Rev. B 48, 13378 91993).
14. H. Ebert, L. Baumgarten, C.M. Schneider, and J. Kirschner, Phys. Rev. B 44, 4406 (1991).
15. B.T. Thole and G. van der Laan, Phys. Rev. Lett. 67, 3306 (1991).
16. B.T. Thole and G. van der Laan, Phys. Rev. Lett: 70, 2499 (1993).
17 G. van der Laan, Int. J. Mod. Phys. B 8, in print (1994).
18. D.M. Brink and G.R. Satchler, *Angular Momentum* (Oxford University Press, London, 1962).
19. A.P. Yutsis, I.B. Levinson, and V.V. Vanagas, *Mathematical Apparatus of the Theory of Angular Momentum* (Israel Program for Scientific Translation, Jerusalem, 1962).
20. D.A. Varshalovich, A.N. Moskalev, and V.K. Khersonskii,*Quantum Theory of Angular Momentum* (World Scientific, Singapore, 1988).
21. B.T. Thole and G. van der Laan, Phys. Rev. B 44, 12424 (1991).
22. G. van der Laan and B.T. Thole, Phys. Rev. B 48, 210 (1993).
23. B.T. Thole and G. van der Laan, Phys. Rev. B 49, 9613 (1994).

SPIN RESOLVED SOFT X-RAY APPEARANCE POTENTIAL SPECTROSCOPY

Volker Dose

Max-Planck-Institut für Plasmaphysik
EURATOM Association
D-85740 Garching bei München, Germany

A solution of the longstanding and persisting problem of band ferromagnetism will finally emerge from an understanding of the spindependent electronic structure of ferromagnetic materials[1]. The prominent difficulty on this way is the rather small energy associated with magnetic ordering which is of the order of KT_c where T_c is the materials Curie temperature. This amount of energy is smaller or at most comparable to the presently achievable precision of band structure theory. The difficulties on the theoretical side have an experimental counterpart, namely the complexity added to standard band structure techniques by the necessity of analysis or proper preparation of electron spin. In fact the limited resolving power of spin detectors and the finite degrees of polarization of electron beams, usually only a fraction of the ideal 100%, enlarge the times for data aquisition sometimes to a prohibitive limit. Consider for example a quantity S which responds with S^+ and S^- counts respectively to a number of N of 100% positively and negatively spin polarized electrons. For the much more realistic situation that only a beam of say p, typically 30%, spin polarization is available one would need at least N/p^2 electrons in order to determine S^+ and S^- with the previous relative precision. Theoretical and experimental approaches to the problem of band ferromagnetism will therefore remain a tedious job.

Electron spectroscopies on ferromagnetic solids may be classified into two categories. The first category aims at the localization of the response, the measured signal, in real space. This class of spectroscopies employs core levels of the material under investigation. Recall, for example, that the diameter of the Fe L shell is of the order of 0.1 Å which is only 3% of the extension of the Fe unit cell in real space. The uncertainty principle, or more generally Fourier transform theory tells immediately that the momentum information extends over several Brillouin zones. Modifying the argument to a reduced zone representation means that spectroscopies involving core levels smear the momentum information throughout the Brillouin zone. They are capable of measuring more or less restricted density of states information.

The second class aims at a localization of the response in momentum space, with an associated loss of real space localization information. The response of such an approach is an energy-versus-momentum dispersion relation $E_\sigma(\underline{k})$. The density of states function $\rho_\sigma(E)$ is of course an integral quantity derived from $E_\sigma(\underline{k})$ dispersion relations and it would seem that a measurement of the latter outweights core level spectroscopies. This holds true in principle.

However, in much the same way, as a measured total scattering cross section has hardly ever been obtained by integration of measured diffential cross sections, an experimental determination of the density of states has to my knowledge never proceeded through experimental $E_\sigma(\underline{k})$ determinations.

Both approaches have therefore their own merits. The dispersion relation $E_\sigma(\underline{k})$ is the far more detailed quantity, and of primary interest for comparison with theory[2]. The spin resolved density of states $\rho_\sigma(E)$ on the other hand provides a clue to the macroscopic magnetic behaviour.

We shall now discuss representative cases of spin resolved core level spectroscopies by reference to fig. 1. As a consequence of core level involvement all of them are in a natural way site/element specific. This is of considerable interest for the investigation of magnetic alloys. Core levels are of course involved in both X-ray emission (SXS) and X-ray absorption (XAS). Structure in X-ray emission bands reflects the density of occupied states. However, as consequence of the dipole selection rules only the partial DOS with symmetry character $l \pm 1$ where l is the core hole angular momentum is projected out of the total DOS. XAS frequent denoted also XANES (X-ray absorption near edge structur) yields the equivalent data for empty electronic states. The symmetry selective property of the two methods is of course a limitation. can turn into an advantage if a more detailed interpretation of available total DOS data required.

Spin resolution capability of SXS would require an analysis of the emitted radiation with respect to its degree of circular polarization. Such an experiment is difficult to do. XAS on the other hand is just a fashion at present. Synchrotron radiation offers linear polarization when taken

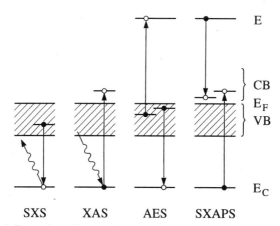

Figure 1. Energy level diagram for selected methods of core level spectroscopies.

inplane and circular polarization when viewed from above or below the synchrotron plane. XAS with appropriately prepared polarization of the incident radiation is now known as <u>M</u>agnetic <u>C</u>ircular (<u>L</u>inear) <u>X</u>-ray <u>D</u>ichroism[3,4].

While the foregoing two methods involve the transition of only one electron and the sprectra obtained contain the DOS information in a linear and therefore rather direct way, life becomes more complicated with Auger electron spectroscopy (AES)[5] and soft X-ray appearance potential spectroscopy (SXAPS)[6]. The intimate relationship between the two approaches is obvious from an inspection of fig. 1. In appearance potential spectroscopy an electron of initial energy E excites a core electron to an empty state above the Fermi level. At threshold for core hole production the final states for both the incident and the excited electron are immediately above the Fermi level. Above threshold, the excess energy $\Delta E = E - |E_c|$ is shared among the two electrons. Assuming energy independent transition matrix elements and sharp initial and final states the rate of core hole production $P(\Delta E)$ is found to be proportional to the autoconvolution of the conduction band density of states, $\rho(E)$

$$P(\Delta E) \propto \int_0^{\Delta E} \rho(E)\rho(\Delta E - \varepsilon)d\varepsilon \qquad (1)$$

P(E) can be monitored by detection of fluorescence[6], Auger electrons, secondary electrons[7] or variation in the primary electron reflectance[8]. An appropriate modification of (1) for experiments with spin polarized incident electrons will follow below.

The above modell for Appearance Potential Spectroscopy is of course the empty band analog of Lander's[9] modell for the line shape of Auger core-valence-valence transitions. Modifications improving over this picture have been developped in the past[10,11,12]. The spin resolution option on Auger spectroscopy is of course obtained by spin analysis after energy selection. This has extensively been used by Landolt[13,14].

Having put the four spectroscopies into proper mutual perspective we shall narrow down our interest in the following on appearance potential spectroscopy, to be used more specific, on soft X-ray appearance potential spectroscopy since this mode offers the best available decoupling between excitation and reorganization[15].

The apparatus employed in this work is shown in fig. 2[16]. Spin polarized electrons are generated by photo excitation of GaAs with circularly polarized laser radiation. Choosing the laser energy to just meet the Γ point excitation in GaAs results in 50% spin polarization of the electrons excited to the conduction band[17]. The crystal surface is covered with a thin layer of

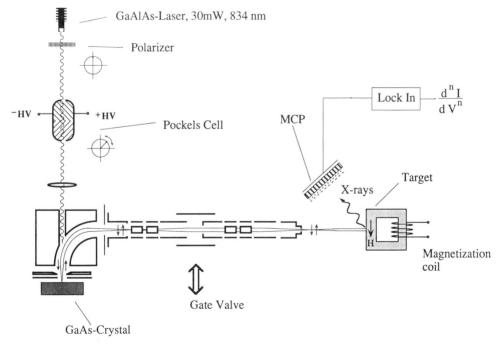

Figure 2. Experimental setup for spinpolarized soft X-ray appearance potential spectroscopy.

cesium oxide in order to provide a negative electron affinity surface such that the conduction band electrons escape into the vacuum. The spin polarization of the emitted electrons is longitudinal and can be reversed by switching the helicity of the laser light with the help of a pockets cell. T[he] spin polarization is subsequently transformed to transverse polarization by 90 degre[e] electrostatic deflection of the beam. An electrostatic transport lens connects the source and targ[et] vacuum chambers. More detailed properties of this system have been described earlier[18].

The targets used in this work were of picture frame shape and monocrystalline with [10[0]] orientation parallel to the electron spin direction. The Fe crystal was stabilized with 3% Si [in] order to enable heat treatment in the process of cleaning without structural phase transition[s]. Target cleaning proceeded via standard techniques of sputtering, gas adsorption and annealing. [In] addition to chemical cleanliness, a proper preparation of the targets magnetization state is [of] special importance in spin polarized work[19]. The targets were magnetized by a current thorough [a] selfsupporting coil wound around one of the legs of the crystals. The magnetization properti[es] were investigated in situ (not shown in fig. 2) with magneto optical Kerr effect (MOKE) and e[x] situ with MOKE microscopy. The Ni crystal showed a behaviour significantly different from th[at] of earlier work with IPE[19,20], presumably because the crystal had undergone mechanical polishir[g] since the previous experiment. While the Ni frame could be cast in a state of remane[nt] magnetization by a short current pulse, measurement with the iron crystal required permane[nt] magnetic excitation, because its surface did not contain an axis of easy magnetization. Core lev[el] excitation in the targets was monitored by fluorescence emission. The X-ray detector consists of [a] CsJ sensitized venetian blind shaped photo cathode, a micro channel plate for photocurre[nt] amplification and a 740Å thick aluminium entrance window to suppress low energ[y] Bremsstrahlung quanta. The performance of the system was shot noise limited for primar[y]

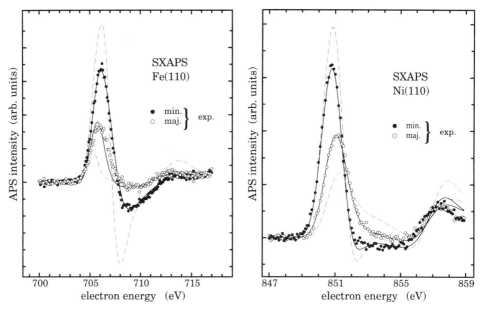

Figure 3. Spinpolarized SXAPS spectra from Fe (L_{III}) compared with model calculations. Data are scaled to a 100% polarization of the incident electron beam. Dashed curves: convolution of minority and majority DOS with total DOS. Solid curves: fit with sum of weighted convolutions of spin- and angular-momentum-resolved DOS subsets.
Figure 4. Spinpolarized SXAPS spectra from Ni(L_{III}). See also caption to Fig.3.

electron currents ≥ 1 µA. Since the fluorescence from core level excitation constitutes only a[ ] minute fraction of the total emitted X-ray flux, potential modulation of the target by 1.5 Vp[p] together with lock-in detection was employed in order to separate the signal from the otherwise overwhelming background.

Figures 3 and 4 show results of spin resolved APS measurements for Fe and Ni near magnetic saturation (T = 150 K)[21]. The spectra correspond to excitation of core electrons from the $2p_{3/2}$ core level above the onsets at about 705eV and 849 eV, respectively. Since spinresolved data are obtained with an electron beam of incomplete polarization, it is usual to project spinresolved data to a hypothetical polarization of 100% of the primary electrons and to complete alignment of the electron polarization vector and the sample magnetization vector. The projected intensities $I_\uparrow$ and $I_\downarrow$ for electrons parallel or antiparallel to the magnetization may then be calculated from the raw data $i_\uparrow$ and $i_\downarrow$ be means of

$$I_{\uparrow(\downarrow)} = \frac{1}{2}\left((i_\uparrow + i_\downarrow) \pm (i_\uparrow - i_\downarrow)\frac{1}{P_{eff}}\right), \quad (2)$$

where P_{eff} is given by the magnitude and direction of the spin polarization vector of the incident electrons relative to the effective sample magnetization determined by the domain structures of the crystals. As a result of this correction, the relative statistical error of the displayed curves is enhanced by a factor 2.5 compared with the raw data. The two spectra shown as open and solid circles correspond to primary electron spins parallel (minority) or antiparallel (majority) to the target magnetization. Simultaneous reversal of both the target magnetization and spin direction was shown to leave the experimental curves unaffected, thus excluding experimental asymmetries and proving the pure magnetic origin of this effect. For iron comparable data in a reduced range from the early work of Kirschner[22] are available. Assuming a single domain structure of the sample in that work and correcting for the different geometry and our domain structure, the results are in excellent agreement.

The experimental data from fig. 3 and fig. 4 will now be compared with simulations on the basis of the simple DOS convolution model. To this end we rewrite equation (1) for the case of spin split densities of states

$$\frac{d}{dE}P_{\uparrow(\downarrow)}(\Delta E) \propto \frac{d}{d\Delta E}\int_0^{\Delta E}\rho_{\uparrow(\downarrow)}(\varepsilon)\rho_{(\uparrow+\downarrow)}(\Delta E - \varepsilon)d\varepsilon \quad (3)$$

For the evaluation of (3) we have used spinresolved DOS based on selfconsistent LRC[23] calculations which employ selfconsistently generated potential for the valence band region and LAPW calculations for higher lying bands using the selfconsistent LRC potentials. The underlying band structure calculations had been previously successfully applied to spinresolved inverse photoemission experiments on Ni[24]. In order to reproduce the measured data in figs. 3 and 4 some broadening has to be imposed on (3). Apparatus broadening resulting from a spread of the primary electron energy and the potential modulation were lumped together with the core level broadening into a single Gaussian with variance $\sigma^2 = (0.51 \text{ ev})^2$. For the final conduction band state an energy dependent Lorentzian lifetime broadening given by $\Gamma(E) = 0.13 \text{x}(E-E_F)$ was used for nickel. This relation had been determined previously by Goldmann et al. for energies $E-E_F \leq 50$ eV[25]. For iron it turned out that the smearing was still unsufficient presumably due to the large DOS resulting from unoccupied d-bands. The discrepancy of experiment and simulation for spinaveraged data at the negative going dip in fig. 3 could be significantly reduced employing a conduction band lifetime $\Gamma_{Fe} = 0.5 \text{x}(E-E_F)$. This choice is in good agreement with recent experience from IPE measurements which demanded a broadening of $0.6 \text{x}(E-E_F)$ for d-bands up to $E-E_F = 2$ eV[26].

Deviations of the simulation from the experimental data are far more striking in spinresolved data than for spinaveraged data. The energetic positions of the main structural features in the signal are quite well described. The discrepancy arises with intensities. As possible reason we considered the influence of the topmost layer of the sample which contributes 20-25% to the observed AP signal. They are known to have DOS differing from the bulk. Layer dependent slab calculations[27] show however, that the surface contribution to the signal would tend to increase spinasymmetries rather than reduce them. The secondary structure at 857 eV for nickel and 714 eV for iron is due to DOS discontinuities deriving from the L_7 critical point in fcc nickel[28] and from the N_1 critical point in bcc iron. The absence of any spinsplitting in these regions comes as a surprise and must remain unexplained at present.

In order to improve the simulations beyond the dashed curve results in figs. 3 and 4 we have considered the possibility of angular momentum dependent transition matrix elements. To this end equation (3) is replaced by a sum over all possible convolution products of the spin and angular momentum resolved DOS. Each of these contributions was weighted with a factor

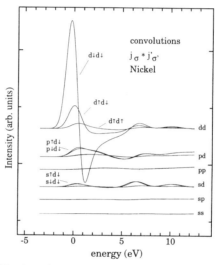

Figure 5. Convolutions of DOS subsets for the case of Ni according to their spin and orbital character. These convolutions are similar in relative magnitude to those of Fe.

adjusted for the best fit between simulation and experiment. We display the different contributions unscaled for nickel in fig. 5. Without any weighting mainly d-d-convolutions, or more precisely convolutions of the $l = 2$ projection onto the atomic sites, contribute to the AP signal. The best fits obtained to the experimental data are shown as solid lines in figs. 3 and 4. The reproduction of the experimental data is now close to perfect, the intensity discrepancy is removed and only slight deviations in the structures deriving from critical point features persist.

The positions of minority and majority peak in the simulation for iron still show a minute deviation from the experimental data. The remaining discrepancy might be due to a spinsplitting of the $2P_{2/3}$ core level which was invoked by Baumgarten et al.[29] from MCXD results. Ebert et al. found such an effect from a fully relativistic theoretical treatment[30]. In the experiment a spinsplitting of 0.5 eV was found while theory found a ground state spinsplitting into four sublevels spread over a range of 0.8 eV. The incorporation of these findings into our simulation had, surprisingly enough, only negligible influence on the quality of the simulation.

In summary we have been able to measure and simulate the line shapes of soft X-ray appearance potential spectra. They exhibit a rather large spinasymmetry in the magnetic ground state. From these results we expect spinpolarized APS to envolve as the empty band counter part of spinpolarized AES. Temperature dependent measurements are now just becoming available and will certainly add to our knowledge of band ferromagnetism.

References

1. Physics Thorough the 1990s, Condensed Matter Physics, National Academy Press, Washington DC, p. 112 (1986).
2. R. Feder (ed), Polarized Electrons in Surface Physics, World Scientific (1986).
3. G. Schütz, W. Wagner, W. Wilhelm, P. Kienle, R. Zeller, R. Frahm and G. Materlik, Phys. Rev. Lett. 58, 737 (1987).
4. J. B. Goedkoop, B.T. Thole, G. van der Laan, G.A. Sawatzky, F.M.F. de Groot and J.C. Fuggle, Phys. Rev. B 37, 2086 (1988).
5. R. Weissmann and K. Müller, Surf. Sci. Rept. 1, 251 (1981).
6. R.L. Park and J.E. Houston, J. Vac. Sci. Technol. 11, 1 (1974).
7. J.E. Houston and R.L. Park, Phys. Rev. B5, 3808 (1972).
8. J. Kirschner and P. Staib, Appl. Phys. 6, 99 (1975).
9. J.J. Lander, Phys. Rev. 91, 1382 (1953).
10. M. Cini, Solid State Commun. 24, 684 (1977).
11. G.A. Sawatzky and A. Lenselink, Phys. Rev. B21, 1790 (1980).
12. W. Nolting, G. Geipel and K. Ertl, Z. Phys. B92, 75 (1993).
13. M. Landolt and D. Mauri, Phys. Rev. Lett. 49, 1783 (1982).
14. M. Landolt, R. Allenspach and M. Taborelli, Surf. Sci. 178, 311 (1986).
15. V. Dose, R. Drube and A. Härtl, Solid State Commun. 57, 273 (1986(.
16. M. Vonbank, K. Ertl, G. Geipel and V. Dose, Symposium on Surface Science Contributions, Obertraun (1991).
17. D.T. Pierce and F. Meier, Phys. Rev. B13, 5484 (1976).
18. U. Kolac, M. Donath, K. Ertl. H. Liebl and V. Dose, Rev. Sci. Instrum. 51, 478 (1988).
19. K. Starke, K. Ertl and V. Dose, Phys. Rev. B46, 9709 (1992).
20. M. Donath, Appl. Phys. A49, 351 (1989).
21. K. Ertl, M. Vonbank, V. Dose and J. Noffke, Solid State Commun. 88, 557 (1993).
22. J. Kirschner, Solid State Commun. 49, 39 (1984).
23. L. Fritsche, J. Noffke and H. Eckardt, J. Phys. F17, 943 (1987).
24. M. Donath, V. Dose, K. Ertl and U. Kolac, Phys. Rev. B41, 5509 (1990).
25. A. Goldmann, W. Altmann and V. Dose, Solid State Commun. 79, 511 (1991).
26. A. Santoni and F.J. Himpsel, Phys. Rv. B43, 1305 (1991).
27. A.J. Freeman and R. Wu, J. Magn. Magn. Mat. 100, 497 (1991).
28. V. Dose and G. Reusing, Solid State Commun. 48, 683 (1983).
29. L. Baumgarten, C.M. Schneider, H. Petersen, F. Schäfers and J. Kirschner, Phys. Rev. Lett. 65, 492 (1991).
30. H. Ebert, L. Baumgarten, C.M. Schneider and J. Kirschner, Phys. Rev. B44, 4406 (1991).

LINEAR MAGNETIC DICHROISM IN DIRECTIONAL PHOTOEMISSION FROM CORE LEVELS AND VALENCE BANDS

Giorgio Rossi, [1,2] Fausto Sirotti, [1,2] and Giancarlo Panaccione [2]

[1]Laboratorium für Festkörperphysik, ETH-Zürich, CH-8093
[2]Laboratoire pour l'Utilisation du Rayonnement Electromagnetique,
CNRS, CEA, MESR F-94305 Orsay

INTRODUCTION

The need for new surface specific probes of magnetism is very high. The ideal surface magnetic probe should be selectively sensitive to surface, subsurface and interface layers, atom-specific and site specific, i.e. sensitive to the local environment of the excited atom, an should provide an absolute measure of the magnetic moment associated to the selected atoms. The field of surface magnetism has been opened by the application of spin-polarimetry to photoelectrons and to the secondary electron yield.[1] The measure of spin polarization (SP) of secondaries is intrinsically surface sensitive due to the short escape depth for low energy photoelectrons in ferromagnets, and can be understood semi-quantitatively.[2] The magnetic resolution of SP is high and can be used for imaging surface magnetic domains[3] and for studying the dynamics of surface magnetism with time resolution in the picosecond range in laser experiments,[4] or in the nanosecond range in synchrotron radiation experiments.[5] SP of secondaries probes the average magnetization and cannot be made atom-specific without independent knowledge on the atomic structure.[6] Spin Polarized Low Energy Electron Diffraction is very surface sensitive,[7] but it is limited to ordered surfaces, and is not atom specific. The magneto-optic Kerr effect [8] is not surface sensitive and can be applied only to the study of ferromagnetic order in monolayers on non magnetic substrates.[9] The related X-ray techniques of circular magnetic dichroism and linear magnetic dichroism [10, 11] have the great advantage to be atom-specific [12] via the characteristic absorption thresholds, and allow for a quantitative derivation of orbital and spin moments via the application of sum rules on the X-ray absorption [13] and emission.[14] X-ray absorption is not surface sensitive: its application is limited to magnetic adlayers or multilayers. The basic spectroscopic tools of surface science, Auger electron spectroscopy and photoelectron spectroscopy of core levels and valence bands, have been implemented in the spin-resolved mode.[15, 16, 17] These are close to ideal tools for surface magnetism, but suffer for a great technical handicap: the low efficiency of spin-detection which is only some 10^{-3} and severely reduces the applications of these techniques. One reasonable approach to overcome these shortcomings is to develop highly efficient spin-detectors.[18, 19] An alternative is to exploit the properties of the dipole matrix elements for directional photoemission in order to probe magnetism via a dichroism approach.

The study of the Fano effect [20] implies the measure of the spin polarization of photoelectrons ejected from unpolarized atoms by circular polarized light. The same information can be obtained by performing a circular dichroism experiment on polarized atoms, with the great practical advantage of measuring an intensity difference, instead of a spin-polarization value.[21] The full angular distribution of photoelectrons ejected from polarized atoms contains all of the information sought in the photoemission experiment, without need of actually measuring the spin polarization of the photoelectrons. Cherepkov et Kuznetsov [22] suggested that the measure of differences of angular distributions, corresponding to polarization changes of the light, or of the initial atomic polarization, are a greatly simplified approach which allows to extract theoretical parameters from the photoemission experiments. Recently a discussion of the photoemission experiments in terms of fundamental spectra and their relationship with difference spectra, has been proposed by van der Laan and Thole.[23, 24] The specific interest in overcoming the need of spin-resolution in electron spectroscopy experiments on magnetic surfaces, and to avoid the use of circularly polarized soft X-rays, has led to the exploitation of the differences of angular distributions of photoelectrons in mirror experiments. Angle resolved experiments with linearly polarized light (as well as with unpolarized light) can be set up in a chiral geometry.[25] Mirror experiments can be obtained, for example, by reversing the initial state polarization of the atoms with an applied magnetic field. The differences in the angular distribution of photoelectrons between two mirror experiments are a special case of magnetic dichroism in photoemission.

PRINCIPLES

Embedding an atom in a magnetic field lowers its symmetry.

The magnetic field has axial symmetry. In an applied magnetic field the interaction between the field and the magnetic moment of the electrons in an atom splits the energy terms which correspond to the different possible values of the magnetic quantum number m. The atoms become polarized. In the limit of the anomalous Zeeman effect (the most general case for applied fields *weak* with respect to spin-orbit interaction of the electrons) the energy splitting of the lines is due to both S and L quantum numbers (LS coupling) i.e. by the total angular momentum quantum number J and are described by the effective gyromagnetic factors g$_j$: $\Delta E_{m,mj-1} = g_j \mu_B H_0$ where μ_B is the Bohr magneton and H_0 is the external field.[26] The light emission/absorption properties (i.e. the dipole transition selection rules $\Delta mj = 0, \pm1$) for polarized atoms are anisotropic: in transversal observation (perpendicular to the direction of the applied magnetic field) three linear polarized lines are emitted (or absorbed) whilst in longitudinal observation (parallel to the field lines) two circular polarized lines are observed (absorbed) with opposite handedness ($\Delta mj = \pm1$). This is shown in the textbook-like figures 1 and 2 for the D_2 transitions of Na atoms in a magnetic field: the magnetic sublevels of atoms embedded in external magnetic fields can be excited by absorption of circular polarized radiation of appropriate frequency in a longitudinal experiment (i.e. photon propagation direction parallel or antiparallel to the magnetization direction) and by linear polarized radiation in a transverse experiment, (i.e. photon propagation direction perpendicular to the magnetization axis).

The photoexcited atom of a ferromagnetic solid is polarized due to its embedment in the magnetic field of the neighboring atoms and in the crystal field. The photoemission final states are split by the exchange and coulomb interactions of the core hole with the exchange-split valence electrons, yielding a multiplet spectrum of magnetic sublevels. Average magnetic fields for bulk atoms are of the order of 10^3 Gauss. This implies a Zeeman splitting between adjacent mj sublevels of the order 10^{-4} eV. This value is small with respect to typical lifetime broadening of core levels and cannot be measured. The exchange interaction, on the other hand is typically four orders of magnitude larger. It induces different combinations of sublevels with respect to the Zeeman effect. LS coupling does not apply and spin-spin interaction dominates over spin-orbit. The interlevel energy splitting between these m sublevels of the order of 10^{-1} eV, which is clearly reflected in the photoemission lineshape.

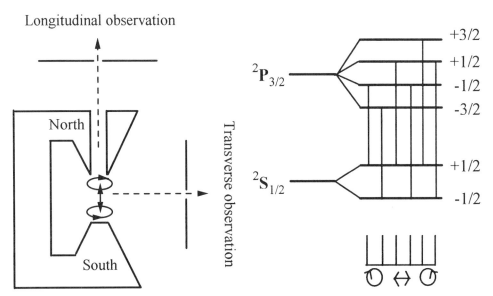

Fig.1: Transversal and longitudinal observation of emission lines from atoms in a magnetic field.

Fig.2: Anomalous Zeeman effect for alkali atoms. The D_2 spectrum is composed of six lines.

EXPERIMENTS

In figure 3 the 3p core level of Fe in bcc-Fe and in semiconducting β-FeSi$_2$ are compared. The silicide spectrum reflects the spin-orbit interaction in the open 3p core, with a splitting of $1.05\pm.05 eV$ between a Fe $3p_{3/2}$ and Fe $3p_{1/2}$ peaks.[27] The spectrum of ferromagnetic bcc-Fe is much broader and asymmetric: it reflects the unresolved fine structure of magnetic and spin-orbit split sublevels. Spin-resolved experiments have shown that the fastest photoelectrons have minority spin character, whilst a peak of majority spin photoelectrons appears corresponding to the shoulder.[28]

In longitudinal photoemission experiments on ferromagnetic surfaces one aligns the propagation direction of circularly polarized radiation (**q**) with the macroscopic magnetization direction (**M**) of the sample surface.[29, 30, 31] The optical orientation of the photohole, due to the interaction of the photon spin with the hole orbit, is modulated by the electrostatic and exchange interactions with the spin-polarized valence band. By reversing the photon helicity the spectrum is dominated by the "other" set of interactions between the polarized hole (opposite spin) and the valence band. If this experiment is angle integrated then it is identical to exciting with a fixed helicity, and reversing the magnetization of the sample. In both cases the spectra corresponding to the interactions between hole and valence band have been separated as a function of core hole spin. This experiment is called Magnetic Circular Dichroism in photoemission, and yields information on the magnetization of the surface.[32, 24] In a transverse experiment one aligns the linearly polarized radiation (**q**) perpendicularly to the magnetization axis, so that also the linear polarization vector **e** is also perpendicular to **M**, therefore exciting all the magnetic sublevels. The spectral resolution of the magnetic sublevels in this case is limited by the magnitude of splitting, lifetime broadening and experimental energy resolution, obtaining the same magnetization averaged spectrum as the one of figure 3 for *bcc*-Fe. In fact these experiments are never exactly performed. This is due to the limited angular acceptance of the electron energy analyzer which operates a selection on the photoelectron emission directions (angle-resolved photoemission).[33] One therefore measures the angle distribution of the photoemission intensity, at the well defined angle set by the experimental geometry.

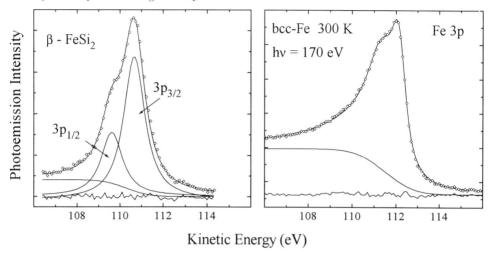

Fig.3: Photoelectron spectra of the Fe 3p core level in bcc-Fe (left) and in the semiconducting phase of iron-disilicide.(right) The semiconductor spectrum is easily decomposed in the two spin-orbit final states.

If the ferromagnetic sample is magnetized by an external field then its atoms have the axial symmetry of the field. The angular distribution of photoelectrons ejected from polarized atoms depends on three vectorial quantities, the direction of the photon beam **q** (or the direction of linear polarization **e**), the direction of photoelectron ejection **k** (or photoelectron momentum), and the direction of atomic polarization which, for a ferromagnetic surface, is given by the magnetization direction **M**. By choosing the photon direction as one axis of the laboratory frame, the angular distribution of the photoelectrons can be represented by a double expansion over spherical functions of two directions **k** and **M**.[34] In an angle resolved experiment the photoemission direction **k** is selected by the analyzer. The directional photoemission spectrum is different from the isotropic spectrum due to the interference between the dipole allowed final states which modify the directional photoemission intensity but not the angle integrated intensity due to complete averaging.[23, 35] The trick for doing a magnetization dependent experiment with angle-resolved photoemission (and without spin-resolution) is to fix the experimental geometry such to have a well defined chirality (i.e. to admit a distinct mirror experiment) and then to perform both the experiment and its mirror experiment obtained by reversing **M**, or by mirroring **k**. The difference of the two experiments is a difference between two angular distributions: it is a special photoemission magnetic dichroism called (Linear) Magnetic Dichroism in the Angular Distribution of photoelectrons [(L)MDAD].[36, 37, 38, 39, 40]

LMDAD experiments on Fe.

Let us consider a photoemission experiment in the geometry sketched in figure 4. It can be seen that two mirror experiments[25] are obtained depending on the magnetization direction oriented "above" the scattering plane (left handed triad, fig 4a) or "below" the scattering plane (right handed triad, fig 4b).

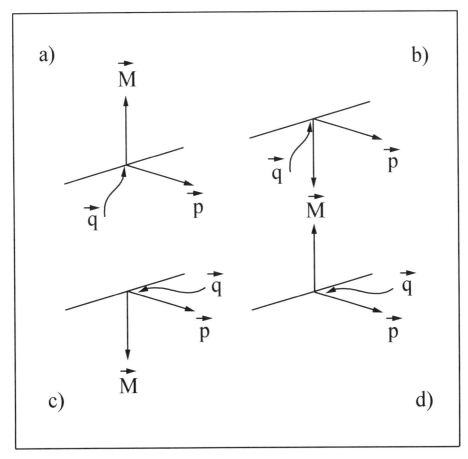

Fig.4: Geometry of the chiral experiment. Mirror experiments are obtained by reversal of M, (a,b) or by mirroring the photon beam direction.(a,d) Experiments (a,c) and (b,d) are identical.

The handedness of the experiment is reversed by reversing the surface magnetization on the sample (along the vertical axis). If the magnetization orientation is kept fixed mirror experiments can be obtained by impinging with the linear polarized photons from the "right" of the pM plane, instead than from the "left" (with the same incidence angle with respect to the photoelectron momentum, as seen in figure 4 a,b and b,c). More complex chiralities occur when more vector quantities enter in the experiment, like crystallographic symmetry directions that could be referred to the sample normal.[35, 41] The three vector chirality of the example is necessary and sufficient to do magnetic dichroism experiments with linear polarized light, and to show their equivalence to longitudinal experiments with circular polarized light.[34, 23] One should note that effects of chirality (even of hidden chiralities) are so large that all photoemission experiments are in fact chiral to some degree, unless the special case of co-linear vectors is approximated.[35] Figure 5 shows core level photoemission data on Fe obtained from experiments with linear polarized undulator radiation from the SuperAco SU7 monochromator in the geometry of figure 4. The magnetization could be imposed parallel or antiparallel to the y-axis direction by applying a current to the magnet coil. The photoelectrons were collected by a MAC-II Riber analyzer whose axis lies along the sample normal. The overall energy resolution was 350 meV for the valence band and 3p spectra and 500 meV for the 3s spectra.[38, 39]

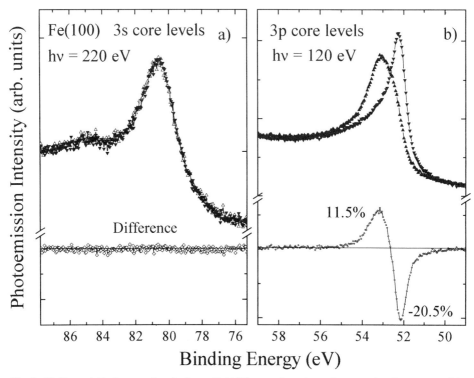

Fig.5: Fe 3s and Fe 3p core levels from Fe(100) as a function of magnetization reversal in a chiral experiment. The difference is zero for l=0 levels.

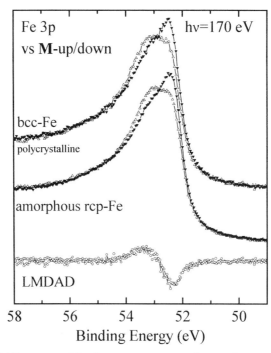

Fig.6: Fe 3p LMDAD measured in the same geometry from polycrystalline Fe and from amorphous rcp-Fe at 30 K.

Two spectra are shown as obtained for opposite directions of **M** and their difference. The left panel shows the 3s core level with its exchange split satellite at $-4.5eV$ [9] no difference is detected between the two magnetization directions. The right panel shows the 3p core levels, which display a large difference on the leading peak. The difference spectra are qualitatively identical for single crystal (100) surfaces of Fe, epitaxial *fcc*-Fe/Cu-Al(100), polycrystalline *bcc*-Fe, amorphous *rcp*-Fe and Fe in $Fe_{.4}Ni_{.4}B_{.2}$. Some spectra are collected in figure 6. These data give directly two informations: a) Fe 3p core levels show magnetic dichroism upon magnetization reversal in chiral experiments. The effect is large giving a maximum asymmetry of 20% of the total intensity. The effect is connected with the axial symmetry of the magnetized sample, and is fully independent on the crystallographic environment or long range order. Finally the LMDAD signal is a difference signal of two Fe 3p photoemission spectra, i.e. it cost a factor of two more measurements than the standard core level photoemission with linearly polarized synchrotron radiation. This is the major advantage with respect to circular magnetic dichroism in photoemission, which requires circular polarized soft X-rays, usually less intense then the linearly polarized sources on synchrotron rings. The advantage over spin-detection is still larger since only photoemission intensity is measured.

b) $l = 0$ core levels, like Fe 3s, do not show LMDAD. This strongly indicates that the LMDAD is determined by the core hole, and not by the final photoelectron states.

LMDAD as a Kerr-like diagnostic tool

The first application of LMDAD is to probe the magnetic order at surfaces in a similar qualitative way as the magneto-optic Kerr effect probes the bulk magnetic order, but with the important advantage of chemical sensitivity. A clear example is given in figures 7 and 8. Fe 3p LMDAD gives a quick and direct evidence of the ferromagnetic order of the Fe(100) substrate surface, and of the ferromagnetic order and magnetization direction of the overlayers.

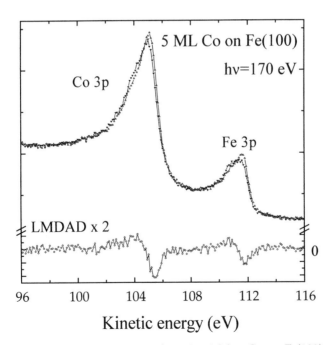

Fig.7: 3p core level spectra and LMDAD for epitaxial bcc-Co on Fe(100). The ferromagnetic coupling through the interface is seen by the same sign of LMDAD for the two elements.

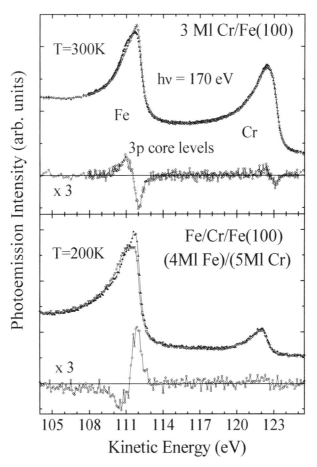

Fig.8: 3p spectra and LMDAD for epitaxial Cr/Fe(100) (top) and for a trilayer structure (bottom). The antiferromagnetic coupling of the Fe overlayer through the Cr spacer is seen by the reversal of the LMDAD asymmetry.

Figure 7 shows the LMDAD signal for 5 monolayers of bcc-Co epitaxially grown onto Fe(100) at 400K. Both Fe 3p and Co 3p show a similar dichroism indicating the ferromagnetic order of the bcc-Co overlayer and the ferromagnetic coupling across the interface. This information is obtained by only doubling the effort of measuring the core level lineshape, relative binding energy and relative intensity, as done in a standard XPS experiment. The LMDAD results for the growth of a Fe/Cr/Fe trilayer structure is shown in figure 8 : the top panel shows that the ferromagnetic order of the Fe(100) substrate is not disturbed by the epitaxial growth of 3 ML of Cr. The Cr 3p also shows LMDAD indicating ferromagnetic order in plane. Cr is known to grow along the [100] direction as a layered antiferromagnet,[43] the measured Cr 3p LMDAD is due to the uncompensated signal from the surface layer. The lower panel shows the completion of the trilayer structure by epitaxial overgrowth of Fe on a 5 ML thick Cr interlayer. The LMDAD signal shows that the top Fe film is antiferromagnetically coupled to the substrate iron across the 5 monolayers of chromium. The surface sensitivity of the photoemission was maximized by tuning the kinetic energy of the photoelectrons to the minimum of the escape depth. The same samples were studied in situ by $h\nu$-dependent SP measurements of the secondary yield. The SP averages over all the contribution to the yield, which include spin-polarized secondaries from the Fe overlayer, polarization-averaged emission from the Cr interlayers, spin filtering of the secondary electrons traversing the Fe overlayer, spin-polarized secondaries from the Fe substrate attenuated through the Cr layers and filtered through the antiferromagnetically coupled Fe overlayer. This means that the SP results can be understood only by integrating independent knowledge on the structure of the trilayer, or with a model.[6]

The LMDAD of Fe 3p core levels is large enough to allow for magnetic hysteresis curves to be measured. This is expected to work on soft magnetic materials where the coercivity is of a few Gauss and the disturbance of the applied field to the photoelectron collection into the analyzer is small.

LMDAD of core levels as a surface magnetometer

The merit of core level LMDAD as a surface magnetometer can be addressed by the study of the temperature dependence of the Fe 3p LMDAD in an epitaxial ultrathin film. Four monolayers of fcc-Fe were epitaxially grown on Cu-$Al(100)$;[42] they could be remanently magnetized in plane. The LMDAD asymmetry as a function of sample temperature is shown in figure 9 where it is also compared with the mean field theory prediction for the behavior of ferromagnetism near the Curie temperature (T_c). By maximizing the linearity of $log[LMDAD]$ vs. $log[1 - T/Tc]$,[44] one finds a value of $T_c = 288$ K and a critical exponent $\beta = 0.212$ which are typical values for ultrathin Fe films.[45] The spread of the LMDAD points above T_c is due to the limited coherence of the ultrathin film, i.e. it represents the density of defects and domain boundaries in the film. LMDAD reflects therefore the average value of the magnetization $<\mathbf{M}>$ and can be used to monitor surface magnetization likewise the SP of secondary electrons and the exchange asymmetry in SPLEED.[45]

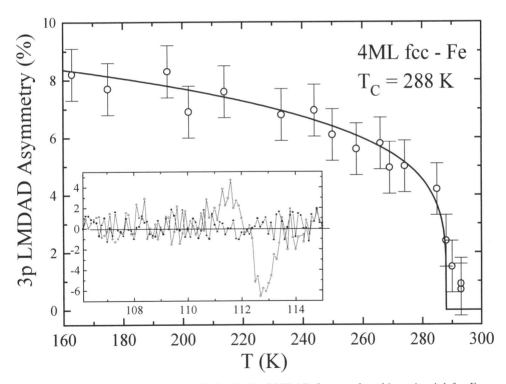

Fig.9: Temperature dependence of the Fe 3p LMDAD for an ultrathin epitaxial fcc-Fe layer grown on Cu-Al(100). The data are fitted by the mean field theory prediction. The LMDAD asymmetry at low and high temperatures are shown in the inset.

The obvious advantage of this probe over the other two since is the atom specificity and the chemical environment specificity via the core level chemical shifts. The surface sensitivity can be tuned by a proper choice of the photon energy. Also the presence of contaminants in the experiment can be easily monitored with photoemission and a reduction of LMDAD due to chemical reaction or contamination can be seen from a change of lineshape and presence of chemical shifts of the peak.

The absolute size of the LMDAD effect is on the other hand not well defined. It depends on the exact geometry of the experiment, the angular aperture of the electron spectrometer and on the photon energy employed and it can only be evaluated in the atomic model, disregarding all the solid state effects. This means that LMDAD cannot be an *absolute* surface magnetometer but its use is restricted to relative measurements of clean surfaces or adsorbates versus thickness, temperature, chemical reaction, etc. The LMDAD asymmetry in a given experiment can be calibrated against independent magnetic measurements, or against standard samples of known surface magnetic moments.

ATOMIC MODEL

Core level splitting

In ferromagnets the (3p,2p) core hole state of total angular momentum J is split by the exchange interaction with the (3d) valence electrons into sublevels with a given projection m_j on the magnetic quantization axis. We maintain this description of magnetic sublevels knowing that it is not an exact picture when considering the exchange interaction for which $J = L + S$ is not a good quantum number, i.e. the magnetic field cannot be treated as a perturbation of the spin-orbit splitting of the J levels, like in the anomalous Zeeman effect. So the m_j sublevels are the components of the splitting due to a *particular* magnetic field, the exchange field acting on spin only. The ordering of the m_j sublevels is different with respect to the Zeeman sublevel order, since exchange dominates over orbital interaction. By disregarding altogether the manybody processes involved in the photoemission from atoms and from solids, as well as the possible splitting of the high energy final states, one expects each m_j sublevel to contribute intensity with a characteristic angular distribution. In this case the core hole multiplet determines completely the intensity and the angular distribution of the photoemission spectrum. The advantage of this atomic model is that it places all of the emphasis on the matrix element for photoionization of a particular m_j core sublevel. This can be calculated within various possible schemes. Testing the atomic model of LMDAD[34] is rather straightforward since the relative variations of lineshape measured in the chiral photoemission experiment and in its mirror experiment, represent the relative variations of intensity from an effective grid of m_j sublevels. This can be compared with the relative magnitude of the calculated intensities for each m_j sublevel, for the specific chirality and photon energy.[39]

Since this model may appear too crude when we deal with photoemission from metals, a few comments are necessary. The photoexcitation of a core hole produces a multiplet of dipole allowed final states which reflect the configurations of the excited atom described by the interaction of the core holes with the conduction band holes. When localized open shells exist in the ground state of the atom, like the d-subshells in transition metals and the f-subshells in the rare earths and actinides, the interaction of the two open subshells can be so strong to spread the transition intensity over very broad multiplets with lots of structures. The intensity from a core level is therefore in general spread over several peaks and satellites. Experimentally one finds a variety of situations. Figure 10 shows extended spectra for Fe 3p and Ni 3p. The Ni 3p peak and satellites correspond to $3p^5d^{10}$, $3p^5d^9$ and $3p^5d^8$ final state configurations. The photoemission intensity of the satellites amounts to 20% of the total intensity. Since every configuration is different, it will display in principle a different magnetic dichroism.

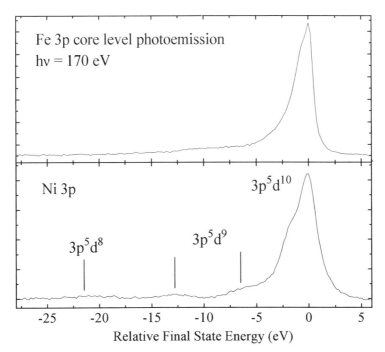

Fig.10: Wide range spectra of Fe 3p and Ni 3p core levels with evidence of satellites. Fe LMDAD (bottom).

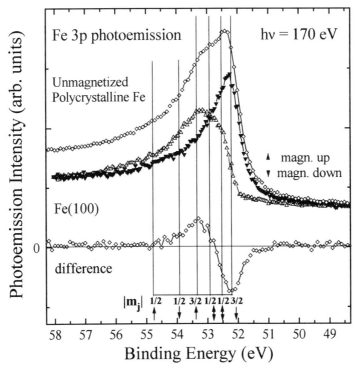

Fig.11: Fe 3p photoemission spectra after substraction of a common amount of unpolarized lineshape. The peaks and shoulders indicate a grid of magnetic sublevels.

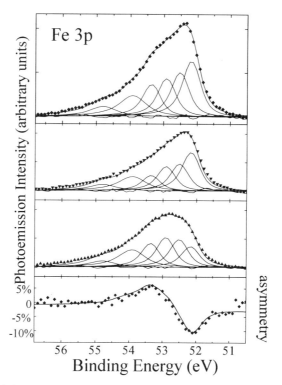

Fig.12: Analysis of the Fe 3p spectra with a sextuplet of Lorentian-Gaussian peaks with the parameters of table 1.

Such result has been suggested by a circular dichroism experiment on Ni.[46] In the case of Fe the measured photoemission intensity in the satellite region of the Fe 3p peak represents less then 10% of the total, and the maximum LMDAD asymmetry in the satellite region is less then 10^{-2}. This result for Fe 3p in bcc-Fe gives an empirical support to the discussion of the Fe 3p intensity as due to one dominant effective atomic configuration. Based on this facts one can attempt an analysis of the Fe 3p lineshape by adjusting a sextuplet of magnetic sublevels to the peaks and shoulders of the magnetization dependent spectra (figure 11). In the fitting procedure, whose results are shown in figure 12, the peak energies and amplitudes were constrained by the condition of simultaneously fitting the unmagnetized as well as the LMDAD spectra. The lineshapes and widths of each component of the sextuplet were kept fixed during the refinement procedure. The fitting parameters, and the results for the peak energies are collected in table 2. Systematically the deeper two states require a lorenzian broadening almost double with respect to the shallower four peaks.[39] One can notice that the spectral weight of the fitted sublevels deviates from a true atomic model. In fact the intensities of the $\pm m_j$ sublevels are not identical. The different intensity of the sublevels and their width (larger for the j=1/2 sublevels) are due to all the solid state effects which we have intentionally disregarded by adopting the atomic-like model. According to the atomic model the *sign and magnitude* of the variation of intensity of each peak as a function of the field reversal reflects the m_j character of each component.

The sublevels fitted to the unmagnetized spectrum are the atomic-like equivalent levels representing the 3p photoemission intensity in bcc-Fe. The goal of this analysis is to test the behavior of the sextet sublevels in the LMDAD experiment by comparing to the results of atomic calculations of the photoemission matrix elements. It is noteworthy that a recent calculation of the density of states of 3p core levels in bcc-Fe gives a very similar sextuplet of sublevels to the one derived with our empirical procedure.[47]

State Multipoles

The angular distribution of photoelectrons can be represented by the density matrices [48] of both photon beam and initial atomic state.[34] A general expression for the angular distribution of photoelectrons ejected from polarized (i.e. axially symmetric) atoms was derived by [49, 22] introducing the atomic state multipoles. The state multipoles [34] describe the polarized core hole states, and also determine the angular distribution of the photoelectrons for a given polarization of the radiation. One consequence of this theory is that the sign and the magnitude of LMDAD for different magnetic sublevels is defined by the sign and the magnitude of the relevant state multipoles. By calculating the 3p state multipoles based on atomic wave functions and by comparing the results to the measured LMDAD asymmetry, a unique value of m_j can be attributed to each sublevel of the experimentally derived sextuplet. Therefore the LMDAD experiment identifies the components of the hole state with different projections m_j i.e. the fine structure of the 3p core holes in the presence of the exchange interaction. The state multipoles ρ_{N0}^n for each magnetic sublevel of p-states are given in Table 2.

For the geometry of the experiment shown in figure 4, the photoelectron current in a given direction, can be written using eq.(10) of ref.[22]:

$$I_j^{LMDAD} = I_j(\mathbf{M}) - I_j(-\mathbf{M}) = \frac{\sigma_{nlj}(\omega)}{2\pi} 3i \, \rho_{10}^n (2j+1)^{1/2} \, C_{221}^j \cdot \sin\vartheta \cos\vartheta$$

where $\mathbf{M}$ is the direction of the sample magnetization, $\sigma_{nlj}(\omega)$ is the photoionization cross section of the nlj subshell, ω is the photon energy, and ϑ is the angle of the grazing incidence of the light. The parameter C_{221}^j depends on the reduced dipole matrix elements for the transitions to the continuum states with $l-1$ and $l+1$ and on the corresponding phase shifts. In this approximation, neither the cross section σ_{nlj} nor the parameters C_{221}^j depend on m_j; therefore the sign and the relative magnitude of LMDAD for different magnetic sublevels is defined exclusively by the sign and the magnitude of the state multipoles ρ_{10}^n.

The comparison of theory and experiment is done in figures 13 and 14 for Fe 3p and Co 3p core levels in the pure metals.

Tab.1: Fitting parameters for the unmagnetized spectra of Fe 3p and Co 3p. The gaussian broadening was 350 meV and 500 meV for Fe and Co respectively.

	Fe − 3p		Co − 3p	
m_j	B.E.(eV)	Lor.	B.E.(eV)	Lor.
+3/2	52.2	0.42	59.6	0.7
+1/2	52.5	0.42	60.0	0.7
−1/2	52.9	0.42	60.5	0.7
−3/2	53.3	0.42	61.2	0.7
−1/2	53.9	0.85	61.8	1.0
+1/2	54.8	0.85	63.0	1.0

Tab .2: State multipoles ρ_{N0}^n for magnetic sublevels of $np_{1/2}$ and $np_{3/2}$ states.

	$j=1/2$		$j=3/2$			
m_j	+1/2	−1/2	+3/2	+1/2	−1/2	−3/2
ρ_{00}^n	$\frac{1}{\sqrt{2}}$	$\frac{1}{\sqrt{2}}$	$\frac{1}{2}$	$\frac{1}{2}$	$\frac{1}{2}$	$\frac{1}{2}$
ρ_{10}^n	$\frac{1}{\sqrt{2}}$	$-\frac{1}{\sqrt{2}}$	$\frac{3}{2\sqrt{5}}$	$\frac{1}{2\sqrt{5}}$	$-\frac{1}{2\sqrt{5}}$	$-\frac{3}{2\sqrt{5}}$
ρ_{20}^n	0	0	$\frac{1}{2}$	$-\frac{1}{2}$	$-\frac{1}{2}$	$\frac{1}{2}$
ρ_{30}^n	0	0	$\frac{1}{2\sqrt{5}}$	$-\frac{3}{2\sqrt{5}}$	$\frac{3}{2\sqrt{5}}$	$-\frac{1}{2\sqrt{5}}$

It allows to conclude that the magnetic sublevel level ordering is reversed with respect to the Zeeman splitting. This is understood as a consequence of the exchange coupling of the 3p core holes with the 3d band. [30, 29]

Fine Structure: Exchange and Spin-Orbit Parameters

From the fine structure of the 3p core levels one can evaluate the exchange splitting of the core hole. The width of the $J = 3/2$ multiplet is found here to be $1.11 \pm .05 eV$ for Fe 3p, while the total width of the sextuplet is $2.6 \pm .1 eV$. The spin orbit splitting of the 3p core holes in Fe and Co is known from silicide data [27] and is respectively $1.04 \pm .05 eV$ and $1.43 \pm .05 eV$ for Fe and Co. The splitting between adjacent m_j sublevels corresponds to a value of exchange field of $3.5 \, 10^7$ Gauss for the 3p core hole in bcc-Fe. It is noteworthy to stress here that chiral experiments including spin-polarization detection can be extended to nonmagnetic samples since the polarization of the initial state is replaced by the spin-polarization of the detected electrons. After the experiments of Schöhnense [50] on rare gases the technique has been applied to the valence band of Pt, measuring the spin-orbit interaction within the bands,[51] and on the 3p core level of fcc-Cu, which confirms exactly the atomic-like behavior.[52]

Spin-Selected Spectra:

With the knowledge of the fine structure of m_j levels projected along the magnetization direction **M**, one can reconstruct the energy distribution of the spin-selected Fe 3p spectra by making the proper linear combination of the individual m_j components of the sextuplet spectrum. The highest kinetic energy 3p sublevel has minority spin character, as it has been confirmed by spin-resolved photoemission experiments.[28, 16]

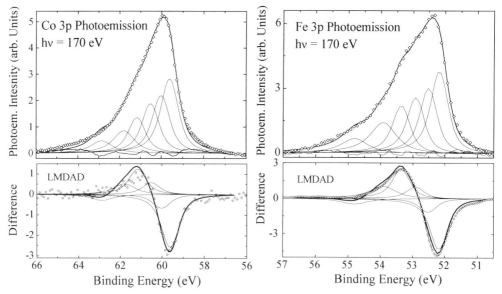

Fig.13: Upper panel: fitting of the Co 3p photoemission spectrum from unmagnetized Co (after integral background subtraction). In the bottom panel are shown the m_j components multiplied by the state multipoles of Table 2 along with the resulting convolution (continuous curve) and the experimental LMDAD spectrum (points).

Fig.14:Upper panel: fitting of the Fe 3p photoemission spectrum from unmagnetized bcc Fe (after integral background subtraction). In the bottom panel are shown the m_j components multiplied by the state multipoles of Table 2 along with the resulting convolution (continuous curve) and the experimental LMDAD spectrum

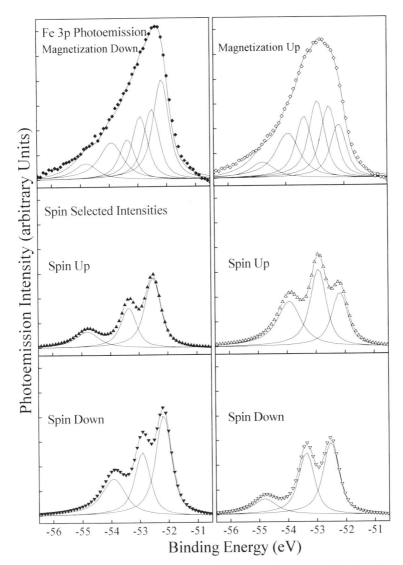

Fig.15:Experimental Fe 3p lineshapes obtained with down and up magnetization directions (full and open diamonds respectively) and the sextuplet fit with the parameters of Table 2. The lower panels show the spin selected lineshapes obtained using the m_j components of the sextuplets according to both spin projection and magnetization direction (full up triangles represent the majority spin spectrum excited with down magnetization, etc.). The curves can be directly compared to the spin-resolved experimental data of ref. 36

We approximate the $J = 3/2$, $m_j = \pm 3/2$ to pure spin states of minority (+) and majority (-) character. The intensity of the $J = 3/2$, $m_j = \pm 1/2$, and the $J = 1/2$, $m_j = \pm 1/2$ undefined spin states contains the linear combination of two spin states,[26] but for the geometry of this experiment only the components with projections of the angular momentum $m = \pm 1$ contribute.

With this procedure it becomes possible to reproduce spin-resolved angle integrated 3p spectra as well as spin-resolved LMDAD spectra based on the experimental intensities and widths of the sextuplet sublevels and on the calculated atomic LMDAD coefficients, as shown in figure 15. The lower panels show the LMDAD spin-selected intensities and can be compared to the experimental results of Roth et al.[36] Apart a sign reversal, which is a mistake, the spectra are well reproduced and explained by our spin selected LMDAD lineshapes. By analyzing figures 15 one notices that a lack of symmetry is present in the spectra. Although the general spectral shape for opposite magnetization and opposite spin compare well, yet a complete symmetry, as expected for a dichroism experiment, is not observed. Figure 16 shows the total majority spin 3p intensity and minority spin 3p intensity for Fe: these spectra represent the spin selected photoelectron intensities as measured in a spin resolved photoemission experiment. The difference curve ($I_{up} - I_{down}$) and the spin-asymmetry ($I_{spin-up} - I_{spin-down}$) / ($I_{spin-up} + I_{spin-down}$) are also plotted. The results for Co are shown in figure 17. The area ratios for the (angle integrated) minority and majority spin 3p spectra are 1.33 for Fe and 1.18 for Co. This lack of symmetry is connected with the net spin polarization of Fe 3p core level photoemission spectra confirmed by several experiments.

These shortcomings are due to the restriction of the analysis only to the main photoemission peak. On the other hand the maximum dichroism measured in the satellite region is less then 10^{-2} i.e. it is very small, and does not allow to conclude about the presence of spin polarization well below the main 3p peak. The spin neutrality of core level photoemission experiments is still an open problem.

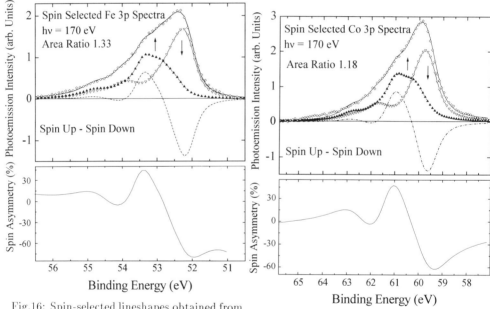

Fig.16: Spin-selected lineshapes obtained from the peaks of figure 15 by selecting all majority and minority spin intensities. Up (down) triangles indicate majority (minority) spins. The sum is compared to the unmagnetized spectrum (diamonds) The spin asymmetry is given at the bottom.

Figs.17: Spin-selected lineshapes of Co 3p core levels. Up (down) triangles indicate majority (minority) spins. The sum is compared to the unmagnetized spectrum (diamonds) The spin asymmetry is given at the bottom.

VALENCE BAND PHOTOEMISSION

LMDAD is also present in valence band photoemission experiments on Fe.[38] Figure 18 presents the 3d and valence band photoelectron energy distribution curves as obtained with excitation energy of $170 eV$ and with the poor angle resolution of a cylindrical electron energy analyzer, i.e in a regime where the spectra represent the one hole density of states. The difference spectrum gives features across the whole d band. We therefore normalized the experimental intensity at 5eV binding energy, just below the bottom of the 3d band, in order to enhance the LMDAD signal from the background difference.

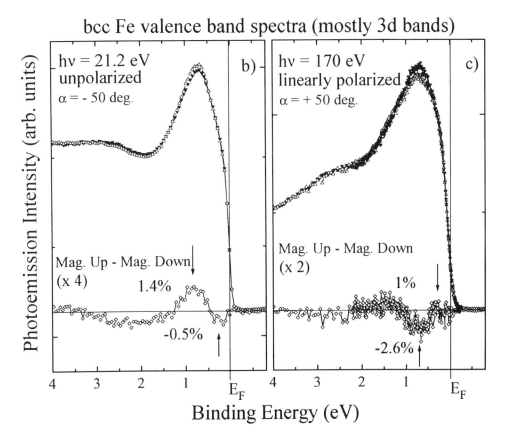

Fig.18: Valence band Fe 3d spectra and LMDAD as measured with linearly polarized synchrotron radiation,(a,b experiment following figure 4) and unpolarized He II radiation in a symmetric experiment.(d,c following figure 4)

The left panel of figure 18 shows a similar experiment performed with unpolarized Helium radiation at $21.2eV$ in the mirror geometry as the previous one, i.e. with the light beam impinging onto the sample from the opposite side of the **pM** plane.[53] The shape of the magnetization dependent spectra is different for $h\nu = 21.2eV$ and for $h\nu = 170eV$. This is due to the changes in the relative values of the photoionization cross sections for the 3d and the 4sp bands, and the different slope of the secondary electron distribution under the spectra. Nevertheless the shape of the difference spectra is very similar but inverted and roughly divided by two in the unpolarized ultraviolet spectrum. A narrow peak appears at $0.25eV$ below the Fermi level, and the main feature appears at about $0.75eV$. The asymmetry between 1.5 and $5eV$ is weak and it is sensitive to the exact alignment of the spectra, i.e .to the background intensity which is strong in the photoemission at the energy of He-I radiation ($h\nu = 21.2eV$). These results are consistent with the atomic theory for MDAD. In fact the sign reversal of the MDAD asymmetry in the two experiments is due to the reversed chirality of the two experiments (the light sources impinge in the same plane but from opposite angles with respect to the photoelectron collection direction). Furthermore the factor of two reduction of the asymmetry in the He-I experiment reflects the use of unpolarized radiation, which can be viewed as the incoherent superposition of s and p linear polarized intensities. The s−polarized component is aligned with the magnetization and gives no MDAD effect,[36] and therefore only half of the unpolarized radiation contributes to the MDAD effect. One should remember, though, that the magnitude of the LMDAD effect also depends on photon energy, as measured and predicted in the soft X-rays range for the 3p subshell.[39] The reduction of LMDAD would be exactly a factor two if the same photon energy was used, and the effects of 100% p−polarized radiation vs. unpolarized radiation were compared. The same spectrum (at the same photon energy) can be obtained in the double-mirror experiment where **M** is inverted and **q** (**e**) is mirrored with respect to the **pM** plane. The experiments were repeated for polycrystalline bcc-Fe yielding the same results. This shows that the selection of special interband-transition and the electron state symmetry arguments put forwards to explain spin polarization in valence band photoemission from centrosymmetric surfaces are not necessary to explain LMDAD.[54] On the other hand the same experiment for amorphous rcp-Fe shows a qualitatively different LMDAD spectrum.[53] This means that the changes in the radial part of the 3d wavefunctions due to different hybridization do affect the LMDAD spectrum. The 3d peak position of bcc-Fe appears shifted by approximately $0.1 eV$ when the magnetization is reversed. This shift is related to the spin-orbit interaction within the 3d band, which is the coupling mechanism for the spin and the photon field, i.e. the necessary interaction for the dichroism to occur.

The basic phenomenon is therefore the same as in core levels, and obeys to the same rules. The difference lies in the initial states which are represented by relativistic bands (in presence of exchange and spin-orbit interactions) instead of atomic-like levels. We note that similar results were obtained for Fe(100) and for fcc Co(001) by angular resolved photoemission with circular polarized synchrotron radiation at far-UV energies.[55, 56] The spectra showed Circular MDAD (CMDAD) asymmetries as a function of magnetization reversal, for a fixed light helicity, or as a function of helicity reversal for a fixed magnetization direction. The CMDAD spectra for Co(100) measured with $h\nu = 23 eV$ [55] are very similar to our LMDAD results. The CMDAD results for Fe(100),[56] obtained with $h\nu = 31 eV$ are also qualitatively the same. These facts are strong hints to the equivalence of LMDAD and CMDAD in valence band photoemission.[53] Experiments on Fe 3p core levels with circular and linear polarized light have been presented at this school by E. Kisker. Sum rules for photoemission with circularly polarized light from open shells have been derived, in analogy with the sum rules for X-ray absorption, but they are only valid for ideal angle integrated experiments.[57]

The valence bands chiral experiments show the generality of (L)MDAD in photoelectron spectroscopy which is not restricted to atomic-like core levels. This fact bears an analogy with the Cooper Minimum effect, which is an atomic property well described in photoionization from atoms, or from core levels, but also measured in hybridized 4d and 5d valence bands of the transition metals, but with intensity changes in magnitude as a function of hybridization.[58]

Valence bands narrow at the surface in connection with the reduced neighbor coordinations of surface atoms. The d bands become therefore more localized and the orbits can partially relax, due to the reduced symmetry with respect to the bulk.The relaxation at the surface of the quenching of the orbital magnetic moment was predicted [59] and recently measured in the case of Ni.[46] The spin-orbit interaction of the 3d states at the surface should therefore in principle be different with respect to the bulk and enhance the LMDAD.

The (L)MDAD results on the valence bands of ferromagnets show that high resolution chiral photoemission experiments with unpolarized ultraviolet laboratory sources can be set up in such a geometry that a magnetic asymmetry can be detected. MDAD on 3p core levels with a laboratory X-ray source has also been demonstrated.[60]

CONCLUSIONS AND OUTLOOK

LMDAD is a large effect in chiral photoemission experiments on $l > 0$ initial states. The measure of LMDAD on core levels is a diagnostic of ferromagnetic order which cost only a double effort with respect to the standard photoemission lineshape inspection. This has, and will have a major impact on the spectroscopy of magnetic surfaces. The LMDAD asymmetry is directly proportional to $<\mathbf{M}>$ and therefore it can be used to study the temperature dependence of ferromagnetic order at surfaces and the surface ferromagnetic hysteresis. The relation between the LMDAD asymmetry and the atomic magnetic moment is complex and, at present the only reliable quantities are the changes of the LMDAD asymmetry. The atomic calculations of LMDAD for the same subshell, say the 3p core level, of different atoms can be easily done and used to evaluate the relative results of LMDAD experiments, strictly done in the same geometry, on Fe, Co, Cr and the other transition metals. This should provide a guideline for the relative magnitudes of the magnetic moments if the solid state and manybody effects are not too different. The empirical transferability of the LMDAD asymmetry/magnetic moment ratio between Fe 3p and Co 3p appears to work, but preliminary results of Ni 3p are at odds, perhaps due to the larger satellite intensity of the Ni 3p spectrum.

Acknowledgements

This work was partially supported by the Swiss National Fund for Research, under program 24. We are in debt with N.A. Cherepkov and F. Combet Farnoux for their collaboration, and to H.C. Siegmann for stimulating comments and continuous support. One of us, G.P. acknowledges the EC for a grant under the Human Capital and Mobility Program.

References

[1] H.C. Siegmann; J. Phys.: Conden. Matter **4**, 8395 (1992) and references quoted therein

[2] H.C. Siegmann, in this book

[3] R. Allenspach, J. Magn. Magn. Materials, **129**, 160 (1994) and references therein

[4] A. Vaterlaus, T. Beutler, and F. Meier, Phys. Rev. Lett. **67**, 3314 (1991)

[5] F. Sirotti, R. Bosshard, G. Panaccione, A. Jucha, and G. Rossi, to be published

[6] F. Sirotti, G. Panaccione, and G. Rossi, J. Physique, to be published

[7] J. Kirschner, in *Polarized Electrons in Surface Physics*, ed. by R. Feder, World Scientific, Singapore, 1985

[8] M.J. Freiser, IEEE Transactions on Magnetics, Mag-4, 152 (1968)

[9] T. Beier, H. Jahrreiss, D. Pescia, Th. Woike, and W. Gudat, Phys. Rev. Lett. **61**, 1875 (1988)

[10] P. Carra and M. Altarelli, Phys. Rev. Lett. **64**, 1286 (1990)

[11] M.Sacchi, F.Sirotti, G.Rossi, Solid State Commun. **81**, 977 (1992); M. Sacchi, O. Sakho, F. Sirotti, and G. Rossi, Appl. Surf. Sci. **56-58**, 1 (1992)

[12] C.T. Chen, Y.U. Idzerda, H.J. Lin, G. Meigs, A. Chaiken, G.A. Prinz, and G.H. Ho Phys. Rev. **B48**, 642 (1993)

[13] B.T. Thole, P. Carra, F. Sette, and G. van der Laan; Phys. Rev. Lett. **68**, 1943 (1992)

[14] J. Stöhr and Y. Wu, in "New Directions in Research with Third Generation Soft X-Ray Synchrotron Radiation Sources " ed by. A.S. Schlachter and F. Wuillemier, 1994 Kluwer Academic Publisher p.221; F. Sette, ibid. p.251

[15] M. Landolt, in *Polarized Electrons in Surface Physics*, ed. by R. Feder, World Scientific Singapore, 1985; R. Allenspach, D. Mauri, M. Taborelli, and M. Landolt, Phys. Rev **B35**, 4801 (1987)

[16] C. Carbone, and E. Kisker; Solid State Commun. **65**, 1107 (1988)

[17] E. Kisker, in *Polarized Electrons in Surface Physics*, ed. by R. Feder, World Scientific Singapore, 1985

[18] D. Tillmann, R. Thiel, E. Kisker, Z. Phys. B-Condensed Matter **77**, 1 (1989)

[19] G. Schönhense and H.C. Siegmann, Ann. Physik **2**, 465 (1993)

[20] U. Fano; Phys. Rev. **178**, 131 (1969)

[21] G. Baum, M.S. Lubell, and W. Raith, Phys. Rev. Lett. **25**, 267 (1970)

[22] N.A. Cherepkov and V.V. Kuznetsov J. Phys. B **22**, L405 (1989)

[23] G. van der Laan and B.T. Thole, Solid State Commun. **92**, 427 (1994

[24] B.T. Thole and G. van der Laan, Phys. Rev. Letters **67**, 3306 (1991); Phys. Rev. **B49**, 9613 (1994)

[25] G. Schönhense, Physica Scripta **T31**, 255 (1990)

[26] L.I. Schiff "Quantum Mechanics ", McGraw-Hill New York (1955).

[27] F. Sirotti, M. De Santis, and G. Rossi, Phys. Rev. **B48**, 8299 (1993)

[28] B. Sinkovic, P.D. Johnson, N.B. Brookes, A. Clarke, and N.V. Smith, Phys. Rev. Lett. **65**, 1647 (1990)

[29] L. Baumgarten, C.M. Schneider, H. Petersen, F. Schafers, and J. Kirschner; Phys. Rev. Letters **65**, 492 (1990)

[30] H. Ebert, L. Baumgarten, C.M. Schneider, and J. Kirschner, Phys. Rev.B **44**, 4406 (1991)

[31] K. Starke, E. Navas, L. Baumgarten, and G. Kaindl, Phys. Rev. **B48**, 1329 (1993)

[32] G. van der Laan, Phys. Rev. Lett. **66**, 2527 (1991)

[33] N.V. Smith and P.K. Larsen; in *Photoemission and the Electron Properties of Surfaces* ed. by Feuerbacher, Fitton and Willis, John Wiley and Sons, New York, 1978.

[34] N.A. Cherepkov, Phys. Rev. B to be published; N.A. Cherepkov, V.V. Kuznetsov, and V.A. Verbitskii; J. Phys. B. to be published

[35] D. Venus; Phys. Rev. **B48**, 6144 (1993), ibid. **49**, 8821 (1994))

[36] Ch. Roth, F.U. Hillebrecht, H. Rose, and E. Kisker, Phys. Rev. Lett. **70**, 3479 (1993)

[37] Ch. Roth, H. Rose, F.U. Hillebrecht, and E. Kisker, Solid State Commun. **86**, 647 (1993)

[38] F. Sirotti and G. Rossi, Phys. Rev. **B49**, 15682 (1994)

[39] G. Rossi, F. Sirotti, N.A. Cherepkov, F. Combet Farnoux, and G. Panaccione, Solid State Commun. **90**, 557 (1994)

[40] E. Kisker, in this book

[41] C.M. Schneider, D. Venus, and J. Kirschner, Phys. Rev. **B45**, 5041 (1992)

[42] W. Macedo, A. Schatz, G. Panaccione, F. Sirotti, and G. Rossi, unpublished results

[43] J. Unguris, R.J. Celotta, and D. Pierce, Phys. Rev. Lett. **69**, 1125 (1992)

[44] D.L. Connelly, J.S. Loomis, and D.E. Mapother, Phys. Rev. **B3**, 924 (1971)

[45] W. Durr, M. Taborelli, O. Paul, R. Germar, W. Gudat, D. Pescia, and M. Landolt, Phys. Rev. Lett. **62**, 206 (1989)

[46] G. van der Laan, M.A. Hoyland, M. Surman, C.F. J. Flipse, and B.T. Thole, Phys. Rev. Lett **69**, 3827 (1993)

[47] E. Tamura, G.D. Waddill, J.G. Tobin, and P.A. Sterne, Phys. Rev. Lett. **73**, 1533 (1994)

[48] K. Blum, *Density Matrix Theory and Application* (Plenum, New York, 1981)

[49] H. Klar and H. Kleinpoppen, J. Phys. B **15**, 933 (1982)

[50] G. Schönhense, Phys. Rev. Lett. **44**, 640 (1980)

[51] N. Irmer, R. David, B. Schmeideskamp, and U. Heinzmann, Phys. Rev. **B45**, 3849 (1992)

[52] Ch. Roth, F.U. Hillebrecht, W.G. Park, H.B. Rose, and E. Kisker, Phys. Rev. Lett. **73**, 1963 (1994)

[53] G. Panaccione, F. Sirotti, and G. Rossi, to be published

[54] E. Tamura and R. Feder, Europhysics Lett. **16**, 695 (1991)

[55] C.M. Schneider, M.S. Hammond, P. Schuster, A. Cebollada, R. Miranda, and J. Kirschner, Phys. Rev. **B 44**, 12066 (1991)

[56] J. Bansmann, C. Westphal, M. Getzlaff, F. Fegel and G. Schonhense, J. Magn.Magn. Mater. **104-107**, 1691 (1992)

[57] B.T. Thole and G. van der Laan, Phys. Rev. Lett. **70**, 2499 (1993)

[58] G. Rossi, I. Lindau, L. Braicovich, and I. Abbati, Phys. Rev. **B 28**, 303 (1983)

[59] O. Eriksson, G.W. Fernando, R.C. Albers, and A.M. Boring, Solid. State Commun. **78**, 801 (1991)

[60] F.U. Hillebrecht, and W.D. Herberg, Z. Phys. **B93**, 299 (1994)

X-RAY RESONANT INELASTIC SCATTERING

Paolo Carra[†], and Michele Fabrizio[‡]

[†] European Synchrotron Radiation Facility,
B.P. 220, F-38043 Grenoble Cédex, France
[‡] Institut Laue-Langevin,
B.P. 156X, F-38042 Grenoble Cédex, France

INTRODUCTION

With the advent of synchrotron radiation sources, characterised by high fluxes, tunability and polarisation control, the inelastic scattering of x rays has become a powerful probe of ground-state properties and elementary excitations in solids.

For energies $\hbar\omega < 100$ keV, the x rays interact weakly with bulk samples, and the scattering may be accurately treated in the lowest Born approximation. The double differential cross-section is then obtained by considering the usual semi-classical coupling of the electrons to the photons:

$$H = \frac{1}{2m} \sum_j \left(\mathbf{p}_j + \frac{e}{c} \mathbf{A}(\mathbf{r}_j) \right)^2,$$

where $\mathbf{A}$ represents the vector potential of the electromagnetic field; m and $\mathbf{p}$ denote the electron mass and momentum, respectively. (Relativistic corrections, which are of order $\hbar\omega/mc^2$, will not be considered in this work.)

Away from resonance, the scattering process is controlled by the $\mathbf{A}^2$ term (treated in first order); in this case, according to the magnitude of energy and momentum transfers, inelastic scattering experiments can roughly be subdivided in three classes:

1. Compton scattering, giving ground-state information in momentum space;
2. scattering by valence electrons, probing the valence-electron dynamics (e.g. plasmons);
3. scattering by collective ion excitations (phonons).

For a complete discussion of non-resonant inelastic x-ray scattering, the reader is referred to the recent review by Schülke [1].

---

[1] Address from January 1, 1995: Scuola Superiore di Studî Avanzati (SISSA), via Beirut 4, I-34014 Trieste, Italy.

Near an absorption edge, the $\mathbf{A} \cdot \mathbf{p}$ term (treated to second order) can give rise to significant contributions to the scattering amplitude; these effects have been thoroughly investigated in the elastic limit (anomalous diffraction), in magnetic and anisotropic systems: starting with the pioneering work of Templeton and Templeton [2] and Gibbs and collaborators [3], various experimental observations have probed charge and magnetic spatial correlations, including the critical region, in transition metals, rare earths and actinides [4, 6, 5]. These experiments can be interpreted using an effective scattering amplitude, derived in terms of simple density operators by applying the methods previously developed for the related processes of x ray absorption and dichroism [7, 8, 9].

Among the resonant inelastic scattering processes, mostly those with a core hole in the final state (Resonant Raman scattering) have been so far considered [10, 11]. Only recently, more attention has been paid to processes where the final state contains an electron-hole excitation in the valence band [12, 13]. The latter appears to be particularly attractive, in view of the possibility of probing elementary excitations in solids, using x rays at resonance. To this end, several high resolution beam lines are currently under construction at third generation synchrotron radiation facilities.

As a result of energy and momentum conservation, the double differential x-ray scattering cross-section is related to the density of states of all possible excitations of given energy: $\hbar\omega = \hbar c(|\mathbf{k}| - |\mathbf{k}'|)$, and momentum: $\hbar\mathbf{q} = \hbar(\mathbf{k} - \mathbf{k}')$; here, $\hbar\mathbf{k}$ and $\hbar\mathbf{k}'$ denote the momentum of the ingoing and outgoing photons, respectively. This applies, in particular, to x-ray resonant inelastic scattering; specifically, if the energy transfer is smaller than the energy required to excite interband transitions, the cross-section will involve valence band excitations only. This may be surprising at first, as the scattering takes place through local transitions, a priori insensitive to the presence of a crystalline lattice; moreover, the propagation of the core states is a rather slow process, so that creation and annihilation of the core-hole should occur on the same atomic site. However, for lattice separations smaller than the photon coherence lenght, resonances occurring on different sites interfere contructively, therefore probing long range (q→0) correlations, when the spatial coherence of the incident photon is large enough. These recent developments provide an opportunity to outline, in this lecture, a symmetry analysis of the correlation functions associated with the resonant inelastic scattering of x rays.

THEORETICAL FRAMEWORK

We consider x-ray resonant inelastic scattering, a process in which, by way of the coupling:

$$\hat{H}_{\text{int}} = \frac{e}{mc}\mathbf{p}\cdot\mathbf{A},$$

a photon of energy $\hbar\omega_\mathbf{k}$, momentum $\hbar\mathbf{k}$ and polarisation ϵ is scattered into a photon with energy $\hbar\omega_{\mathbf{k}'}$, momentum $\hbar\mathbf{k}'$ and polarisation ϵ'. An energy $\hbar\omega = \hbar(\omega_\mathbf{k} - \omega_{\mathbf{k}'})$ and a momentum $\hbar\mathbf{q} = \hbar(\mathbf{k} - \mathbf{k}')$ are transferred to the scattering system.

The scattering rate can be determined using Fermi's golden rule:

$$P(i \rightarrow f) = \frac{2\pi}{\hbar}|U_{i\rightarrow f}|^2 \delta(E_f + \hbar\omega_{\mathbf{k}'} - E_i - \hbar\omega_\mathbf{k}),$$

with the resonant transition amplitude $U_{i\rightarrow f}$ given by (second order perturbation theory):

$$U_{i\rightarrow f} = \sum_n \frac{\langle f|\hat{H}_{\text{int}}|n\rangle\langle n|\hat{H}_{\text{int}}|i\rangle}{E_i + n_{\epsilon\mathbf{k}}\hbar\omega_\mathbf{k} - \mathcal{E}_n + i\eta} \quad (1)$$

The notation is as follows: $|i\rangle = |\psi_i; n_{\epsilon\mathbf{k}}, n_{\epsilon'\mathbf{k}'} = 0\rangle$ represents the initial state, with the scattering system in a quantum state ψ_i, in the presence of $n_{\epsilon\mathbf{k}}$ photons, and $|f\rangle = |\psi_f; n_{\epsilon\mathbf{k}} - 1, n_{\epsilon'\mathbf{k}'} = 1\rangle$ denotes a final state induced by $\hat{H}_{\text{int}}$. E and $\mathcal{E}$ stand for the energy of the electron system and electron system + e.m. field, respectively. Furthermore, in the second quantisation formalism:

$$\hat{H}_{int} = \frac{e}{mc} \int d\mathbf{r} \mathbf{A}(\mathbf{r}) \cdot \Psi^\dagger(\mathbf{r}) \mathbf{p} \Psi(\mathbf{r})$$

with $\Psi(\mathbf{r})$ the electron field (to be defined below), and

$$\mathbf{A}(\mathbf{r}) = \sum_{\mathbf{k}\epsilon} \sqrt{\frac{2\pi\hbar c^2}{V\omega_\mathbf{k}}} \left(\epsilon a_\mathbf{k} e^{i\mathbf{k}\mathbf{r}} + \text{H.c.}\right).$$

The total scattering rate, obtained by summing $P_{i \to f}$ over all final states allowed by energy conservation, is proportional to the dynamic structure factor (also known as the scattering law):

$$S(\mathbf{q}, \omega) = \int_{-\infty}^{\infty} dt e^{i\omega t} \langle i | \hat{O}^\dagger(\mathbf{q}, t) \hat{O}(\mathbf{q}) | i \rangle,$$

with

$$\hat{O}(\mathbf{q}) = \sum_n \frac{\hat{H}_{int} |n\rangle \langle n| \hat{H}_{int}}{E_i + n_{\epsilon \mathbf{k}} \hbar \omega_\mathbf{k} - \mathcal{E}_n + i\eta},$$

and $\hat{O}(\mathbf{q}, t) = exp(-i\hat{H}t/\hbar) \hat{O}(\mathbf{q}, 0) e(i\hat{H}t/\hbar)$.

The above expressions are general and yield the response function associated with a given x-ray resonant inelastic scattering process; however, they are not of great help interpreting experiments, as the operator $\hat{O}(\mathbf{q})$ (as defined above) has, in general, complex dynamics.

In this lecture it will be shown that under particular assumptions, which correspond to realisable experimental conditions, $\hat{O}(\mathbf{q})$ can be expressed in terms of elementary spin and charge density operators, and the correlation function, appearing in the definition of the dynamic structure factor, can be given a standard two-particle form.

When $\hbar \omega_\mathbf{k}$ is close to an absorption threshold, the transition amplitude $U_{i \to f}$ is dominated by those processes in which the ingoing photon is first absorbed by promotion of a core electron into the conduction band; the core-hole is then annihilated by creation of a hole in the conduction band and emission of a photon of momentum $\mathbf{k}'$ and polarization ϵ'. In such a case, a simple expression for the electron field operator is obtained by considering core and conduction electron states only:

$$\Psi(\mathbf{x}) = \sum_{\boldsymbol{\kappa}}^{BZ} \left[\sum_{\lambda\sigma} l_{\lambda\sigma, \boldsymbol{\kappa}} \psi_{\lambda\sigma, \boldsymbol{\kappa}}(\mathbf{x}) + \sum_m c_{jm, \boldsymbol{\kappa}} \psi_{cjm, \boldsymbol{\kappa}}(\mathbf{x}) \right]$$

with the Bloch waves expanded using Wannier's functions:

$$\psi_{\lambda\sigma, \boldsymbol{\kappa}}(\mathbf{x}) = \frac{1}{N^{\frac{1}{2}}} \sum_\mathbf{R} e^{i\boldsymbol{\kappa} \cdot \mathbf{R}} \varphi_{\lambda\sigma}(\mathbf{x} - \mathbf{R})$$

and

$$\psi_{cjm, \boldsymbol{\kappa}}(\mathbf{x}) = \frac{1}{N^{\frac{1}{2}}} \sum_\mathbf{R} e^{i\boldsymbol{\kappa} \cdot \mathbf{R}} \varphi_{cjm}(\mathbf{x} - \mathbf{R}).$$

Here $\mathbf{R}$ is a lattice site label and N the number of sites; $l_{\lambda\sigma, \boldsymbol{\kappa}}$ and $c_{jm, \boldsymbol{\kappa}}$ denote the annihilation operators for conduction and core electrons, respectively. The core electron Wannier functions have been labelled using atomic quantum numbers: c (orbital angular momentum), $j = c \pm \frac{1}{2}$ (total angular momentum) and m (its projection on the quantisation axis); the spin-orbit coupling of the conduction band has been neglected, thus justifying the labelling by a set of orbital quantum numbers λ and spin σ. In this case, $\hat{H}_{int}$ takes the form:

$$\begin{aligned}\hat{H}_{int} &= \frac{e}{mc}\frac{1}{N}\sum_{\epsilon\mathbf{k}}N_{\mathbf{k}}a_{\epsilon\mathbf{k}}\sum_{\kappa\kappa'}\sum_{\mathbf{R}\mathbf{R}'}\sum_{\lambda\sigma m}e^{-i\boldsymbol{\kappa}\cdot\mathbf{R}+i\boldsymbol{\kappa}'\cdot\mathbf{R}'}l^{\dagger}_{\lambda\sigma,\boldsymbol{\kappa}}c_{jm,\boldsymbol{\kappa}'}\cdot\\ &\quad \cdot\int d\mathbf{x}\varphi^*_{\lambda\sigma}(\mathbf{x},\mathbf{R}\ )e^{i\mathbf{k}\cdot\mathbf{x}}\boldsymbol{\epsilon}\cdot\mathbf{p}\varphi_{cjm}(\mathbf{x},\mathbf{R}\ ')+\text{H.c.} \quad (2)\\ &= \sum_{\epsilon\mathbf{k}}N_{\mathbf{k}}a_{\epsilon\mathbf{k}}\hat{H}_{\epsilon\mathbf{k}}+\text{H.c.},\end{aligned}$$

with $N_{\mathbf{k}} = \sqrt{2\pi\hbar c^2/V\omega_{\mathbf{k}}}$. Making use of the properties of Wannier's functions, the operators $\hat{H}_{\epsilon\mathbf{k}}$, defined by expression (2), can be written as:

$$\hat{H}_{\epsilon\mathbf{k}} = \sum_{\boldsymbol{\kappa}\mathbf{G}}\sum_{\lambda\sigma m}M_{\lambda\sigma;jm}(\epsilon;\mathbf{k}\ ,\boldsymbol{\kappa})l^{\dagger}_{\lambda\sigma,\boldsymbol{\kappa}+\mathbf{k}+\mathbf{G}}c_{jm,\boldsymbol{\kappa}},$$

with $\mathbf{G}$ a reciprocal lattice vector, and

$$M_{\lambda\sigma;jm}(\epsilon,\mathbf{k}\ ,\boldsymbol{\kappa}) = \frac{e}{mc}\sum_{\mathbf{R}}e^{-i(\boldsymbol{\kappa}+\mathbf{k})\mathbf{R}}\int d\mathbf{x}\varphi^*_{\lambda\sigma}(\mathbf{x},\mathbf{R}\ )e^{i\mathbf{k}\cdot\mathbf{x}}\boldsymbol{\epsilon}\cdot\mathbf{p}\varphi_{cjm}(\mathbf{x},0)\ .$$

By inserting (2) into (1), we find:

$$U_{i\to f} = \sum_{I}N_{\mathbf{k}}N_{\mathbf{k}'}\sqrt{n_{\epsilon\mathbf{k}}}\frac{\langle\psi_f\mid\hat{H}^{\dagger}_{\epsilon'\mathbf{k}'}\mid I\rangle\langle I\mid \hat{H}_{\epsilon\mathbf{k}}\mid\psi_i\rangle}{E+\hbar\omega_{\mathbf{k}}-E_I+i\Gamma_I/2}, \quad (3)$$

where the intermediate states $|I\rangle$ are eigenstates of the system Hamiltonian with energies E_I; they are assumed to be of the form: $|I\rangle = l^{\dagger}c|\psi_i\rangle$. This amounts to neglecting many core-hole decay processes, e.g. non-radiative (Auger) decays, which might occur before the core-hole is filled by a conduction electron. In order not to completely disregard these events, a finite core-hole lifetime: $\tau_I = \hbar/\Gamma_I$ has been introduced. The final states $|\psi_f\rangle$ differ from the initial one by the presence of a particle-hole excitation in the conduction band.

To obtain a suitable form for the operator $\hat{O}(\mathbf{q})$, further approximations are required. Firstly, we assume that the intermediate state lifetime, determined by the minimum between τ_I and $\hbar/|\hbar\omega_{\mathbf{k}}-E_I+E_i|$, is so small that photon absorption and emission are practically simultaneous (no core-hole propagation). Experimentally, this amounts to having: $Max(|\hbar\omega_{\mathbf{k}}-E_I+E_i|,\Gamma_I) \gg D$, with $D = \langle(E_I-\langle E_I\rangle)^2\rangle^{\frac{1}{2}}$ of the order of the conduction bandwidth. In a metal, the propagation of the core hole would give rise to the well-known infrared singularities [14]; such effects can be disregarded, if the above condition is fulfilled. The above assumption is equivalent to neglecting the dispersion of the intermediate states; E_I and Γ_I can be taken as constants, and the expansion for the resonant denominator:

$$\begin{aligned}(E_I - E_i - \hbar\omega - i\tfrac{\Gamma_I}{2})^{-1} &= (\langle E_I\rangle - E_i - \hbar\omega - i\tfrac{\langle\Gamma_I\rangle}{2})^{-1}\cdot \quad (4)\\ &\quad \cdot\sum_{n=0}^{\infty}\left(\frac{\langle E_I-i\tfrac{\Gamma_I}{2}\rangle - E_I + i\tfrac{\Gamma_I}{2}}{\langle E_I\rangle - E_i - \hbar\omega - i\tfrac{\langle\Gamma_I\rangle}{2}}\right)^n,\end{aligned}$$

truncated at $n=0$. The sum over the intermediate states $|I\rangle$ in Eq. (3) can then be performed by completeness. We have:

$$\begin{aligned}U_{i\to f} &= N\sum_{\boldsymbol{\kappa}\boldsymbol{\kappa}'}^{BZ}\sum_{\lambda\sigma m}\sum_{\mathbf{G}\mathbf{G}'}\sum_{\lambda'\sigma'm'}M_{\lambda\sigma;jm}(\epsilon,\mathbf{k}\ ,\boldsymbol{\kappa})M^*_{\lambda'\sigma';jm'}(\epsilon',\mathbf{k}\ ',\boldsymbol{\kappa}')\\ &\quad \cdot\langle\psi_f|c^{\dagger}_{jm',\boldsymbol{\kappa}'}l_{\lambda'\sigma',\boldsymbol{\kappa}'+\mathbf{k}'+\mathbf{G}'}l^{\dagger}_{\lambda\sigma,\boldsymbol{\kappa}+\mathbf{k}+\mathbf{G}}c_{jm,\boldsymbol{\kappa}}|\psi_i\rangle,\end{aligned}$$

with
$$\mathcal{N} = \frac{N_\mathbf{k} N_{\mathbf{k}'} \sqrt{n_{\epsilon \mathbf{k}}}}{E_i + \hbar\omega_\mathbf{k} - \langle E_I \rangle + i\langle \Gamma_I \rangle/2}.$$

As the core band is full in both the initial and final states, the expression for the transition amplitude can be further simplified:

$$U_{i \to f} = \mathcal{N} \sum_{\boldsymbol{\kappa} \mathbf{G}} \sum_{\lambda \sigma \lambda' \sigma' m} M_{\lambda \sigma; jm}(\epsilon, \mathbf{k}, \boldsymbol{\kappa}) M^*_{\lambda' \sigma'; jm}(\epsilon', \mathbf{k}', \boldsymbol{\kappa}) \cdot$$
$$\cdot \langle \psi_f | l_{\lambda' \sigma', \boldsymbol{\kappa}} l^\dagger_{\lambda \sigma, \boldsymbol{\kappa}+\mathbf{q}+\mathbf{G}} | \psi_i \rangle,$$

with $\boldsymbol{\kappa} + \mathbf{q} + \mathbf{G} \in BZ$. The double differential cross-section is obtained by multiplying the total scattering rate by the density of states of the outgoing photon: $V\omega_{\mathbf{k}'}^2/\hbar c^3 (2\pi)^3$, and dividing by the incident flux: $cn_{\epsilon \mathbf{k}}/V$.

X-ray resonant scattering is controlled by electric dipole and quadrupole transitions; an appropriate formulation is obtained by performing a coupled-multipolar expansion of the electron-photon interaction [15] and retaining electric transitions (EL) only. This is achieved by considering the standard expansion:

$$\mathbf{p} \cdot \mathbf{A} = \mathbf{p} \cdot \mathbf{e}_i \, g_l(k_i r) \sum_m Y^{l*}_m(\hat{\mathbf{k}}_i) Y^l_m(\hat{\mathbf{r}}),$$

and then recoupling $\mathbf{p}$ and $Y^l(\hat{\mathbf{r}})$ to a total L, giving the term [16]:

$$\sum_L \left[[\mathbf{e}_i, Y^l(\hat{\mathbf{k}}_i)]^L [\mathbf{p}, Y^l(\hat{\mathbf{r}})]^L \right]^0 g_l(k_i r).$$

In the limit $k_i r \ll 1$, $g_l(k_i r) \sim (k_i r)^l$, one finds

$$\langle \psi_2 | \mathbf{p} \cdot \mathbf{A} + \mathbf{A} \cdot \mathbf{p} | \psi_1 \rangle \sim \frac{2m(E_2 - E_1)}{i\hbar \sqrt{L(2L+1)}} \langle \psi_2 | r^L Y^L_M(\hat{\mathbf{r}}) | \psi_1 \rangle,$$

i.e. the general electic multipole matrix element.

The double-differential scattering cross-section can then be given the form:

$$\frac{d^2\sigma}{d\Omega d\hbar\omega_{\mathbf{k}'}} = \frac{\omega_{\mathbf{k}'}}{\omega_\mathbf{k}} \sum_f \left| \frac{\sum_{z=0}^{2L} \left[T^{(z)}(\epsilon'^*, \mathbf{k}'; \epsilon, \mathbf{k})_{EL} \langle \psi_f | F^{(z)}(k, k')_{EL} | \psi_i \rangle \right]^0}{\langle E_I \rangle - E_i - \hbar\omega_\mathbf{k} - i\langle \Gamma_I \rangle/2} \right|^2$$
$$\cdot \delta(E_f + \hbar\omega_{\mathbf{k}'} - E_i - \hbar\omega_\mathbf{k}), \qquad (5)$$

with

$$T^{(z)}_\zeta(\epsilon'^*, \mathbf{k}'; \epsilon, \mathbf{k})_{EL} = \frac{2L+1}{L+1} \left[[\epsilon' Y^{L-1}(\mathbf{k}')]^L [\epsilon^* Y^{L-1}(\mathbf{k})]^L \right]^z_\zeta,$$
$$F^{(z)}_\zeta(k, k')_{EL} = \sum_{\boldsymbol{\kappa} \mathbf{G}} \sum_{\lambda\sigma, \lambda'\sigma' m} \left[A^{L^\dagger}(k'\lambda'\sigma') A^L(k\lambda\sigma) \right]^z_\zeta l_{\lambda'\sigma', \boldsymbol{\kappa}} l^\dagger_{\lambda\sigma, \boldsymbol{\kappa}+\mathbf{q}+\mathbf{G}}, \qquad (6)$$

and

$$A^L_q(k\lambda\sigma) = \frac{4\pi e(ik)^L}{(2L+1)!!} \sqrt{\frac{L+1}{L}} \cdot$$
$$\cdot \sum_\mathbf{R} e^{-i(\boldsymbol{\kappa}+\mathbf{k})\mathbf{R}} \int d\mathbf{x} \varphi^*_{\lambda\sigma}(\mathbf{x}, \mathbf{R}) r^L Y^L_q(\mathbf{x}) \varphi_{cjm}(\mathbf{x}, 0);$$

the tensor couplings are defined by:

$$[T^{L'} F^{L''}]^L_l = \sum_{l'l''} T^{L'}_{l'} F^{L''}_{l''} \langle L'L''l'l'' | Ll \rangle.$$

Expressions (5) and (6) show that, when the fast collision approximation holds, the scattering is described by standard particle–hole creation operators. Their full expression is obtained by analysing the relations between hole and particle quantum numbers, as determined by the tensors $\left[A^{L^\dagger} A^L \right]^z$; this will be discussed in the next section.

CORRELATION FUNCTIONS

The double differential scattering cross-section can be expressed in terms of the dynamic structure factor, $S(\mathbf{q}, \omega)$, once we make the identification:

$$\hat{O}(\mathbf{q}) = \sum_{z=0}^{2L} [T^z(\epsilon, \mathbf{k}; \epsilon'^*, \mathbf{k}')_{EL} F^z(k, k')_{EL}]^0$$

$$= \sum_{z=0}^{2L} \hat{O}_z(\mathbf{q}),$$

write the δ-function, which appears in (5), as:

$$\delta(E_f + \hbar\omega_{\mathbf{k}'} - E_i - \hbar\omega_{\mathbf{k}}) = \frac{1}{2\pi\hbar} \int_{-\infty}^{\infty} dt\, e^{i(E_f + \hbar\omega_{\mathbf{k}'} - E_i - \hbar\omega_{\mathbf{k}})/\hbar};$$

and perform the sum over the final states $|\psi_f\rangle$ by completeness. This yields:

$$\frac{d^2\sigma}{d\Omega' d\hbar\omega_{\mathbf{k}'}} = \frac{\omega_{\mathbf{k}'}}{\omega_{\mathbf{k}}} \frac{1}{2\pi\hbar} \int_{-\infty}^{\infty} dt\, e^{i\omega t} \langle \psi_i | \hat{O}^\dagger(\mathbf{q}, t) \hat{O}(\mathbf{q}, 0) | \psi_i \rangle.$$

To obtain a simple form for the operator $\hat{O}(\mathbf{q})$ we make a further approximation regarding the range of the matrix elements that enter the definition of F^z. As we expect the Wannier functions to be localised about the corresponding lattice sites, we assume that the integral:

$$\sum_{\mathbf{R}} e^{-i(\boldsymbol{\kappa}+\mathbf{k})\mathbf{R}} \int d\mathbf{x}\, \varphi^*_{\lambda\sigma}(\mathbf{x}, \mathbf{R})\, r^L Y^L_q(\mathbf{x})\, \varphi_{cjm}(\mathbf{x}, 0)$$

is non-neglegible only for $\mathbf{R}=0$.

So far, the form of the valence band Wannier functions, $\varphi_{\lambda\sigma}(\mathbf{x})$, has not been specified. In what follows, we will assume that, as for the core functions, their symmetry properties are atomic-like at each lattice site; i.e., they coincide with those of the atomic orbitals whose angular symmetry is expected to be the dominant one in the conduction band (orthogonal tight binding approximation). Consequently, the quantum numbers l and λ are identified with an orbital angular momentum and its projection along the quantisation axis, respectively. The resonant scattering amplitude involves an overlap integral between conduction and core electron wave functions where only the local (atomic) properties of the Wannier functions are expected to be of importance, thus justifying the above approximation.

Non-neglegible crystal fields are accounted for by the transformation: $\varphi_\lambda = \sum_m U_{\lambda m} Y^l_m$, with $U_{\lambda m}$ a unitary matrix, yielding a linear combination of spherical harmonics of equal rank l. Similarly, when atomic orbitals with different angular momenta are expected to be equally important, φ_λ is rewritten as a linear combination of these orbitals., For simplicity, the extension of our formalism to cover these effects will not be considered in this lecture.

Finally, the expression: $\sum_{\lambda\sigma,\lambda'\sigma'm} \left[A^{L\dagger}_{k'\lambda'\sigma'} A^L_{k\lambda\sigma} \right]^z_\zeta l_{\lambda'\sigma',\boldsymbol{\kappa}} l^\dagger_{\lambda\sigma,\boldsymbol{\kappa}+\mathbf{q}+\mathbf{G}}$, is transformed by applying the Wigner-Eckhart theorem and recoupling the angular part using standard methods of angular momentum theory [17]. This is a rather technical part of our derivation; a sketc of the method is given in the Appendix.

In the important case of electric dipolar (E1) transitions (for which $z = 0, 1, 2$), the operators $\hat{O}_z(\mathbf{q})$ can be given the form:

$$\hat{O}_0(\mathbf{q}, 0) = R_{cj;1;l} \sum_{\boldsymbol{\kappa}\mathbf{G}} \left[\delta_0(c, l) \sum_{\lambda\sigma} l_{\lambda\sigma,\boldsymbol{\kappa}} l^\dagger_{\lambda\sigma,\boldsymbol{\kappa}+\mathbf{q}+\mathbf{G}} \right.$$

$$\left. + \delta_1(c, l) \sum_{\lambda\sigma,\lambda'\sigma'} (l_{\lambda\lambda'} \cdot s_{\sigma\sigma'}) l_{\lambda'\sigma',\boldsymbol{\kappa}} l^\dagger_{\lambda\sigma,\boldsymbol{\kappa}+\mathbf{q}+\mathbf{G}} \right] \epsilon'^* \cdot \epsilon$$

$$\begin{aligned}
\hat{O}_1(\mathbf{q},0) &= R_{cj;1;l}\sum_{\boldsymbol{\kappa}\mathbf{G}}\left[\rho_0(c,l)\sum_{\lambda\lambda'\sigma}l_{\lambda\lambda'}l_{\lambda\sigma,\boldsymbol{\kappa}}l^\dagger_{\lambda'\sigma,\boldsymbol{\kappa}+\mathbf{q}+\mathbf{G}}\right.\\
&\quad + \rho_1(c,l)\sum_{\lambda\sigma\sigma'}s_{\sigma\sigma'}l_{\lambda\sigma,\boldsymbol{\kappa}}l^\dagger_{\lambda\sigma',\boldsymbol{\kappa}+\mathbf{q}+\mathbf{G}}\\
&\quad \left.+ \rho_2(c,l)\sum_{\lambda\lambda'\sigma\sigma'}t_{\lambda\lambda'\sigma\sigma'}l_{\lambda\sigma,\boldsymbol{\kappa}}l^\dagger_{\lambda'\sigma',\boldsymbol{\kappa}+\mathbf{q}+\mathbf{G}}\right]\cdot\boldsymbol{\epsilon}'^*\times\boldsymbol{\epsilon},
\end{aligned}$$

with $t = s - \frac{3}{2l(2l+1)}[s\cdot l, l]_+$, $v_{\mu\mu'} = \langle\mu'|v|\mu\rangle$; $\delta_i(c,l)$ and $\rho_i(c,l)$ are coefficients, which depend on core (c) and valence (l) electron orbital angular momenta only (see Appendix). $R_{cj;1;l} \sim |\langle R_{ncj}(r)|r|R_{n'l}(r)\rangle|^2/(\langle E_I\rangle - E_i - \hbar\omega - i\frac{\langle\Gamma_I\rangle}{2})$ denotes the "reduced resonance scattering amplitude". The operator $\hat{O}_2(\mathbf{q},0)$ can be obtained in a similar way. It describes the quadrupolar moments of the valence electron distribution, and will not be treated in this lecture; its explicit form can be found in Ref. [18].

The $\sigma \to \sigma$ channel (ingoing and outgoing photon polarisation perpendicular to the scattering plane) selects charge scattering, as determined by the operator: $\hat{O}_0$. Neglecting the spin-orbit interaction, and defining charge density fluctuations at wavevector $\mathbf{q}$ as:

$$\rho_\mathbf{q} = \sum_{\boldsymbol{\kappa}\mathbf{G}}\sum_{\lambda\sigma}l_{\lambda\sigma,\boldsymbol{\kappa}}l^\dagger_{\lambda\sigma,\boldsymbol{\kappa}+\mathbf{q}+\mathbf{G}},$$

we obtain:

$$\left.\frac{d^2\sigma}{d\Omega'd\hbar\omega_{\mathbf{k}'}}\right|_{\text{charge}} = \frac{\omega_{\mathbf{k}'}}{\omega_\mathbf{k}}\frac{1}{2\pi\hbar}R^2_{cj;1;l}\delta^2_0(c,l)\int_{-\infty}^\infty dt e^{i\omega t}\langle\psi_i|\rho_{-\mathbf{q}}(t)\rho_\mathbf{q}|\psi_i\rangle, \quad (7)$$

expressing the cross-section in terms of the charge density-density correlation function.

The $\sigma \leftrightarrow \pi$ channel (one polarisation perpendicular to the scattering plane) selects magnetic scattering, as described by the operator: $\hat{O}_1$. Neglecting the orbital and magnetic dipole contributions (i.e. assumme, for simplicity, a negligible spin-orbit coupling, a strong orbital quenching, and a cubic crystal field), and defining spin density fluctuations at wavevector $\mathbf{q}$ as:

$$\mathbf{s}_\mathbf{q} = \sum_{\boldsymbol{\kappa}\mathbf{G}}\sum_{\lambda\sigma\sigma'}s_{\sigma\sigma'}l_{\lambda\sigma,\boldsymbol{\kappa}}l^\dagger_{\lambda\sigma',\boldsymbol{\kappa}+\mathbf{q}+\mathbf{G}}$$

we have:

$$\left.\frac{d^2\sigma}{d\Omega'd\hbar\omega_{\mathbf{k}'}}\right|_{\text{spin}} = \frac{\omega_{\mathbf{k}'}}{\omega_\mathbf{k}}\frac{1}{2\pi\hbar}R^2_{cj;1;l}\rho^2_1(c,l)\int_{-\infty}^\infty dt e^{i\omega t}\langle\psi_i|\mathbf{s}_{-\mathbf{q}}(t)\mathbf{s}_\mathbf{q}|\psi_i\rangle, \quad (8)$$

that is, the spin-density correlation function.

CONCLUDING REMARKS

The previous sections have discussed general features of the resonant inelastic scattering of x rays from valence (conduction) electrons. It has been shown that, under particular assumptions, the double-differential cross-section can be expressed in terms of elementary two-particle charge and spin-density correlation functions; these, in turn, can be independently probed by selecting different scattering geometries (polarisation channels).

Formally the results are rather similar to those obtainable for inelastic neutron scattering [19]. There is, however, an important difference: on account of electic dipole selection rules, $\hat{O}_z(\mathbf{q})$ is a *projected* operator; consequently, $\langle\psi_i|\hat{O}^\dagger(\mathbf{q},t)\hat{O}(\mathbf{q},0)|\psi_i\rangle$ describes the correlations among electrons of a given angular symmetry l (shell selectivity).

Our derivation applies to fast collisions, which amounts to assuming a dispersionless set of intermediate states. When this approximation breaks down [5, 16], $n > 0$ terms in expansion (4) have to taken into account; they yield higher order correlation functions.

Expressions (7) and (8) can provide useful information on charge and spin collective excitations: plasmons and spin waves. The former are characterised by an energy scale of ~ 10 eV, and they should be observable with the energy resolution currently attainable at synchrotron sources. The latter on the contrary, with an energy scale of ~ 20 meV, appear to be at the limit of high resolution beam lines.

APPENDIX

In this Appendix we provide a few technical details about the recoupling of the scattering amplitude. Consider the tensor $F^{(z)}(k)$, as defined by expression (6). One has:

$$F_\zeta^{(z)}(k) = |R_{cLl}|^2 \sqrt{\frac{2z+1}{2L+1}} \sum_{MM'} C_{LM';z\zeta}^{LM}$$
$$\sum_{m,all\lambda\sigma\gamma} C_{c\gamma';\frac{1}{2}\sigma'}^{jm} C_{c\gamma;\frac{1}{2}\sigma}^{jm} C_{c\gamma';LM'}^{l\lambda'} C_{c\gamma;LM}^{l\lambda} l_{\lambda'\sigma',\kappa} l_{\lambda\sigma,\kappa+q+G}^\dagger .$$

Neglecting unessential (for our purposes) phase factors, the term:

$$\sum_{m,all\gamma} C_{c\gamma';\frac{1}{2}\sigma'}^{jm} \ldots C_{c\gamma;LM}^{l\lambda}$$

is represented, graphically, by:

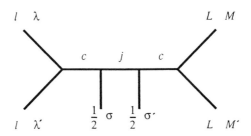

Applying the theorem:

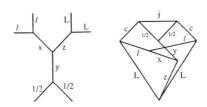

(α and β denote arbitrary block-diagrams, i.e. diagrams with no external lines) yields:

where the external lines have been connected to obtain "logical" pairs. The "bubble" diagram can be further decomposed using the theorem:

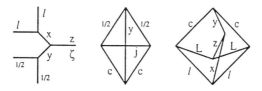

Together, the above transformations give:

$$F_\zeta^{(z)} \sim \sum_{\kappa G} \sum_{\lambda\sigma\lambda'\sigma'} l_{\lambda'\sigma',\kappa} l^\dagger_{\lambda\sigma,\kappa+q+G}$$

The "tree" diagram (i.e. that with external lines) determines the symmetry properties of the operatot $ll^\dagger$; the remaining ones (a 6j and a 9j symbol, respectively) enter the definition of the coefficients $\delta(c,l)$ and $\rho(c,l)$.

REFERENCES

[1] W. Schülke, Inelastic scattering by electronic excitations, in: "Handbook on Synchrotron Radiation", G. Brown and D.E. Moncton eds., Elsevier Science Publishers, Amsterdam (1991).

[2] D.H. Templeton, and L.K.Templeton, *Acta Crystallogr. Sect. A* 36:237 (1980); *Ibidem*, 38:62 (1982).

[3] D. Gibbs, , D.R.Harshman, E.D.Isaacs, D.B.McWhan, D.Mills, and C.Vettier, *Phys. Rev. Lett.* 61:1241 (1988).

[4] E.D.Isaacs, E.D., D.B.McWhan, C.Peters, G.E.Ice, D.P.Siddons, J.B.Hastings, C.Vettier, and O.Vogt, *Phys. Rev. Lett.* 62:1671 (1989).

[5] K.D. Finkelstein, Q.Shen, and S.Shastri, Phys. Rev. Lett. 69, 1612 (1992).

[6] T.R. Thurston, G. Hegelsen, D. Gibbs, J.P. Hill, B.D. Gaulin, and G. Shirane, *Phys. Rev. Lett.* 70:3151 (1993).

[7] B.T. Thole, P.Carra, F.Sette, and G. van der Laan, *Phys. Rev. Lett.* 68:1943 (1992).

[8] P. Carra, B.T.Thole, M.Altarelli, and X.Wang, *Phys. Rev. Lett.* 70:694 (1993).

[9] J. Luo, G.T.Trammell and J.P.Hannon, *Phys. Rev. Lett.* 71:287 (1993).

[10] K. Hämäläinen, D.P. Siddons, J.B. Hastings and L.E. Berman, *Phys. Rev. Lett.* 67:2850 (1991).

[11] P. Carra, M. Fabrizio, and B.T. Thole, unpublished.

[12] Y. Ma, *Phys. Rev. B* 49:5799 (1994).

[13] P.D. Johnson, and Y. Ma, *Phys. Rev. B* 49:5024 (1994).

[14] G.D. Mahan, "Many-Particle Physics," Plenum Press, New York (1991).

[15] A.I. Akhiezer, and V.B. Berestetsky, "Quantum Electrodynamics", Consultants Bureau, New York (1957).

[16] P. Carra, and B.T. Thole, *Rev. Mod. Phys.*, to appear.

[17] D.M. Brink, and G.R. Satchler, "Angular Momentum", Oxford University Press, London (1962).

[18] P. Carra, H. König, B.T. Thole and M. Altarelli, *Physica B* 192:182 (1993).

[19] W. Marshall and S. Lovesey, "Theory of Thermal Neutron Scattering", Oxford University Press, Oxford (1971).

INVESTIGATION OF LOCAL ELECTRONIC PROPERTIES IN SOLIDS BY TRANSMISSION ELECTRON ENERGY LOSS SPECTROSCOPY

Christian Colliex

Laboratoire de Physique des Solides associé au CNRS
Bâtiment 510, Université Paris Sud, 91405 Orsay, France

INTRODUCTION

High energy electron energy-loss spectroscopy (EELS) has been used over the last decades for the investigation of the local electronic properties in solids, as an alternative to optical techniques, with the advantage of covering a very large spectral range (from the visible to the x-ray domain) in a single experiment. On the low energy side it has provided a direct insight into the physics of plasmons and interband transitions as summarized in the textbooks by Raether[1,2], Schnatterly[3], Schattschneider[4], Fink[5] and Colliex[6.].

On the high energy side atomic core excitations which had been observed as early as 1942 by Ruthemann[7] have been recognized as useful probes of electronic states in molecules and crystals only around 1970 following the pioneering work of Colliex and Jouffrey[8,9] and of Isaacson[10]. The present review describes recent experimental and theoretical developments in this field with a special emphasis on the perspectives offered by the latest improvements in energy and in spatial resolution.

In a transmission scattering experiment, a beam of monochromatic particles is incident on the specimen and one measures the changes in energy and momentum which they suffer while propagating through the thin foil (see fig.1). Inelastic scattering data are therefore related to the excitation spectra of the different electron populations in the target. If the used geometry offers a good angular resolution, the study of dispersion curves is potentially accessible. On the other hand, if the incident beam is focussed into a probe of small size on the specimen, the momentum distribution is partially lost, but one can select the information from a reduced area on the specimen, which can be as small as a fraction of a nanometre. It offers unique possibilities for correlating the spectroscopic information to a detailed structural and topographical knowledge of the specimen.

Over the last few years a number of spectacular magnetic properties have been discovered in the case of thin film, multilayer, granular structures (see other contributions in this volume). The nanostructure of these heterogeneous objects, their local chemical composition constitute important parameters which have to be known to understand the measured macroscopic magnetic properties. Because of its momentum dependence and its high spatial resolution, it is believed that the information provided by EELS studies on magnetic materials will help predicting and tailoring the magnetism of ultra-small objects, with great improvements of the recording media properties in perspective.

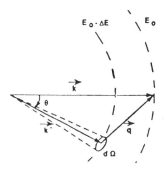

Fig. 1 : Scattering geometry of an electron energy loss experiment. A primary electron of energy E_0 and wavevector **k** is inelastically scattered into a state of energy $E_0 - \Delta E$ and wavevector **k'**. The energy loss is ΔE and the momentum transfer is $\hbar\mathbf{q}$. The scattering angle is θ and the scattered electrons are collected within an aperture of solid angle $d\Omega$ to be analysed by the spectrometer.

ANATOMY OF AN ENERGY LOSS SPECTRUM

The EELS spectrum for a beam contained within a given angular acceptance around the direction of transmission displays the characteristic features illustrated in fig.2.

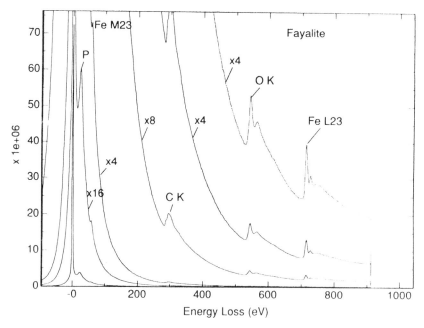

Fig.2 : Electron energy loss spectrum through a thin foil of fayalite mineral of nominal composition $SiFe_2O_4$, typically 70 nm thick, recorded with primary electrons of 100 keV. Total acquisition time is about 30 s in several times in order to provide a satisfactory signal-to-noise ratio over the whole energy loss range from 0 to about 1000 eV

Apart from the zero-loss peak, one identifies from lower to higher energy losses, the dominant contribution of the plasmon peak (P) at about 23 eV and a series of edges, corresponding to the excitation of electrons from deeper and deeper atomic levels, superposed over a continuously decreasing background. It is important to note the large difference in intensities between the low-loss structures, plasmons and interband transitions lying typically between 5 and 30 eV, and the core-loss features between 200 and 1000 eV. The involved dynamical range encompasses several orders of magnitude.

The major contribution to the low loss spectrum comes from the excitation of electrons initially lying in the valence bands and is generally dominated by the collective behaviour of this electron gas, giving rise to intense plasmon peaks (P in fig. 2).

The probability for exciting these plasmon modes is rather high. It can be expressed in terms of inelastic mean free path λ_{in}, corresponding to the average travel distance of the primary high energy electron between two successive inelastic events. This distance obviously depends on the nature of the material but lies in the range 60-150 nm for 100 keV incident electrons. As the statistical occurrence of these events obeys a Poisson law ($P_n = \frac{1}{n!}(t/\lambda_{in})^n \exp(-t/\lambda_{in})$) as a function of the specimen thickness t, it implies that t must be smaller than a few tens of nanometers in order to keep negligible the contribution of multiple scattering in the EELS spectrum. For higher thickness, deconvolution procedures have been developed to extract single scattering profiles from the experimental data.

At higher energy losses, core-level excitations appear as edges superposed over a monotonously decreasing background (Fe-M_{23}, C-K, O-K, Fe-L_{23} designed with the commonly used spectroscopy nomenclature). If the edges due to iron, oxygen and contamination carbon are quite clearly visible as sharp saw-teeth profiles, it is much more difficult to pick up at first sight the Si L_{23} edge at about 100 eV on such a wide range spectrum, because it appears as weak oscillations over a steep slope background. This example demonstrates that all edges do not display similar shapes, the general behaviour being imposed by atomic considerations.

The non-characteristic background in EELS is made of the superposition of different contributions: the high energy tail of the valence electron excitations, the tails of core edges with lower binding energies, bremstrahlung losses, plural scattering. It is therefore a complex function of composition and thickness, which it is rather difficult to model a-priori for every point of analysis on the specimen.

Basically core-level EELS spectroscopy investigates transitions from one well defined atomic orbital to a vacant state above the Fermi level: it is a probe of the energy distribution of unoccupied states in a solid. As the excited electron is promoted from a given atomic orbital on a well defined site, the corresponding edge must first satisfy the dipolar selection rules valid for small scattering angle, so that s-type electrons are promoted to p-type states, p-type electrons predominantly go to d-type vacant states and so on. Then it provides the local atomic point of view on the concerned site. Finally one must not neglect completely the existence of the created hole which can be more or less screened by the surrounding population of electrons in the solid. The distribution of vacant states revealed by the excitation of a core hole may consequently differ from the unperturbed one. All these aspects will be more thoroughly discussed in a next paragraph.

INSTRUMENTATION

In a machine specially dedicated to the investigation of inelastic scattering processes, one aims at the best momentum and energy resolution and uses a well collimated and monochromatized primary beam at the cost of limited spatial resolution. The illumination optics includes a monochromator between the electron source and the specimen at ground potential, and delivers a parallel beam on the specimen, covering areas of the order of 100 nm in diameter. Scan coils (or plates) after the specimen are used to investigate the angular (or momentum-dependence) of the inelastic process. The energy analyser between the specimen and the detector is generally of the same type as the monochromator: these are electrostatic hemispherical spectrometers in the group in Karlsruhe[5] and of the mixed electrostatic-magnetic Wien filter type in the group in Sendai[11]. Typical energy resolution of 0.1 eV and momentum transfer resolution of 0.05 - 0.1 Å^{-1} have been demonstrated.

However EELS studies have obtained a wide increase in popularity when they have become currently available from an attachment on an electron microscope (EM) column. This configuration offers access on the same instrument to the standard EM capabilities for image and diffraction characterization. A detailed discussion on EELS spectrometers coupled to electron microscopes can be found in the textbook by Egerton [12].

Among many types of spectrometers investigated over decades, the homogeneous magnetic prism with a design involving second-order aberrations correction has proved to be most popular, the more as it has become commercially available by Gatan (US) and can be fitted to any type of microscope [13]. When attached on the latest generation of dedicated scanning transmission microscopes (Vacuum Generators HB series), it offers a quite useful association of energy resolution of the order of 0.4 eV, mostly limited by the natural energy distribution of the field emitted primary beam, and of ultra-high spatial resolution defined by the diameter of the focussed beam issued from the field emission gun. In the 100 kV STEM instrument in operation in Orsay [14] (see fig.3), a probe of 0.5 nm in diameter on the entrance surface of the specimen can carry a few tenths of nA, quite sufficient to provide spectra of good signal-to-noise ratio within a short acquisition time. Recent advances in technology made by VG (primary voltage up to 300 kV, improved design of the focussing objective lens) have further reduced the diameter of this incident probe to 0.2 nm, which is typically the interatomic distance within a crystal. It is the reason why there has been at the end of 1993 a burst of papers announcing atomic resolution EELS studies [15-17].

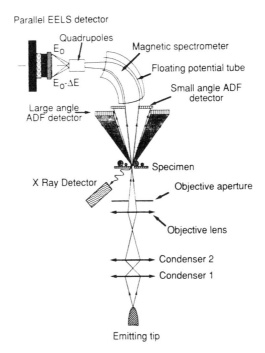

Fig.3 : Principle and basic components of the dedicated STEM equipped with an EELS spectrometer and a parallel detector array, in operation in Orsay. It offers the simultaneous capability of visualizing the object topography through the annular detectors (small angle for diffraction contrast, large-angle for "Z-contrast") and of analyzing with the magnetic spectrometer the energy loss distribution of forwardly transmitted electrons.

An alternative solution to the magnetic sector is the Wien filter adapted on one of these STEM instruments by Batson [18]. It offers a very good energy resolution down to 0.15 eV, nearly equivalent to that of dedicated EELS instruments. The detailed shape of the energy distribution of the electrons emitted from the field emission source can then be taken into account to extract a refined threshold shape on selected core edges by deconvolution techniques (see fig. 4). However, in most core-loss spectra shown further on, relative to oxygen-K, transition-metal L_{23} and rare-earth M_{45} edges, the practical energy resolution is of the order of 0.5 to 0.7 eV, providing the possibility of resolving fine structures separated by 0.8 to 1.0 eV.

The major step forward made by EELS studies over the last few years relies largely on progress in detection. The formerly used sequential acquisition mode in which the dispersed beam was scanned in front of a narrow slit located in the spectrometer dispersion plane, has been replaced by an array of diodes which offer parallel acquisition of all energy loss channels[19]. In most cases an auxiliary electron optics matches the required dispersion to the size of the individual detection cells. This new generation of detectors has led to significant improvements : enhanced detection limits, reduced beam damage in beam-sensitive materials, data of improved quality both in terms of SNR and resolution, access to time-resolved spectroscopy and to image spectrum-mode.

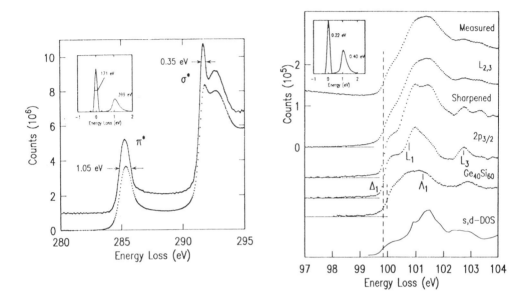

Fig. 4 : Background subtraction and spectral sharpening of a carbon 1s and of a silicon 2p edges to obtain respectively 0.17 and 0.22 eV resolution. The insets show the used field emission source distribution and the resulting symmetric spectral resolution obtained through deconvolution. Furthermore for the Si case, the L_2 part is stripped away by deconvoluting a spin-orbit function made of 2 δ peaks separated by the spin-orbit term of 0.608 eV. From Batson[20-21].

PEELS THEORY: A SUMMARY

The interpretation of the various features observed in an EELS spectrum has been gradually elaborated over three decades by many different authors and one can refer to a number of references [1-6,12,22] for complete surveys. In the present text, we shall only summarize a few major aspects. Practically an EELS experiment measures the partial differential cross section $d^2\sigma / d\Omega.d(\Delta E)$ corresponding to the scattering geometry of fig. 1.

Using the dielectric formalism, this cross section can be written as :

$$\frac{d^2\sigma}{d\Omega.d(\Delta E)} = \frac{1}{(\pi e a_o)^2} \cdot \frac{1}{q^2} \cdot \operatorname{Im}\left\{\frac{-1}{\varepsilon(\mathbf{q},\omega)}\right\}$$

where $\text{Im}\left\{\dfrac{-1}{\varepsilon(q,\omega)}\right\}$ is the so-called energy-loss function, depending on the variation of the macroscopic dielectric function $\varepsilon(q,\omega)$ with a momentum transfer $\hbar q$ and the energy loss $\Delta E = \hbar\omega$. This function varies only slowly with q, so that the cross section is roughly proportional to the inverse square of the transferred wave vector. Practically $\dfrac{1}{q^2} \propto \dfrac{1}{\theta^2 + \theta_E^2}$ where the inelastic characteristic angle $\theta_E = \dfrac{\Delta E}{2E_0}$ in the non relativistic case (and $\dfrac{\Delta E}{\gamma m_0 v^2}$ in the relativistic case). For 100 keV incident electrons, θ_E amounts respectively to $\cong 10^{-4}$ for plasmon losses and to $\cong 3 \times 10^{-3}$ rad for oxygen K losses. In the incident parallel beam mode one can then investigate the inelastic scattering as a function of **q** from **q** parallel to the incident direction at low θ values, to **q** perpendicular to the incident direction at high θ values. However the detection of anisotropy effects associated to the orientation of the foil with respect to the incident electrons (see ref. 23) requires a very good angular resolution as available with the parallel beam illumination mode. In the convergent beam mode, the momentum transfer is dominantly perpendicular to the incident electron direction and anisotropy effects are partially smeared out.

1) Energy loss function in the low-energy loss range ($\cong$ 1 - 50 eV)

This description in terms of the dielectric function is most useful in the low-loss region. In the small scattering angle limit which is valid when the collection angle for transmitted electrons is of the order of 1 - 2 Å^{-1}, the energy-loss function can be written as :

$$\text{Im}\left\{-\dfrac{1}{\varepsilon(\omega)}\right\} = \dfrac{\varepsilon_2}{\varepsilon_1^2 + \varepsilon_2^2}$$

It displays a dominant peak for $\varepsilon(\omega) = 0$, which practically corresponds to $\varepsilon_1 = 0$ and ε_2 small. Since ε describes the screening of an external field E_{ext} inside the solid, the condition $\varepsilon = 0$ indicates the instability of the system against very small external perturbations. These are the collective excitations known as plasmons, the frequency and dispersion curves of which being given by $\varepsilon_1(q,\omega) = 0$.

Practically the maxima of the energy loss function correspond to collective excitations through the zero values of ε_1 as well as to interband transitions through the maxima of ε_2, however strongly screened by $(\varepsilon_1)^2$. Simple models, such as the Drude one for metals and the Lorentz one for insulators, can provide useful rules for understanding the general behaviour of this dielectric function.

For a more satisfactory description of the relationship between the energy-loss function and the optical absorption function $\omega\,\varepsilon_2(\omega)$ in real solids, one has to use the Kramers-Kronig transformation between the imaginary and the real parts of the dielectric function :

$$\text{Re}\left\{\dfrac{1}{\varepsilon(\omega)}\right\} = 1 - \dfrac{2}{\pi}\,\text{PP}\int_0^\infty \text{Im}\left\{-\dfrac{1}{\varepsilon(\omega')}\right\}\cdot\dfrac{\omega'\,d\omega'}{(\omega'^2 - \omega^2)}$$

where PP denotes the principal part of the integral. This processing step is necessary before further correlating the EELS results to band structure data, such as expressed in :

$$\varepsilon_2(\omega) = \dfrac{A}{\omega^2}\left|M_{jj'}\right|^2 \cdot \text{JDOS}(\hbar\omega)$$

where $M_{jj'}$ is the matrix element between occupied Bloch wave states j in the valence band and unoccupied Bloch wave states j' in the conduction band, and JDOS the joint density of states between the two bands. Extracting useful information from this comparison is rather difficult except in the neighbourhood of band gaps[24,25].

Another important feature visible in this low energy loss range is the occurrence of surface or interface modes imposed by the existence of boundary conditions between two homogeneous media (A) and (B) :

$$\varepsilon_A(\omega) + \varepsilon_B(\omega) = 0$$

From this equation one can deduce the existence of charge fluctuations in the boundary surface, giving rise to fluctuations of the electrostatic potential normal to these surfaces, which can be coupled to the external incident electrons. The eigenfrequencies of these localized excitations depend on the dielectric properties of the adjacent media and of the geometrical shape (planar, spherical, cylindrical) of the interface. In a real solid, one searches for the roots of the equation $\varepsilon(\omega_s) = -1$ for the solid-vacuum mode and of $\varepsilon(\omega_s) = -\varepsilon_d$ for a solid covered with a layer of dielectric constant ε_d.

Figure 5 shows the existence of such a surface mode when the electron beam crosses the outer surface of a ruthenium metal particle.

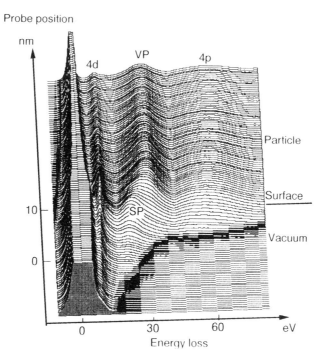

Fig.5 : Plot of a sequence of spectra acquired while scanning a 0.6 nm probe with 0.3 nm steps from vacuum into a metallic Ru particle, across its external surface. Most visible is the shift of the surface plasmon at 21 eV into a bulk plasmon at 26 eV, over a width of about 2.5 nm.

2) Energy loss function for core losses

Far above the plasmon frequency, ε_2 is small and ε_1 close to unity. Consequently the relationship between the energy-loss function and the imaginary part of the dielectric function is more straightforward because in this case :

$$\text{Im}\left\{-\frac{1}{\varepsilon(q,\omega)}\right\} \rightarrow \varepsilon_2\{q,\omega\}$$

Moreover for small momentum transfers $\hbar q \rightarrow 0$ and EELS measures the same solid state properties as in far UV or soft x-ray absorption spectroscopy, i.e. $\varepsilon_2(0,\omega)$ in the equivalent range of frequencies. Consequently, the detailed interpretation of fine structures on EELS core-edges heavily relies on theoretical developments elaborated for the interpretation of data recorded on synchrotron radiation lines.

Using the Fermi golden rule, the cross section can be written in terms of matrix elements :

$$\frac{d^2\sigma}{d\Omega.d(\Delta E)} = \frac{4\gamma^2}{a_o^2 q^4}|<f|e^{i\mathbf{q}.\mathbf{r}}|i>|^2 \rho_f(\Delta E)$$

between the initial $|i>$ and the final $|f>$ states of the transition, $\rho_f(\Delta E)$ designing the density of accessible final states for the energy loss ΔE. The dipole approximation consists in developing the interaction potential in the matrix element as :

$$e^{i\mathbf{q}.\mathbf{r}} \cong 1 + i\mathbf{q}.\mathbf{r} - \frac{(\mathbf{q}.\mathbf{r})^2}{2} + ...$$

It is valid if $|\mathbf{q}.\mathbf{r}| \ll 1$, i.e. if $q \ll \frac{1}{R_c}$ where R_c is an average extension of the promoted core electron. An equivalent expression for the validity of this approximation is $\theta \ll \left(\frac{\Delta E}{E_o}\right)^{1/2}$. Considering the values of collection angle and typical energy losses mentioned above, this condition is verified for all experiences described thereafter.

The transition matrix element can then be rewritten as :

$$|<f|e^{i\mathbf{q}.\mathbf{r}}|i>|^2 \rightarrow q^2 |<f|\mathbf{e_q}.\mathbf{r}|i>|^2$$

where $\mathbf{e_q}$ is the unit vector in the direction of the momentum transfer. It plays a role equivalent to $\mathbf{E_q}$ (polarization vector of the electric field) in x-ray absorption studies. This dependence can then be used to investigate the transitions allowed along given reciprocal space directions.

EELS CORE EDGES: A SPECTROSCOPY OF UNOCCUPIED ELECTRON STATES

Energy loss spectroscopy of core-edges offers a variety of advantages for the investigation of unoccupied electron states :

i) they can be recorded on many types of specimens as well natural as synthetic, originating from quite different fields. They are prepared as for any electron microscopy observation. The recorded data originate from selected areas, which can otherwise be characterized by imaging and diffraction, reducing the possibilities of errors on the specimen nature. As a consequence of the projection geometry, they are more representative of the properties of the bulk of the material than of its surface. It is clearly an elemental and chemical specific technique.

ii) A wide energy loss range can easily be investigated within an experimental session, offering the possibility of comparing different edges originating from different atoms. This constitutes a quite unique feature of EELS because different probes of the accessible electron configuration (i.e. seen by the promotion of 1s electrons on the anion or of the 2p electrons on the cation in a transition metal oxide) provide complementary information.

The total intensity contained in each edge identified in one spectrum can be used for determining the elemental composition of the irradiated volume. Furthermore different types of information on the unoccupied electron states can be made accessible through a detailed analysis of the edge fine structures.

1) Quantitative elemental analysis [26]

Two methods are now generally used for chemical analysis purposes, as illustrated on fig. 6. In the first case the characteristic signals (S_A) and (S_B) are measured after subtraction of an extrapolated background. This procedure relies on the possibility of modeling the background with a reasonable power-law behaviour over an energy window Γ preceding the edge. It then implies some calculated values of the cross sections (σ_A and σ_B) for the concerned edges using the relevant experimental conditions (β = collection angle of the analysed electrons, Δ = energy window of integration of the measured signal), because :

$$\frac{N_A(\beta,\Delta)}{N_B(\beta,\Delta)} = \frac{S_A(\beta,\Delta)}{S_B(\beta,\Delta)} \times \frac{\sigma_B(\beta,\Delta)}{\sigma_A(\beta,\Delta)}$$

There exists now in the literature several sets of calculated cross sections to be used in this context. They are all atomic-type calculations from a core orbital towards empty continuum states with the required angular symmetry. When integrated over energy widths (Δ) of typically 100 eV above threshold, these calculated cross sections provide a ± 10 % accuracy when compared to experimental measurements, at least for the K and L_{23} edges[27]. For more complex edges such as M_{23}, M_{45}, N_{45}, the agreement with the values measured on standards has not yet reached this level.

In many cases, such as for overlapping edges, an alternative approach is in rapid development. It consists in finding with a MLS fitting procedure the combination of reference signals which best reproduce the experimental curve, encompassing as well the background as the edges. This procedure is quite versatile: it can apply to any type of spectrum recorded in the normal or in a difference mode. It implies reference data, either calculated or previously obtained on standards. The thickness effect and its associated multiple scattering events can be accounted for by convolution techniques. The major limitation is due to the fact that the detailed edge shape depends on the atom environment as it will be demonstrated in a next paragraph. It is therefore not possible to use any type of reference in this fitting process and care must be devoted to the choice of the most satisfactory ones.

The table compares data obtained by the two methods on different iron oxides. If both of them provide satisfactory results (which is normal because there is no special background extrapolation problem), the statistical dispersion is reduced in the second case.

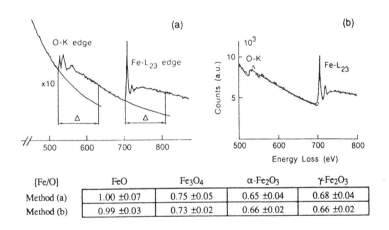

[Fe/O]	FeO	Fe_3O_4	α-Fe_2O_3	γ-Fe_2O_3
Method (a)	1.00 ±0.07	0.75 ±0.05	0.65 ±0.04	0.68 ±0.04
Method (b)	0.99 ±0.03	0.73 ±0.02	0.66 ±0.02	0.66 ±0.02

Fig. 6 : Two different ways of using core-loss excitations for the quantitative elemental analysis of two components, applied to the iron-oxygen system: a) conventional background stripping method; b) spectrum modelling method. Bottom: comparison of the results [28].

2) Limits of detection.

Test experiments have been realized in order to evaluate the detection limits corresponding to quite different situations. The sensitivity has been estimated on a reference glass containing 66 elements including transition metals and rare earths at trace concentrations[29]. This study has shown that the detection limits are not only governed by statistical considerations but that the shape of the edge is also an important factor. In particular, those elements which display clear white lines between 300 and 1200 eV, can be detected at lower levels, because of their characteristic signature in the difference modes: 3d transition elements and preceding alkaline earth L_{23} edges (Ca to Ni) as well as 4f-lanthanides and preceding alkaline earth M_{45} edges (Ba to Yb) should be detected at the 10 to 100 ppm atomic concentration when using a high total current over a large specimen area.

On the other hand, the single atom detection capability has been demonstrated on suitable specimens made of clusters of thorium atoms scattered on an ultra-thin carbon foil[30]. This has been made possible by reducing as much as possible the analysed volume containing the atom to identify. This performance is a test of a good signal-to-noise ratio but not of spatial resolution and has been possible by using a very intense primary flux of electrons on the specimen.

The relationship between these complementary aspects of detection capabilities (concentration, resolution, number of detectable atoms) as a function of the primary dose has already been discussed by Colliex[31].

3) Edge fine structures

The same EELS core-loss edge recorded on various compounds exhibits different fine structures extending from threshold up to a few tens (ELNES) and even hundreds of eV (EXELFS) above (see fig. 7). High energy resolution EELS data demonstrate their full potentialities as a spectroscopical tool when the specific information carried by these fine structures can be identified. This detailed information on electronic, structural, optical or eventually magnetic nature has to be extracted from different features: the number of peaks or steps, their position and widths, the relative intensity of different components such as prepeaks on oxygen 1s edges or white lines on transition metal 2p edges. This step requires the support of the different theoretical tools which have been developed in parallel for the interpretation of the fine structures (XANES or EXAFS) on x-ray absorption edges.

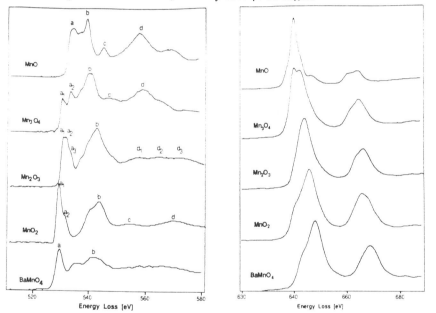

Fig. 7 : Oxygen K-edges (left) and manganese L_{23}-edges (right) for five different oxides after deconvolution of multiple losses and background subtraction [32].

In the one-electron approximation, the relaxation or rearrangement of the other atomic or valence electrons following the creation of the hole on a core orbital, is neglected. Within the dipole approximation, the EELS core level transitions follow selection rules imposed by the transition matrix elements between the initial and final states, as in x-ray spectroscopy. Their quantum numbers (l' and j') are deduced from those of the initial state (l and j) by the rules :

$$l' = l \pm 1 \text{ and } j' = j \text{ or } j \pm 1$$

To determine the accessible final states, which depend on the atomic site and quantum character of the promoted electron, and if possible to calculate the corresponding transition rates, several methods are available :

i) A *molecular orbital (MO)* description : the final state is formed by the interaction of atomic orbitals on the central atom in a polyatomic cluster with those from the first coordination shell. The SCF-Xα multiple scattered wave method used in particular by Tossell et al.[33], has proved its efficiency to assign the various near-edge features to specific molecular orbitals governed by crystal field symmetry considerations. It constitutes the first step to attribute an ELNES profile to a given coordination fingerprint (see Brydson[34]). As an example (see fig. 8), the different splittings of the first peak on the O-K edge in TiO_2 and CeO_2 can be qualitatively interpreted in terms of difference in site symmetry and nature of vacant orbitals, using MO calculations in a TiO_6^{8-} cluster (rutile symmetry) and in a CeO_8^{12-} cluster (fluorite symmetry). The position in energy of the t_{2g} and e_g orbitals corresponding to peaks b and c in the experimental spectra is reversed as a consequence of symmetry considerations and the existence of unoccupied f-state on the Ce ion introduces extra vacant states at threshold which could explain the existence of peak a.

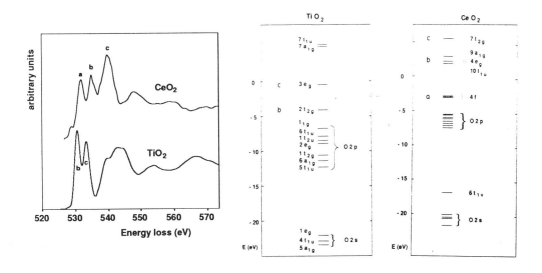

Fig. 8 : Experimental EELS oxygen K near-edge structures in TiO_2 and CeO_2 (left, courtesy of E. Lefèvre) and comparison with the calculated energy distribution of molecular orbitals in the relevant clusters from Tossell et al.[33] (centre) and from Ryzshkov et al.[35] (right).

ii) An extension of the previous approach is the *multiple scattering description in real space (XANES type calculation)* which allows the cluster size, and consequently the number of shells, to be progressively increased. In particular the importance of specific scatterers and scattering paths in determining the resultant near-edge features can be estimated. These oscillations are rather sensitive to coordination numbers and interatomic distances as discussed by Kurata et al. in the case of transition metal oxides[36].

Figure 9 shows how the ELNES peaks b, c and d in the O-K edge of MnO are best reproduced in a multiple scattering description by considering the successive shells of strongly backscattering oxygen atoms (i.e. 2, 4 and 6 nearest neighbours). The resonance peak d is mostly due to single scattering by the nearest oxygen neighbours and its position is quite sensitive to the bond length as already noticed by Stohr et al.[37] on the carbon K edge in different gaseous hydrocarbons. The presence of the intermediate peak c requires the introduction in the calculation of more distant neighbours.

iii) A *band theory calculation* of the unoccupied density of states resolved into angular momentum components at a particular atomic site, i.e. $\rho_{l'}(E, \mathbf{r})$. This description generally attributes spectral features to strong contributions in the density of states and assumes that the matrix elements vary slowly over the concerned energy loss range. The maxima close to an edge are then mostly related to portions of the band structure where the band is flat in the reciprocal space. Useful band structure calculations to be compared with experimental data, must therefore provide the required description in angular momentum and site, and are then generally derived from combinations of atomic orbitals, such as tight-binding or pseudo-atomic-orbital methods (see Rez[38] for a review). These methods have proved to account satisfactorily for a selection of features in the vicinity of the threshold, corresponding to the bottom part of the conduction band, as demonstrated by Weng et al.[39] for the Si L_{23} and the diamond carbon K edge.

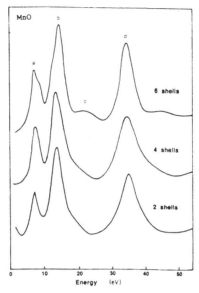

Fig. 9 : Multiple scattering calculation of the oxygen K-edge ELNES structures in MnO for increasing numbers of shells surrounding the excited atom [36].

Basically the multiple scattering and the band structure theory are equivalent to each other in their exact forms. Differences come from the shape of potential used and the approximations employed. The multiple scattering method emphasizes the scattering processes, and the results can then be best interpreted in terms of bond lengths and bond angles between the scattering centers. On the other hand the band theory points out the accessible orbitals of certain symmetry, corresponding most often to singular points in the band structure.

However when there is a strong interaction between the core-level and the valence electrons, which also means a strong overlap between the initial and final state wave functions in the transition matrix element, the one-electron approximation fails. It is the case of the *white lines* arising from transitions to atomic-like bound characters such as occurring on the 2p edges of transition metal atoms or on the 3d edges of rare-earth atoms. The final state (respectively of dominant 3d or 4f character) is screened and the associated spectra exhibit atomic like multiplet structures which are fingerprints of the ground state spin and angular momentum characters. This situation can aso be described in the real space in terms of a criterion R<<d, where R is the radial extent of the free atom wave function for the final state and d the interatomic spacing in the solid. The on-site Coulomb interactions are then large or at least comparable to band widths so that correlation effects are of great importance (see Sawatzky[40] and references therein).

An important family of edges exhibiting such features are the 2p spectra of transition metal ions, which cannot be described in terms of a one-particle density of states. The localization of the 3d wavefunction in the final state imposes to take into account the strong Coulomb and exchange interactions 2p-3d and 3d-3d in the calculation of the $2p^6 3d^n$ to $2p^5 3d^{n+1}$ transition matrix elements. This is first achieved in a spherical atomic model, on which the crystal field effects are superposed, leading to splittings of both the initial- and final-state multiplets. Complete sets of calculations of the 2p edges of first-row transition metal ions in tetrahedral and octahedral crystal field symmetry by de Groot et al. [41] and by van der Laan and Kirkman[42] demonstrate that the detailed spectral shapes provide a valence- and symmetry-selective information. One approach would then consist in searching for the best fit between the experimental profiles such as shown in fig. 7 and the calculated ones in the above references 41 or 42, convoluted with a Lorentzian broadening to simulate life-time effects. For instance an excellent agreement has been demonstrated between the Mn L_{23} edge in MnO and the Mn^{2+} atomic multiplet spectrum calculated in ref. 41 with 10 Dq equal to 0.6 eV.

As a matter of fact the detailed features of the multiplet configuration are grouped in two major components, the L_3 and L_2 lines, separated by the spin-orbit splitting of the inner 2p shell, which is slightly larger than the other energies involved in the calculation of the multiplet splitting. It has been noticed for many years that the intensity ratio between these white lines is noticeably different from the statistical ratio of 2:1 which would result from the degeneracy of the initial state. The origin of this discrepancy has been shown to be another consequence of the importance of multiplet effects, see Thole and van der Laan [43]. Different methods have then been elaborated to measure the relative weight of these transitions to bound states, individually and relatively to the whole intensity of the edge including the higher energy transitions to continuum states[44].

The branching ratio $I(L_3)/[I(L_3) + I(L_2)]$ in the 3d transition metals has been shown to vary from 0.4 at the beginning of the series up to 0.7 to 0.8 in the middle of the series. Figure 10 shows a more detailed comparison of measurements achieved on various Mn and Fe oxides with branching ratios of the divalent 3d transition metal ions predicted from a ligand field multiplet model.

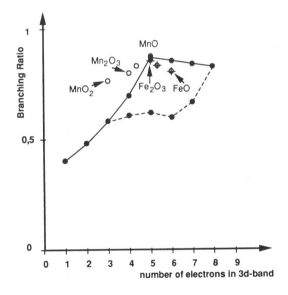

Fig. 10 : Measured [28, 32] and calculated [45] branching ratios for 3d transition metal ions. Solid line : high spin state. Dashed line : low spin state.

The comparison of the branching ratio in different compounds can be used to discriminate high-spin and low-spin configurations. Several authors [46-49] have established a relation between the branching ratio and the local magnetic moment.

An extended quantitative analysis of the various fine structures on core edges therefore constitutes a powerful route to extend the field of information of EELS analysis far beyond the plain elemental analysis. In particular, it will undoubtly with the support of more refined theoretical descriptions, give access to electronic, structural, bonding and eventually magnetic parameters.

SPATIALLY RESOLVED EELS

A major step forward has been accomplished with the introduction and implementation of the image- or line-spectrum techniques. An image-spectrum, as defined by Jeanguillaume and Colliex[50] in EELS, is a 3D-ensemble of numbers : the first two axes correspond to the x-y coordinates on the specimen, as for any image. The third axis is associated with the energy-loss spectrum. The line-spectrum (see fig.11) is a simpler version in which case the probe is scanned along a 1D-line on the specimen across a specific object of interest (interface, surface, precipitate). The major advantage of this technique is to map with a spatial resolution defined by the size of the primary probe of electrons, i.e. of the order of 0.5 to 0.8 nm in our instrument shown in fig.3, the information contained within the EEL spectrum [51-52]. Using the high detection efficiency of parallel detectors, complete spectra can be recorded within times as short as 25 ms (it will be reduced with the availability of detector arrays with faster read-outs), before moving the probe to its next position defined with a typical 0.3 nm accuracy. These are the typical conditions for recording the sequence of spectra displayed in fig. 5. When interested by fine structures on core edges, good signal-to-noise values obviously require longer dwell times per spectrum, varying from a few hundreds of milliseconds to a few seconds.

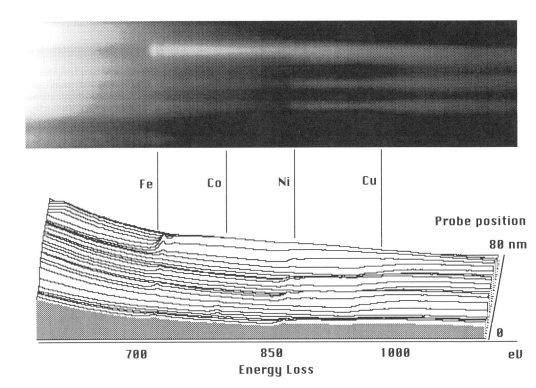

Fig. 11 : Example of a line-spectrum acquired across a magnetic multilayer made of a sequence of NiFe permalloy, Co and Cu layers oriented with interfaces parallel to the incident electron beam (specimen courtesy of P. Galtier, Thomson and data recorded by M. Tencé and M. Walls). In (a) the sequence of spectra acquired while scanning the probe along a profile across the specimen is displayed as a normal micrograph with black and white contrast along two axes, the first one being scaled in terms of position across the specimen, the second one in terms of electron energy loss. In (b) the same data are viewed in a 3D perspective with the intensity displayed along the z-axis.

One interesting possibility of this computer-controlled data acquisition mode, besides access to spatially resolved EELS, is to open the way to time resolved spectroscopy. In that case which does not require the DAC governing the probe scanning, the incident beam is maintained fixed on the specimen and a time sequence of spectra is recorded. Up to now this possibility has been mostly used for monitoring beam-induced spectral changes as a function of the incident electron dose and dose rate.

The real output of the technique relies on the development of data processing and theoretical modelling tools to extract from the recording of huge data sets the relevant information to be mapped, i.e. elemental distribution or any type of low-loss or core-loss fine structure, with a sub-nanometre spatial resolution. For PEELS compositional profiling and mapping, different types of routines have been implemented to extract quantitative chemical maps from spectrum-images[53]. They extend the use of the methods elaborated for processing one single spectrum and illustrated in figure 6, to series of 10^2 to 10^5 spectra. It has proved to be quite efficient in a series of examples involving nanometre dimensions, such as carbon nanotubes filled with metallic manganese[54], dielectric SiO_2-TiO_2 multilayers for optical coatings[55], Si nanowires in type-n nanoporous Si[56].

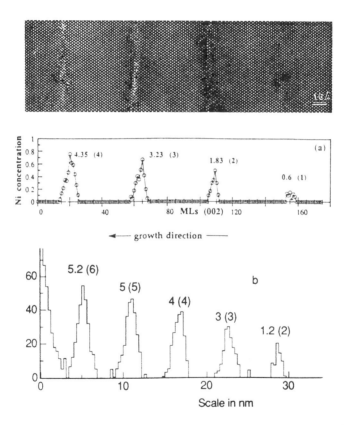

Fig. 12 : Atomic structure and chemistry of a Ni/Au multilayer cross section containing Ni layers from 1 to 6 atomic layers thick. (a) HREM image. (b) Comparison of the Ni chemical profiles dedu-ced from HREM strain profile and PEELS modelling in the second derivative mode. Nomi-nal thicknesses are indi-cated between brackets.

Of particular interest in the context of this school is the case of metallic Ni/Au MBE grown multilayers[57]. The individual Ni layer thicknesses range from 1 to 6 atomic layers, then requiring high sensitivity of detection. Consequently the useful characteristic signal is the Ni M_{23} edge at 68 eV of high cross section, which however suffers from a strong overlap with the Au O_{23} and N_{67} edges at 54 and 83 eV. Using reference edges recorded during the same line-spectrum acquisition from thicker areas of spacer Au and test Ni introduced during the growth sequence, it has nevertheless been possible with the simulation technique in the second difference mode, to measure quantitatively the Ni local atomic concentration for each pixel separated by 0.3 nm[53]. For Ni layers thinner than 5 monolayers, we have shown that the measured total amount of Ni in each layer is in good agreement with the nominal Ni amount estimated during the MBE deposition sequence. Furthermore the Ni chemical profile extracted from the PEELS measurements is quite similar to that deduced from a detailed analysis of HREM strain profiles (see fig. 12). The large misfit between the Ni and the Au lattices in the plane of growth would be accomodated by a compositional ramp with an interplanar distance taking the combined effect of composition and strain in account, as confirmed quite independently by molecular dynamics simulations [58].

Beyond elemental mapping, the greatest potential use of such a spatially resolved spectroscopy lies in its ability of displaying with a subnanometre spatial resolution the distribution of all the different physical properties which can be identified through an extended analysis of the fine structures appearing both in the low-loss range and on the core-edges. As illustrated in the previous paragraph, these spectral features concern quite different aspects of the electronic structure. For instance the presence of a prepeak on the onset of the oxygen K edge reflects the existence of unoccupied hybridized states between the O 2p states and the transition metal 3d states. Furthermore the fact that these states are investigated from the anion point of view strongly reduces the dominant intra-atomic multiplet splitting effects which dominate the final state distribution when investigated from the cation point of view. It seems therefore possible to extract from a quantitative analysis of the weight of different components in the fine structure distribution an information concerning the bonding character, the local site symmetry, the nearest neighbour distance, and eventually the magnetic moment, all quantities which could thus be mapped with a subnanometre resolution, when using the EELS spectrum-image techniques.

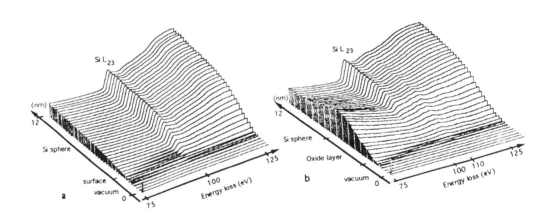

Fig. 13 : Line spectra acquired while scanning the electron probe across the outer surface of oxide-free and oxide-covered silicon spheres, demonstrating the different Si L_{23} edge behaviours.

As an example, figure 13 demonstrates how the local environment of Si atoms can be analysed with step increments of 0.3 nm while scanning a primary probe of electrons of about 0.6 nm in diameter across the outer surface of small silicon spheres, one being free of oxide coverage, the other one being covered with a 3 nm thick oxide layer. The recorded signal is the Si 2p edge which has been thoroughly investigated by different authors on test Si, Si oxides and other compounds. The pure silicon case with a threshold at 99.9 eV and a first maximum at 101.0 eV has been interpreted by Batson[19] in terms of transitions to s-d unoccupied states at the bottom of the conduction band, i.e. to Δ_1, L_1, Λ_1 and Λ_3 states in the Brillouin zone. In amorphous silica the same Si L_{23} edge displays three peaks at 106.0, 108.5 and 115.5 eV with no absorption over the 100-105 eV range. The first peaks have been interpreted as due to a Si-2p core exciton at 105.7 eV followed by the L_3 absorption edge at 107.1 eV [59]. The major advantage of the present technique is to provide sequences of spectra unambiguously related to the local topography, giving access, through a-posteriori processing, to the variation of the environment of the Si atoms (number of Si and O atoms on the surrounding tetrahedra).

CONCLUSION

The major step forward recently made in the development and application of EELS techniques is the close association of the spectroscopy with the microscopy. Mapping or profiling spectra is particularly suited to the investigation of heterogeneities in the specimen. As many new functional materials or designs with original properties are nowadays elaborated with typical nanometre dimensions in one direction at least, the control of their structural, chemical and electronic properties at this level of spatial resolution is essential. The techniques and methods described in the present review constitute best candidates to provide this information.

Using the most recent STEM instruments, several groups at Oak Ridge, Cornell and IBM Yorktown[15-17] have demonstrated atomic column-by-column imaging, accompanied by electron energy-loss spectrometry from a specified column. It means that from now the ultimate goal in analysing solid specimens, i.e. identifying the component atoms, one by one, has been reached, at least within the limit of the 2D projection of the real 3D structure[60-61]. In the meantime the access to the 3D chemical tomography of solids, atom by atom, has been provided by the latest developments in field ion microscopy and time-of-flight mass analysis techniques[62].

However, as extensively illustrated thereabove, the electron energy-loss techniques extend the information beyond the identification of the atoms present along the trajectory of the incident electrons, towards the determination of their bonding and coordination within a short range neighbourhood. But to extract full benefit from such investigations, a theoretical effort has to be devoted to the description of the modifications of the electron states (involved in EELS measurements) within confined materials or in the close vicinity of the heterogeneities, boundaries and interfaces of interest.

When considering more explicitly the impact of EELS mapping on magnetic materials, it must be recognized that relatively few studies have been accomplished up to now except those using the relationship between the branching ratio of white lines and the local magnetic moments[46-48]. In one case it must be pointed out that this relation has been used to investigate the magnetism of nm-sized iron crystallites[49].

I wish to conclude this review by suggesting a few routes of potentially high interest in the field of EELS applications to magnetic material studies : improve our understanding of the determination of the magnetic moments from EELS fine structures, use this information to investigate dimension or proximity effects in magnetic nanoclusters, close to magnetic-non magnetic interfaces or across non magnetic spacers in a multilayer, finally explore the possibility of linear magnetic dichroism studies in thin foils by using the equivalency between the direction of polarization of the linearly polarized photon in an x-ray absorption experiment and that of the momentum transfer in an electron inelastic scattering experiment.

Acknowledgements :

This review has been prepared at the request of the organizers of the International School of Solid State Physics which has taken place in Erice from 15 to 26 May 1994. Thanks are due to them for their stimulating invitation and for their patience afterwards. All my colleagues at the Laboratory of Solid State Physics in Orsay who have collaborated over years to develop the instrument and the methods, to use them in many practical situations and to promote the technique through its successes, have to be warmly acknowledged.

References

1. H. Raether, "Electron Energy Loss Spectroscopy", in Springer Tracts in Modern Physics, 38:85, Springer Berlin (1965).

2. H. Raether, "Excitation of Plasmons and Interband Transitions by Electrons", Springer Tracts in Modern Physics, Vol. 88, Springer Berlin (1980).

3. S.E. Schnatterly, Inelastic electron scattering spectroscopy, *Solid State Physics* 14:275 (1979).

4. P. Schattschneider, "Fundamentals of Inelastic Electron Scattering", Springer Verlag, Wien New York (1986).

5. J. Fink, Transmission electron energy-loss spectroscopy, in "Unoccupied Electronic States", J.C. Fuggle and J.E. Inglesfield eds., Topics in Applied Physics, 69:203 Springer Berlin (1992).

6. C. Colliex, Electron energy loss spectroscopy in the electron microscope, in "Adv. in Optical and Electron Microscopy", R. Barer and V.E. Cosslett eds., 9:65 Academic Press London (1984).

7. G. Ruthemann, *Naturwissenschaften* 30:145 (1942)

8. C. Colliex and B. Jouffrey, Contribution à l'étude des pertes d'énergie dues à l'excitation de niveaux profonds, *C.R. Acad. Sci. Paris* 270:144 (1970).

9. C. Colliex and B. Jouffrey, Diffusion inélastique des électrons dans un solide par excitation de niveaux atomiques profonds : Spectres de pertes d'énergie, *Phil. Mag.* 25:491 (1972).

10. M. Isaacson, Interaction of 25 keV electrons with the nucleic acid bases adenine, thymine and uracyl. I. Outer shell excitations. II. Inner shell excitation and inelastic scattering cross section. *J. Chem. Phys.* 56:1803 and 1813 (1972).

11. M. Terauchi, R. Kuzuo, F. Satoh, M. Tanaka, K. Tsuno and J. Ohyama, Performance of a new high-resolution electron energy-loss spectroscopy microscope, *Microsc. Microanal. Microstruct.* 2:351 (1991).

12. R.F. Egerton, "Electron Energy Loss Spectroscopy in the Electron Microscope", Plenum Press, New York (1986).

13. O.L. Krivanek and P.R. Swann, An advanced electron energy loss spectrometer, in "Quantitative microanalysis with high spatial resolution", 136, The Metals Society London (1981).

14. D. Bouchet, C. Colliex, P. Flora, O.L. Krivanek, C. Mory and M. Tencé, Analytical electron microscopy at the atomic level with PEELS, *Microsc. Microanal. Microstruct.* 1:443 (1990).

15. N.D. Browning, M.F. Chisholm and S.J. Pennycook, Atomic-resolution chemical analysis using a scanning transmission electron microscope, *Nature* 366:143 (1993).

16. D.A. Muller, Y. Tsou, R. Raj and J. Silcox, Mapping sp^2 and sp^3 states at sub-nanometre spatial resolution, *Nature* 366:725 (1993).

17. P.E. Batson, Simultaneous STEM imaging and electron energy-loss spectroscopy with atomic-column sensitivity, *Nature* 366:727 (1993).

18. P.E. Batson, High resolution electron energy-loss spectrometer for the STEM, *Rev. Sci. Instr.* 57:43 (1986).

19. O.L. Krivanek, C.C. Ahn and R.B. Keeney, Parallel-detection electron energy-loss spectrometer using quadrupole lenses, *Ultramicroscopy* 22:103 (1987).

20 P.E. Batson, Electronic structure in confined volumes using spatially resolved EEL scattering, *Materials Science and Engineering* B14:297 (1992).

21. P.E. Batson, Carbon 1s near-edge absorption fine structure in graphite, *Phys. Rev. B* 48:2608 (1993).

22. C. Colliex, Electron energy loss spectroscopy on solids, in "International Tables for Crystallography", Volume C Mathematical, Physical and Chemical Tables, 338, International Union of Crystallography, Kluwer Academic Press, Dordrecht (1992).

23. R.D. Leapman, P.l. Fejes and J. Silcox, Orientation dependence of core edges from anisotropic materials determined by inelastic scattering of fast electrons, *Phys. Rev. B* 28:2361 (1983).

24. P.E. Batson, K.L. Kavanagh, J.M. Woodall and J.W. Mayer, Observation of defect electronic states associated with misfit dislocations at the GaAs/GaInAs interface, *Phys. Rev. Lett.* 57:2729 (1986).

25. P. Schattschneider, The dielectric description of inelastic electron scattering, *Ultramicroscopy* 28:1 (1989).

26. R. D. Leapman, EELS quantitative analysis, in "Transmission Electron Energy Loss Spectrometry in Materials Science", M.M. Disko, C.C. Ahn and B. Fultz eds. p.47, TMS Warrendale, Pa 15086 (1992).

27. F. Hofer, Determination of inner-shell cross-sections for EELS quantification, *Microsc. Microanal. Microstruct.* 2:215 (1992).

28 C. Colliex, T. Manoubi and C. Ortiz, EELS near-edge fine structures in the iron-oxygen system, *Phys. Rev. B* 44:11402 (1991).

29. R.D. Leapman and D.E. Newbury, Trace analysis of transition elements and rare earths by PEELS, *Proc. EMSA meeting* 1250 (1992).

30. O.L. Krivanek, C. Mory, M. Tencé and C. Colliex, EELS quantification near the single-atom limit, *Microsc. Microanal. Microstruct.* 2:257 (1991).

31. C. Colliex, The impact of EELS in materials science, *Microsc. Microanal. Microstruct.* 2:403 (1991).

32. H. Kurata and C. Colliex, Electron-energy-loss core-edge structures in manganese oxides, *Phys. Rev. B* 48:2102 (1993).

33. J.A. Tossell, D.J. Vaughan and K.H. Johnson, The electronic structure of rutile, wustite and hematite from MO calculations, *Am. Mineralogist* 59:319 (1974).

34. R. Brydson, H. Sauer and W. Engel, Electron energy-loss near-edge fine structure as an analytical tool: The study of minerals, In "Transmission Electron Energy Loss Spectrometry in Materials Science", M.M. Disko, C.C. Ahn and B. Fultz eds., p. 131, TMS Warrendale, Pa 15086 (1992).

35. M.V. Ryzhkov, V.A. Gubanov, Yu.A. Teterin and A.S. Baev, Electronic structure, chemical bonding and x-ray photoelectron spectra of light rare-earth oxides, *Z. Phys. B Cond. Matt.* 59:1 (1985).

36. H. Kurata, E. Lefèvre, C. Colliex and R. Brydson, EELS near-edge structures in the oxygen K-edge spectra of transition-metal oxides, *Phys. Rev. B* 47:13763 (1993).

37. J.Stohr, F. Sette and A.L. Johnson, Near-edge x-ray absorption fine structure studies of chemisorbed hydrocarbons: bond lengths with a ruler, *Phys. Rev. Lett.*. 53:1684 (1984).

38. P. Rez, Energy loss fine structure, in "Transmission Electron Energy Loss Spectrometry in Materials Science", M.M. Disko, C.C. Ahn and B. Fultz eds. TMS, Warrendale, Pa 15086 (1992).

39. X.Weng, P. Rez and P.E. Batson, Single electron calculations for the Si L_{23} near edge structure, *Sol. State Comm.* 74:1013 (1990).

40. G.A. Sawatzky, Theoretical description of near edge EELS and XAS spectra, *Microsc. Microanal. Microstruct.* 2:153 (1991).

41. F.M.F. de Groot, J.C. Fuggle, B.T. Thole and G.A. Sawatzky, 2p x-ray absorption of 3d transition-metal compounds : an atomic multiplet description including the crystal field, *Phys. Rev. B* 42:5459 (1990).

42 G. van der Laan and I.W. Kirkman, The 2p absorption spectra of 3d transition metal compounds in tetrahedral and octahedral symmetry, *J. Phys. : Condens. Matter* 4:4189 (1992).

43 B.T. Thole and G. van der Laan, Branching ratio in x-ray absorption spectroscopy, *Phys. Rev. B* 38:3158 (1988).

44. C. Colliex, Spatially-resolved electron energy-loss spectrometry, in "Transmission Electron Energy Loss Spectrometry in Materials Science", M.M. Disko, C.C. Ahn and B. Fultz eds. p.85, TMS Warrendale, Pa 15086 (1992).

45 F.M.F. de Groot, X-ray absorption and dichroism of transition metals and their compounds, *to be published in Journal of Electron Spectroscopy and Related Phenomena* (1994).

46. T.I. Morrison, M.B. Brodsky, N.J. Zaluzec and L.R. Sill, Iron d-band occupancy in amorphous Fe_xGe_{1-x}, *Phys. Rev. B* 32:3107 (1985).

47. D.M. Pease, S.D. Bader, M.B. Brodsky, J.I. Budnick, T.I. Morrison and N.J. Zaluzec, Anomalous L_3/L_2 white line ratios and spin pairing in 3d transition metals and alloys : Cr metal and $Cr_{20}Au_{80}$, *Phys. Lett.* 114A:491 (1986).

48. T.I. Morrison, C.L. Foiles, D.M. Pease and N.J. Zaluzec, Relationships between local order and magnetic behavior in amorphous $Fe_{0.3}Y_{0.7}$: Extended x-ray absorption fine structure and susceptibility, *Phys. Rev. B* 36:3739 (1987).

49. H. Kurata and N. Tanaka, Iron L_{23} white line ratio in nm-sized γ-iron crystallites embedded in MgO, *Microsc. Microanal. Microstruct.* 2:283 (1991).

50. C. Jeanguillaume and C. Colliex, Spectrum-image: the next step in EELS digital acquisition and processing, *Ultramicroscopy* 28:252 (1989).

51. C. Colliex, M. Tencé, E. Lefèvre, C. Mory, H. Gu, D. Bouchet and C. Jeanguillaume, Electron energy-loss spectrometry mapping, *Mikrochim. Acta* 114/115:71 (1994).

52. C. Colliex, E. Lefèvre and M. Tencé, High spatial resolution mapping of EELS fine structures, *Inst. Phys. Conf. Ser.* 138:25 (1993).

53. M. Tencé, M. Quartuccio and C. Colliex, PEELS compositional profiling and mapping at nanometer spatial resolution, *to be published in Ultramicroscopy* (1994).

54. P. Ajayan, C. Colliex, J.M. Lambert, P. Bernier, L. Barbedette, M. Tencé and O. Stephan, Growth of manganese filled nanofibers in the vapor phase, *Phys. Rev. Lett.* 72:1722 (1994)

55. K. Yu-Zhang, G. Boisjolly, J. Rivory, L. Kilian and C. Colliex, Characterization of TiO_2/SiO_2 multilayers by high resolution transmission electron microscopy and electron energy loss spectroscopy, *to be published in the Proceedings of Int. Conf. Metallurgical Coatings and Thin Films, San Diego* (1994)

56. A. Albu-Yaron, S. Bastide, D. Bouchet, N. Brun, C. Colliex and C. Lévy-Clément, Nanostructural and nanochemical investigation of luminescent photoelectrochemically etched porous n-type silicon *J. Phys. I France* 4:1181 (1994).

57. P. Bayle, T. Deutsch, B. Gilles, F. Lançon, A. Marty, J. Thibault, C. Colliex and M. Tencé, HREM observations, PEELS analysis and numerical simulations of Au/Ni MBE multilayers, *to be published in Proc. MRS Society*, Boston (Dec 1993).

58. P. Bayle, T. Deutsch, B. Gilles, F. Lançon, A. Marty and J. Thibault, Au/Ni MBE multilayers : quantitative analysis at the atomic scale of the deformation and of the chemical profiles, *to be published in Ultramicroscopy* (1994).

59. V.J. Nithianandam and S.E. Schnatterly, Soft x-ray emission and inelastic electron scattering study of the electronic excitations in amorphous and crystalline silicon dioxide, *Phys. Rev. B* 38:5547 (1988)

60. L.M. Brown, The ultimate analysis, *Nature* 366:721 (1993)

61. C. Colliex, Energy-loss spectroscopy looks up, *Physics World* Vol. 7, N°5, 27 (1994)

62. D. Blavette, A. Bostel, J.M. Sarrau, B. Deconihout and A. Menand, An atom probe for three-dimensional tomography, *Nature* 363:432 (1993).

d-LIKE QUANTUM-WELL STATES AND INTERFACE STATES OF PARAMAGNETIC OVERLAYERS ON Co(0001)

D. Hartmann, A. Rampe, W. Weber*, M. Reese, and G. Güntherodt

2. Physikalisches Institut
RWTH Aachen
D-52056 Aachen
Germany

INTRODUCTION

Quantum-well states (QWS) are well known from semiconductor heterostructures and have recently also been discovered in metallic overlayers on metal substrates[1]. Such metallic overlayers with thicknesses up to 30 atomic layers (AL) form a quantum well for the electrons inside them. One barrier of the quantum well is made up by the potential step due to the work function at the vacuum/overlayer interface and the other barrier by the potential step at the overlayer/substrate interface due to their different crystal potentials. Of special interest are overlayers on ferromagnetic substrates, because then the potential step at the overlayer/substrate interface can be spin-dependent leading to a quantum well of different depth for majority- and minority-spin electrons. The corresponding QWS therefore are spin-polarized. Furthermore, the binding energy of the QWS varies as a function of the overlayer thickness, following the dispersion of the overlayer bulk band from which they are derived[2]. For (100)-oriented noble metal overlayers this bulk band was identified as a sp-like band[2-4]. The spin-polarized QWS[4-6] are considered as mediators of the oscillatory interlayer exchange coupling[7,8]. This description is equivalent to the RKKY-like coupling model of two ferromagnetic layers through a noble metal interlayer[9]. The connection between the two models is that the oscillation period is determined by extremal wave vectors of the Fermi surface, which are the corresponding wave vectors of the QWS. The present understanding is that the spin-dependent reflection coefficients of the electrons in the interlayer lead to dominantly minority-spin QWS and to a spin-density modulation. Whenever with increasing interlayer thickness such a minority-spin QWS crosses the Fermi level, it negatively polarizes the conduction electrons of the interlayer and the interlayer coupling switches from ferromagnetic to antiferromagnetic coupling. If the QWS has crossed the Fermi level, the spin polarization of the conduction electrons again reverses to positive values thereby mediating a ferromagnetic coupling. Such an oscillation of the spin polarization of the conduction electrons between positive and negative values has been observed by measuring the spin polarization of secondary electrons of Au on Fe(110). With increasing Au thickness this spin polarization oscillates[10]. An oscillatory behavior has recently also been observed in the Kerr rotation of bcc-Fe(100) layers on Au(100) with increasing Fe layer thickness[11] and in the Kerr rotation of Au/Co(0001)/Au(111) as a function of the Au overlayer thickness[12].

So far QWS have mainly been investigated in (100)-oriented overlayers on ferromagnetic substrates and only such QWS which are derived from sp-states of the overlayer. In our work we focussed our attention on QWS derived from d-states of the overlayer. We chose (111)-oriented noble metal overlayers because for the (111)-orientation the interlayer coupling is still discussed quite controversially[13-17]. Up to now it is not unambiguously proven whether the interlayer coupling in Co/Cu/Co(111) multilayers oscillates or not. In particular, there is so far no evidence for QWS for this orientation.

Besides QWS, interface states play an important role in thin film magnetism. Such states result from the hybridization of electronic states of the paramagnetic overlayer and of the ferromagnetic substrate. According to layer-projected spin-density functional calculations[18], magnetic moments can be induced by this hybridization. If the interface state possesses a significant spin-orbit interaction it may contribute to the magnetic anisotropy or via matrix elements of magnetooptic transitions to the Kerr rotation. We looked in Pd, Rh and Ru overlayers on Co(0001) for such interface states. First attempts have been undertaken to identify the spin-orbit interaction of QWS or interface states by the linear magnetic dichroism in the angular distribution of photoelectrons (LMDAD).

APPARATUS AND SAMPLE PREPARATION

The experiments were performed with two different experimental setups. The apparatus used for spin-resolved photoemission has been described elsewhere[19]. Unpolarized vacuum ultraviolet (VUV) light of a noble gas resonance lamp with photon energies of 21.2eV, 16.85eV, and 11.83eV is used. The energy and angle resolution of the system are 200meV and ±3°, respectively. Spin analysis is carried out with a 100-keV Mott detector. The base pressure of the chamber ($1*10^{-10}$mbar) rose to $5*10^{-10}$mbar during electron beam evaporation. The film quality and the growth mode is examined by low-energy electron diffraction (LEED) and Auger electron spectroscopy (AES). In addition, spin-integrated photoemission experiments and the magnetic dichroism (LMDAD) measurements were performed at the TGM 3 beamline at BESSY, Berlin. The combined energy resolution of the monochromator and the hemispherical energy analyzer was 200meV. The angle resolution was ±2°.

Following previous work, thick Co films were evaporated onto a W(110) single crystal held at 400K with deposition rates of 2.0Å/min. The film thickness was measured with a calibrated quartz microbalance. Co grows in the hcp(0001) orientation[20]. Onto this ferromagnetic "substrate" the paramagnetic overlayers were evaporated at a rate of 0.5Å/min with the substrate held at room temperature to avoid interdiffusion with the ferromagnetic substrate.

LEED AND AUGER

The LEED patterns of the overlayers on Co(0001) indicate the fcc(111) orientation for Au, Pd, Pt, Rh and the hcp(0001) orientation for Ru. The LEED pattern of 1AL Au on Co

(14.9% lattice misfit) displays a linear superposition of the Co(0001) and Au(111) surface nets. The in-plane lattice constant of 1AL Au on Co is slightly smaller (2%) than the corresponding bulk lattice spacing which is attained at 8AL Au. 1AL of Pd and of Rh on Co(0001) (9.6% (7.0%) lattice misfit for Pd (Rh)) show a (1 x 11) and (1 x 10) superstructure along the Co[-1,2,-1,0] direction, respectively, but grow pseudomorphically along the Co[1,0,-1,0] direction. This is analogous to the Nishiyama-Wasserman orientational relationship[21]. 1AL of Pt on Co (10.4% lattice misfit) grows pseudomorphically. Ru on Co(0001) (7.6% lattice misfit) grows with its bulk lattice constant in the Ru(0001) orientation. All overlayers attain their bulk lattice constants at 5 to 8AL thickness.

The growth modes of Au, Pd, Pt, Rh and Ru have been studied by AES. The corresponding Co Auger lines at 53eV, 775eV, 775eV and 775eV have been measured to analyze the growth mode of Au, Pt, Rh and Ru, respectively. In each case the intensity of the Co Auger line decreases corresponding to $\exp(-d/\lambda)$ with decay lengths of $\lambda = 3.5$Å, 11.0Å, 12.4Å, and 12.5Å, respectively, when Co is covered by the overlayer of thickness d. Since this decay length has approximately the same value as the inelastic mean free path of electrons of the corresponding kinetic energy[22], the observed exponential decay is at least consistent with a layer-by-layer growth. For Pd/Co(0001) the 330-eV Pd Auger line has been recorded. Its intensity exhibits kinks which are characteristic of the layer-by-layer growth mode.

Au/Co(0001)

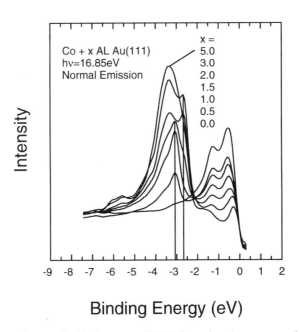

Figure 1. Spin-integrated energy distribution curves (EDCs) for various Au coverages (in atomic layers (AL)) on Co(0001) plotted on the same absolute intensity scale.

Figure 1 shows the photoemission spectra of Co(0001) (x=0.0AL), which is successively covered by up to 5AL Au. All spectra were taken in normal emission for a photon energy of hv=16.85eV and under identical conditions to get comparable absolute intensities. At 0.5AL coverage a first peak evolves at -3.1eV binding energy and saturates in intensity at 1.5AL Au. With further coverage this peak shifts to -2.7eV and saturates in intensity at 2-3AL coverage. Such a thickness-binding energy dependence for the two new peaks is typical of a quantum-well state[23]. At a coverage of 5AL Au the spectrum resembles that of a Au(111) single crystal.

Upon varying the photon energy the binding energy of the new peak at -3.1eV remains constant. This is shown in Fig. 2 for 1AL of Au on Co(0001). We conclude that the peak observed at -3.1eV does not show any dispersion with the electron wave vector component perpendicular to the layer. The same refers to the peak at -2.7eV. Thus the state to which the two peaks belong is a two-dimensional electronic state, i.e. a QWS.

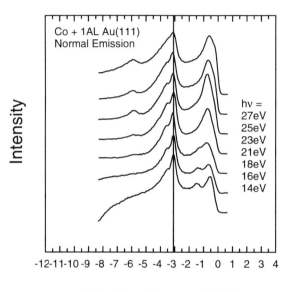

Figure 2. Spin-integrated EDCs for 1AL of Au on Co as a function of the photon energy.

Because the binding energies of the QWS lie in the energy interval within which the density of Au-5d states is high and because the atomic photoionization cross section of Au-5d states is by more than a factor of 100 larger than that of s-states, we conclude that it is derived from d-states of the overlayer and thus is a d-like QWS.

It is interesting to inspect the bulk band structure of Co(0001)[24] and Au(111)[25]. Obviously, the binding energies of the QWS lie in an absolute energy gap of Co(0001) between -3.8eV and -2.7eV. Therefore the Au/Co interface totally reflects the Au electrons at these energies. On the other hand, the Au electrons are totally reflected at the Au/vacuum interface due to the work function. Thus the Au overlayer forms an ideal quantum well. If in first-order approximation the Au electronic bands of the ultrathin overlayers are approximated by the bulk bands of Au(111), we would assume the QWS to be derived from the Au Λ_{4+5} band. Because of a weaker dispersion of this band compared to the Λ_6 band, it yields a higher density of states at the binding energies at which the QWS appears. If the overlayer thickness is increased to 4AL or 5AL the QWS in first-order approximation would shift from -2.7eV to -2.3eV, which lies outside the absolute energy gap and close to the cut-off of the Λ_{4+5}-band dispersion. Therefore at coverages more than 4AL the QWS disappears and the electronic structure of the Au overlayer converges to that of the bulk.

Figure 3 shows the spin-resolved photoemission spectrum of 1AL of Au on Co. The QWS shows neither a spin splitting nor a spin polarization after subtraction of the Co background spin polarization. This can be understood, since because of the absolute energy gap the reflection coefficients are not spin-dependent and thus the QWS shows a spin degeneracy. In other words, there is no hybridization between Au-5d and Co-3d electrons. The same holds for the QWS at 2AL of Au on Co.

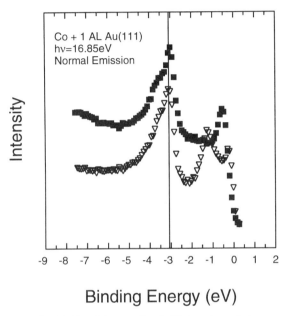

Figure 3. Spin-resolved EDC of 1AL of Au on Co. Solid (open) symbols correspond to the majority- (minority-) spin direction.

Because the QWS is not spin-polarized and also does not cross the Fermi level, it cannot polarize Au conduction electrons. Therefore this d-like QWS does not contribute to the interlayer coupling in corresponding magnetic multilayers. Up to now no QWS derived from sp-states of the Au(111) overlayer have unambiguously been observed in off-normal photoemission at the $\bar{M}$-point of the surface Brillouin zone, although at the $\bar{M}$-point QWS with an extremal wave vector near the (1,1,-1) and (-1,-1,1) neck of the Au Fermi surface are theoretically predicted[8].

Pd/Co(0001) AND Pt/Co(0001)

Overlayers of Pd(111) on Co(0001) behave quite similar to the Au/Co system as can be seen in Fig. 4. Analogously, at a coverage of 1AL Pd a QWS appears at -1.2eV, saturates at 1-1.5AL and then shifts to -0.6eV at 2-3AL Pd. The two-dimensional character of this QWS is confirmed by the independence of its binding energy (-1.2eV) upon varying the photon energy as shown in Fig. 5. The photoemission intensity of the QWS has been followed up to 200eV photon energy. It shows a minimum at 110eV, which coincides with the Cooper minimum of the atomic photoionization cross section for the Pd-4d subshell. Therefore this QWS is a d-like QWS[23,26,27]. Another feature appears in Fig. 5 for $h\nu \geq$ 24eV at -2.3eV binding energy, which will be discussed later.

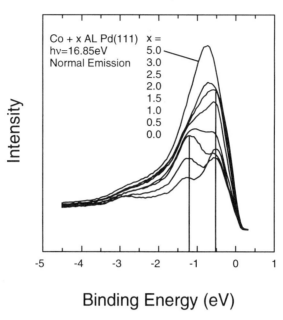

Figure 4. Spin-integrated EDCs for various Pd coverages (in atomic layers (AL)) on Co(0001) plotted on the same absolute intensity scale.

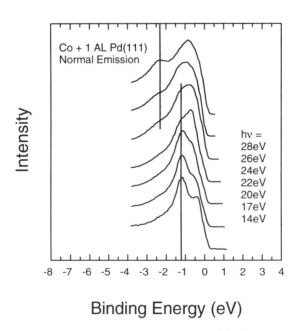

Figure 5. Spin-integrated EDCs for 1AL of Pd on Co as a function of the photon energy. The lines mark the QWS at -1.2eV and the interface state at -2.3eV.

Fig. 6 shows the spin-resolved spectrum of 1AL of Pd on Co(0001). The QWS of 1AL Pd reveals a spin splitting of about 180meV, which is inverted compared to the exchange splitting of Co. For 2AL Pd the QWS does not show a measureable spin splitting[27]. Pd on Fe(100) has been studied by Rader et al. and shows similar results[28].

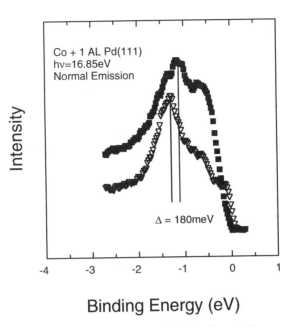

Figure 6. Spin-resolved EDC of 1AL of Pd on Co. Solid (open) symbols correspond to the majority- (minority-) spin direction.

Two conclusions can be drawn from the data: 1) The spin splitting of the QWS proves that the Pd monolayer is magnetically polarized and probably possesses an induced magnetic moment. A hybridization of a Pd-4d with a Co-3d state should be the reason for that. 2) Because the QWS for 2AL Pd is not spin-polarized, the induced magnetic moment must rapidly decrease with increasing distance from the Co substrate and should be confined mainly to the first AL of Pd. This does not seem to hold for the conduction electrons of Pd. They, in contrast, can be long-range spin-polarized and thereby mediating a long-range ferromagnetic interlayer coupling as will be seen in the section below on the interlayer coupling.

A d-like QWS has also been observed for Pt(111) overlayers on Co(0001). Its binding energies are -1.3eV and -0.6eV for 1-1.5AL and 2-3AL Pt, respectively. The QWS is spin-polarized only for the first AL[27,23].

Rh/Co(0001) AND Ru/Co(0001)

In the previous section it is proven that the Pd monolayer is magnetically polarized by hybridization with the ferromagnetic substrate. Besides Pd there are still other elements, such as Ru or Rh, which might be magnetically polarized by a ferromagnetic substrate or even are supposed to order ferromagnetically as monolayer on a Ag(001) substrate[18]. Therefore we studied Ru and Rh monolayers on Co(0001). Fig. 7 shows the results for Ru on Co. The spectra are taken with unpolarized VUV light of hv=21.2eV in off-normal geometry ($k_{||}$=1.19Å^{-1} along $\overline{\Gamma M}$), because in off-normal geometry the reversal of the spin polarization (see below) is most pronounced. Furthermore they are recorded with constant photon flux to get comparable absolute intensities. Fig. 7 a) displays the spectra of Co and Co+1AL Ru separated into the majority- and the minority-spin channel. In the majority-spin channel the coverage of Co with 1AL Ru leads to an increase in the intensity between the Fermi level and -0.3eV due to a Ru-4d interface state. The intensity in the minority-spin channel essentially remains constant. Therefore the spin polarization (Fig. 7 b) in this binding energy interval reverses its sign from -40% to +5% . Especially the conduction electrons are positively spin-polarized. This is an unambiguous proof for an induced spin polarization of Ru, i.e. for a spin-polarized electronic structure, which probably corresponds to an induced magnetic moment of Ru. Analogous results have been obtained for Rh on Co which also confirm a spin-polarized electronic structure of the Rh monolayer on Co[29].

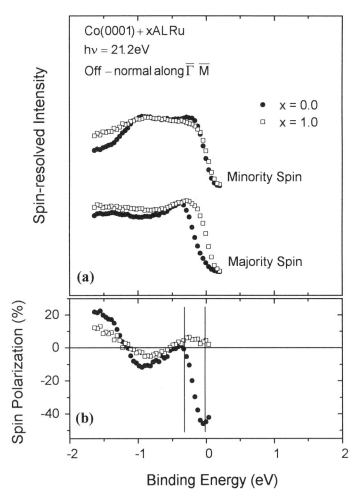

Figure 7. a) Spin-resolved EDCs for 0 and 1AL of Ru on Co(0001) plotted on an absolute intensity scale and b) corresponding spin polarization. The vertical bars in b) are guide lines marking the binding energy interval within which the spin polarization reverses its sign.

Two mechanisms can lead to such a spin-polarized electronic structure: 1) The number of nearest neighbors is reduced in a monolayer compared with the bulk crystal. Consequently, this results in a weaker d-d-hybridization between the overlayer atoms and therefore in a narrower band width of the overlayer d-electrons than in the bulk crystal. This causes a higher density of states at the Fermi level, the Stoner criterion may be fulfilled and ferromagnetism may be favored. In a previous theoretical work Blügel has attributed this mechanism to be responsible for the ferromagnetism of 1AL of Rh and Ru on a Ag(001) substrate[18]. 2) Also a hybridization between d-states of the overlayer and the ferromagnetic substrate could lead to a spin-polarized electronic structure. This results in a complex spin-polarized electronic band structure, which is only accessible by layer-projected spin-density functional calculations.

Spin-polarized Rh and Ru monolayers on Fe(100) have already been found in previous experiments. Kachel et al.[30] found a spin-polarized electronic structure for 1AL of Rh on Fe(100) using spin-resolved photoemission, which was analyzed by ab-initio calculations. Totland et al.[31] measured spin-polarized Auger electrons of Ru on Fe(100) and an exponential decay of the spin polarization of the Ru Auger electrons from 10% to 0% in going from 1AL to 8AL Ru. For Rh and Ru an induced magnetic moment of $0.82\mu_B$[30] and $0.7\mu_B$[31], respectively, was found.

INTERLAYER COUPLING ACROSS Rh, Ru AND Pd

In this section our results of the (oscillatory) interlayer exchange coupling of two Co layers across Rh, Ru and Pd are presented. For the experiments we have prepared trilayers of the following sequence: 12AL Co / xAL interlayer / 7AL Co. Due to its larger thickness the 12-AL base layer was always remanently magnetized in field direction. Thus, as a function of the interlayer thickness the thinner Co top layer couples parallel (ferromagnetically) or antiparallel (antiferromagnetically) to the Co base layer according to the interlayer coupling. The spin polarization of the secondary electrons (SPSE) of the top Co layer which is proportional to its magnetization has been measured. In Fig. 8 this SPSE is plotted as a function of the Rh-, Ru-, and Pd-interlayer thickness. Positive (negative) values correspond to ferromagnetic (antiferromagnetic) coupling. The period lengths of the oscillations observed for Ru and Rh are determined by taking twice the difference of the second and third zero crossing point and amount λ_{Ru} = 8AL and λ_{Rh} = 9AL. The coupling across the Pd(111) interlayer was ferromagnetic up to 16AL thickness.

The only ferromagnetic coupling across Pd can be explained by the Stoner-enhanced Pauli spin susceptibility of Pd which causes a long-range (>10AL) positive spin polarization of the Pd conduction electrons[32]. The interlayer coupling across Rh and Ru has an antiferromagnetic onset already for an interlayer thickness of 2.5AL and 1.5AL, respectively. Especially the Ru case is surprising, since we proved the Ru monolayer to be positively spin-polarized. Moreover the spin polarization of the conduction electrons at the investigated **k**-point is positive. Because the first AL of Ru is positively spin-polarized by the Co base layer and the second AL of Ru should also be positively spin-polarized by the Co top layer, a ferromagnetic coupling across 2AL Ru should be favored. This raises the question whether the first AL of Ru couples antiferromagnetically to the second AL of Ru. But this question cannot be answered by our data and should be addressed by layer-projected spin-density functional calculations.

The observed period lengths (λ_{Ru} = 8AL and λ_{Rh} = 9AL) strongly differ from those found for sputtered samples λ_{Ru} = 5AL and λ_{Rh} = 4AL[33] and that of MBE-grown samples on mica with λ_{Ru} = 5AL[34]. This may be attributed to the different preparation conditions.

At least our value for the Ru-period is consistent with the theoretically predicted periods as determined by the extremal wave vectors of the Ru-Fermi surface. They range from 4-20AL for Ru(0001) interlayers[35]. It is an open question why in our experiment the extremal wave vector which corresponds to a period of 8AL is preferred. Perhaps our results can motivate ab-initio calculations of the absolute contribution of the different extremal **k**-points to the interlayer coupling energy.

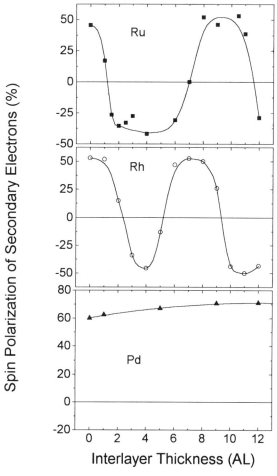

Figure 8. Spin polarization of the secondary electrons of the Co top layer of the Co / (Rh, Ru or Pd) / Co trilayer as a function of the interlayer thickness (in atomic layers (AL)).

LMDAD

In order to gain more information about the symmetry or spin-orbit interaction of the QWS the linear magnetic dichroism in the angular distribution of photoelectrons (LMDAD)[36,37] has been investigated for the valence band electrons of the Pd(111)/Co(0001) system. Our aim is to elucidate the role of such QWS in the magnetic interface anisotropy or magnetooptic transitions in Pd/Co multilayers. Besides the QWS the Pd-derived interface state of 1AL of Pd on Co at -2.3eV (see Fig. 5) is investigated, which is visible for photon energies higher than 24eV.

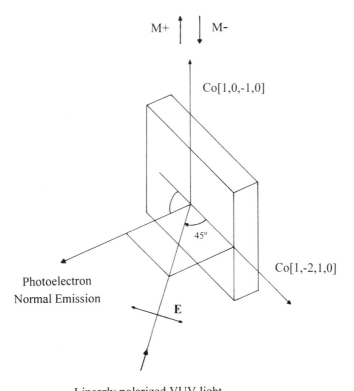

Figure 9. Geometry used to investigate the LMDAD. M+ (M-) corresponds to the magnetization vector parallel (antiparallel) to the Co(1,0,-1,0) direction.

Figure 9 displays the geometrical arrangement used for the LMDAD measurements. Linearly polarized VUV light impinged at an angle of 45° onto the surface of the sample. The **E**-vector of the VUV light lay in the reaction plane spanned by the photon and electron **k**-vectors. The magnetization vector is perpendicular to this reaction plane. The photoemission spectra were taken in normal emission. Photoemission spectra recorded with the magnetization vector parallel (M+) or antiparallel (M-) to the Co[1,0,-1,0] direction are termed I^+ or I^-, respectively.

Fig. 10 shows the LMDAD photoemission spectra of Co and Co+1AL Pd and the corresponding dichroism asymmetry, $(I^+ - I^-) / (I^+ + I^-)$. The spectra are normalized such that I^+ and I^- coincide in the secondary electron peak (not shown here). The Pd-derived state at -2.3eV exhibits a small dichroism asymmetry of +1.5%, whereas the pure Co at this binding energy shows a negative asymmetry of about -2%. This LMDAD proves that the Pd-derived state is magnetically polarized and that it possesses a spin-orbit interaction. It is probably derived from a Pd-4d state which hybridizes with a Co-3d state and which for higher Pd coverages evolves into a state of the lower Λ_3-band of bulk Pd at about -2.5eV[38]. Because of the hybridization this state should be termed interface state. It does not shift with further Pd coverage as the QWS does. In contrast to this state the Pd-QWS itself does not show any linear magnetic dichroism as has been confirmed by LMDAD measurements at lower photon energies (17eV), where the QWS reveals a pronounced peak (see Figs. 4 and 5).

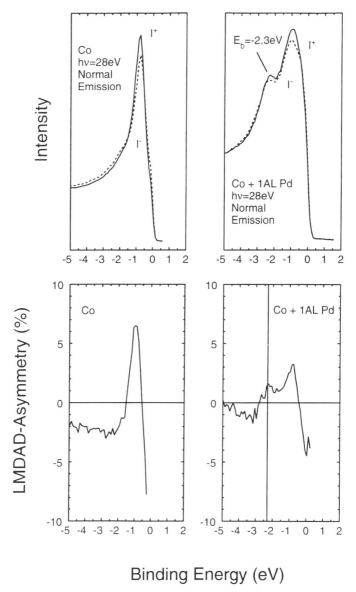

Figure 10. LMDAD spectra of Co and Co+1AL Pd and corresponding LMDAD asymmetry, $A = (I^+ - I^-) / (I^+ + I^-)$.

SUMMARY

d-like quantum-well states (QWS) have been found in (111)-oriented overlayers of Au, Pd, and Pt on Co(0001) for thicknesses up to 3 atomic layers (AL). Au/Co is a textbook-type example of a d-like QWS, because it lies in an absolute energy gap of Co(0001) along the $\Gamma A \Gamma$-direction of the first Brillouin zone. This QWS exhibits no spin polarization and does not cross the Fermi level. Therefore it cannot mediate the interlayer coupling in corresponding magnetic multilayers. The QWS of Pd (Pt) on Co is spin-polarized only for the first AL caused by hybridization of Pd-4d (Pt-5d) states with Co-3d states, whereas for 2AL the QWS is not spin-polarized. Thus the induced magnetic moment should be confined mainly to the first AL of Pd (Pt). Analogously 1AL of Ru(0001) and Rh(111) on Co are hybridized with the substrate, therefore exhibiting an induced positive spin polarization. The oscillatory interlayer coupling across Ru, Rh, and Pd has been studied. The oscillation periods amount λ_{Ru} = 8AL and λ_{Rh} = 9AL. The interlayer coupling across Pd(111) is ferromagnetic up to 16AL interlayer thickness, indicating a long-range positive spin polarization of its conduction electrons. Of special interest is the onset of the antiferromagnetic coupling across already 1.5AL Ru, despite the Ru monolayer showing an induced positive spin polarization. Furthermore first attempts to investigate the linear magnetic dichroism in the angular distribution of photoelectrons (LMDAD) of the Pd QWS and the Pd interface state have been undertaken. This interface state of 1AL Pd on Co(0001) at -2.3eV shows a positive LMDAD which proves that it is spin-polarized and possesses a significant spin-orbit interaction. In contrast, the QWS does not show a linear magnetic dichroism.

ACKNOWLEDGEMENT

The spin-integrated measurements at BESSY, Berlin, have been supported by the Bundesministerium für Forschung und Technologie (BMFT), Project No. 05 5PCFXB 2. Funding for the spin-resolved measurements in Aachen has been provided by the Deutsche Forschungsgemeinschaft (DFG) through SFB 341.

REFERENCES

*Present address: IBM Research Division, Zürich Research Laboratory, CH-8803 Rüschlikon, Switzerland

1. T. Miller, A. Samsavar, G. E. Franklin, T.-C. Chiang, *Phys. Rev. Lett.* 61:1404 (1988)
2. J. E. Ortega, F. J. Himpsel, G. J. Mankey, R. F. Willis, *Phys. Rev. B* 47:1540 (1993)
3. N. B. Brookes, Y. Chang, P. D. Johnson, *Phys. Rev. Lett.* 67:354 (1991)
4. F. J. Himpsel, *Phys. Rev. B* 44:5966 (1991); J. E. Ortega, F. J. Himpsel, *Phys. Rev. Lett.* 69:844 (1992)
5. K. Garrison, Y. Chang, P. D. Johnson, *Phys. Rev. Lett.* 71:2801 (1993)
6. C. Carbone, E. Vescovo, O. Rader, W. Gudat, W. Eberhardt, *Phys. Rev. Lett.* 71:2805 (1993)
7. D. M. Edwards, J. Mathon, R. B. Muniz, M. S. Phan, *Phys. Rev. Lett.* 67:1927 (1991)
8. P. Lang, L. Nordström, P. H. Dederichs, *Phys. Rev. Lett.* 71:1927 (1993), and private communication
9. P. Bruno, C. Chappert, *Phys. Rev. Lett.* 67: 1602 (1991); and *Phys. Rev. B* 46:261 (1992)
10. K. Koike, T. Furukawa, G. P. Cameron, Y. Murayama, *Colloquium Digest of the 14th International Colloquium on Magnetic Films and Surfaces, Düsseldorf 1994*, p. 788
11. Y. Suzuki, P. Bruno, W. Geerts, T. Katayama, *Colloquium Digest of the 14th International Colloquium on Magnetic Films and Surfaces, Düsseldorf 1994*, p. 777
12. P. Beauvillain, A. Bounouh, C. Chappert, R. Mégy, P. Veillet, *Colloquium Digest of the 14th International Colloquium on Magnetic Films and Surfaces, Düsseldorf 1994*, p. 712

13. V. Grolier, D. Renard, B. Bartenlian, P. Beauvillain, C. Chappert, C. Dupas, J. Ferré, M. Galtier, E. Kolb, M. Mulloy, J. P. Renard, P. Veilleit, *Phys. Rev. Lett.* 71:3023 (1993) and references therein.
14. W. F. Egelhoff Jr., M. J. Kief, *Phys. Rev.* B 45:7795 (1992)
15. M. T. Johnson, R. Coehoorn, J. J. de Fries, N. W. E. McGee, J. aan de Steege, P. J. H. Bloemen, *Phys. Rev. Lett.* 69:969 (1992)
16. J. Kohlhepp, S. Cordes, H. J. Elmers, U. Gradmann, *J. Magn. Magn. Mater.* 111:L231 (1992)
17. A. Schreyer, K. Bröhl, J. F. Akner, Th. Zeidler, P. Bödecker, N. Metoki, C. F. Majkrzak, H. Zabel, *Phys. Rev. B* 47:15334 (1993)
18. S. Blügel, *Phys. Rev. Lett.* 68:851 (1992) and references therein
19. R. Raue, H. Hopster, E. Kisker, *Rev. Sci. Instrum.* 55:383 (1984)
20. B. G. Johnson, P. J. Berlowitz, D. W. Goodman, C. H. Bartholomew, *Surf. Sci.* 217:13 (1989)
21. A. Zangwill, Chapter 16, in: "Physics at Surfaces", Cambridge University Press (1988)
22. P. Seah, W. A. Dench, *Surf. Interface Anal.* 1:2 (1979)
23. D. Hartmann, W. Weber, A. Rampe, S. Popovic, G. Güntherodt, *Phys. Rev. B* 48:16837 (1993)
24. F. J. Himpsel, D. E. Eastman, *Phys. Rev. B* 21:3207 (1980)
25. R. Courths, H.-G. Zimmer, A. Goldmann, H. Saalfeld, *Phys. Rev. B* 34:3577 (1986)
26. W. Weber, D. A. Wesner, G. Güntherodt, U. Linke, *Phys. Rev. Lett.* 66:942 (1991)
27. W. Weber, D. A. Wesner, D. Hartmann, G. Güntherodt, *Phys. Rev. B* 46:6199 (1992)
28. O. Rader, C. Carbone, W. Clemens, E. Vescovo, S. Blügel, W. Eberhardt, *Phys. Rev. B* 45:13823 (1992)
29. A. Rampe, D. Hartmann, W. Weber, S. Popovic, M. Reese, G. Güntherodt, accepted for publication in *Phys. Rev. B* (1994)
30. T. Kachel, W. Gudat, C. Carbone, E. Vescovo, S. Blügel, U. Alkemper, W. Eberhardt, *Phys. Rev. B* 46:12888 (1992)
31. K. Totland, P. Fuchs, J. C. Gröbli, M. Landolt, *Phys. Rev. Lett.* 70:2487 (1993)
32. P. H. Dederichs, private communication
33. S. S. P. Parkin, N. More, K. P. Roche, *Phys. Rev. Lett.* 67:2304 (1990)
34. K. Ounadjela, A. Arbaoui, A. Herr, R. Poinsot, D. Dinia, D. Müller, P. Panissod, *J. Magn. Magn. Mater.* 106:1896 (1992)
35. M. D. Stiles, *Phys. Rev. B* 48:7238 (1993)
36. Ch. Roth, F. U. Hillebrecht, H. B. Rose, E. Kisker, *Phys. Rev. Lett.* 70:3479 (1993)
37. D. Venus, *Phys. Rev. B* 49 (1994) 8821
38. B. Schmiedeskamp, B. Kessler, N. Müller, G. Schönhense, U. Heinzmann, *Solid State Commun.* 65:665 (1988)

ELECTRONIC CORRELATIONS IN THE 3s PHOTOELECTRON SPECTRA OF THE LATE TRANSITION METAL OXIDES

Fulvio Parmigiani [1] and Luigi Sangaletti [2]

[1] C.I.S.E. Tecnologie Innovative S.p.A., Materials Division,
P.O.Box 12081, 20134, Milano, Italy
[2] Dipartimento di Fisica A.Volta, Università di Pavia,
Via A.Bassi, 6, 27100, Pavia, Italy

INTRODUCTION

The X-Ray Photoemission (XP) core level spectra of transition metal (TM) compounds have been intensively investigated with the aim to obtain appropriate parameters to characterise their electronic properties [1-15]. In particular, the charge transfer (CT) energy, Δ, which describes the charge fluctuation from the ligand to the metal according to the mechanism $d^n \rightarrow d^{n+1}\underline{L}$ and the Mott-Hubbard energy U, which accounts for the electronic charge fluctuation between two transition metal ions in the crystal lattice, $d_i^n d_j^n \rightarrow d_i^{n-1} d_j^{n+1}$, are basic concepts to understand the electronic correlations of TM compounds [7-9].

Ionic TM compounds are characterised both by polar type fluctuations (U) and by charge transfer type (Δ) fluctuations. These mechanisms affect the XP spectra and a proper interpretation of the photoemission data leads to information on Δ and U. Indeed, charge fluctuations introduce peak asymmetries and satellites structures in XP spectra. Furthermore, the presence of charge fluctuations contributes to entangle the relationship between the density of states DOS and the photoelectron spectral distribution $P(E,\omega)$ since it might be difficult to relate the DOS, which refers to the neutral N-electron system, to the $P(E,\omega)$, which refers to the final state of the excited N-1 electron system. The complexity of the spectral features arises from the interaction between the core hole created upon the photoemission process and the electrons in the outer open shell that, in this case of Mn, Fe, Co and Ni divalent compounds, contains a number n = 5,6,7,8 of 3d electrons.

Nevertheless, the presence of satellite features brings, besides a complexity and difficulty in the data analysis, further information on the electronic structure of the compounds under exam.

THE TRANSITION METAL OXIDES

Within the framework so far outlined, a study of the 3s XP spectra in MnO, FeO, CoO and NiO has been performed. The study of the electronic properties of TM compounds is deeply rooted in solid state physics (see, e.g., Ref. 10 and references therein). These compounds were and still are a cross-roads for testing band structure calculations applied to correlated systems; their optical properties have been interpreted through crystal field and ligand field theories; direct and inverse photoemission studies have shown the features of their occupied and empty electronic states.

The four compounds under exam represent an homogeneous group of ionic insulators with the NaCl crystal structure where the TM ion is octahedrally coordinated with six oxygen atoms. The cell size ranges from 4.445 Å (MnO) down to 4.1769 Å (NiO). In addition, all these divalent oxides are antiferromagnets with a Néel temperature T_N ranging from 120 K (MnO) to 523 K (NiO) (Table 1). These common features make this group of compounds a suitable sample for an experimental and theoretical study where trends in the parameter values as well as in the change of spectral features must be identified without ambiguities.

Table 1. Néel temperatures and lattice parameters of late transition metal oxides

	T_N (K)	Lattice parameter (Å)
NiO	523	4.1769
CoO	271	4.260
FeO	198	4.307
MnO	120	4.445

The choice of the TM 3s levels, instead of the most studied 2p TM levels, was made since it is basically simpler to treat the photoemission problem with the multiplet structure induced by the 3s core-hole creation than treating the problem with the multiplet structure induced by a 2p core-hole.

EXPERIMENTAL DATA

The 3s XP spectra were measured, on commercially available high-quality NiO, CoO, FeO, and MnO single crystals, using a Perkin-Elmer Mod. 5400 spectrometer, equipped with a monochromatic Al-K_α source. The experimental details are reported elsewhere [16].

The XP spectra are shown in Fig.1. Usually three features, labelled A, B and C, can be detected in each spectrum except for NiO where the main line, A, appears structured and an additional feature, A', is detected. Going from MnO to NiO, it is possible to notice that (i) the FWHM of the A and B structures increases, (ii) their separation decreases and (iii) for Co and Ni a fine structure appears in peak B (CoO), which considerably broadens with respect to peak B in MnO and FeO, and in peak A (NiO).

EARLY STUDIES

The 3s core-level XP spectra, mostly those of Mn and Fe compounds, received a lot of attention in the past owing to the attempt to estimate the magnitude of exchange energies (EX) through the measure of the spectral components splitting [1-6,15]. As regards the exchange splitting, the Mn^{2+} ion is perhaps the most studied case. The doublet A,B shown in Fig. 1 corresponds to the exchange splitted 3s level of Mn^{2+}, peak A being the high spin contribution and peak B the low spin contribution.

The terminology relative to the exchange interaction studies can be usefully introduced discussing the case of Mn^{2+} compounds such as MnO and MnF_2 (see also Ref. 15). In these compounds the ground state configuration is $3s^2 3p^6 3d^5$ (^6S). The five 3d electrons have parallel spins. After the creation of the core hole two configuration may arise, $3s^1 3p^6 3d^5$ (^7S) and $3s^1 3p^6 3d^5$ (^5S) where the core-hole spin is, respectively, parallel and anti parallel to the 3d electrons spin. The energy separation between these two configurations is given by the exchange energy and can be calculated according to the Van Vleck's rule as:

$$\Delta E = (2S+1) K_{ns,n'l'} \qquad (S \neq 0) \qquad (1)$$

where S is the ground state spin,

$$K_{ns,n'l'} = \frac{e^2}{2l'+1} \iint \frac{r_<^{l'}}{r_>^{l'+1}} P_{ns}(r_1) P_{n'l'}(r_2) P_{ns}(r_2) P_{n'l'}(r_1) dr_1 dr_2 \qquad (2)$$

and the quantum numbers ns represent the core-hole (e.g. 3s) while the quantum numbers $n'l'$ represent the open shell electrons (e.g. 3d). The intensity of the multiplet peaks in the final state, denoted by the angular and spin quantum numbers S_f, L_f (e.g. ^7S or ^5S), can be calculated according to the intensity rules introduced by Cox and Orchard:

$$I(L^f, S^f) \propto (2S^f + 1)(2L^f + 1) \qquad (3)$$

giving for the intensity ratio between the high spin and the low spin spectral structures the following relation:

$$\frac{I(S+1/2)}{I(S-1/2)} = \frac{2S+2}{2S} \qquad (4)$$

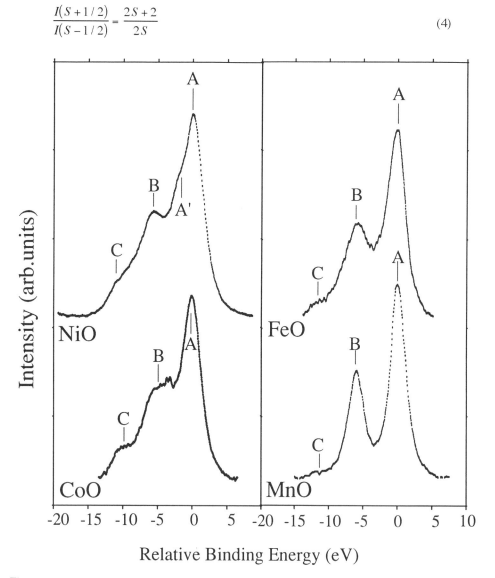

Figure 1. X-ray photoemission spectra of 3s levels in NiO, CoO, FeO, and MnO. To allow comparison, the main line energy of each spectrum has been set at zero eV.

However, the calculated intensity ratio does not correspond to the measured one and the experimental energy separation between the two exchange splitted contributions is roughly one half of the calculated ΔE.

To explain the discrepancy between the theoretical and the measured main line to satellite intensity ratio, as well as the presence of further satellites in the spectra at high binding energies, interpretations of the experimental results were carried out by configuration interaction (CI) approaches where intra-atomic correlation effects, involving redistribution of electrons among the TM atomic 3s, 3p and 3d shells, were included [4,5]. These studies contributed mainly to explain the origin of the broad structure on high binding energy side (~30 eV from the main line) of the 3s spectrum in, e.g., MnO. It has been shown that these broad satellites, whose intensity decreases going from Mn to Ni compounds, receive a considerable contribution from configurations where the 3s core hole is filled by a 3p electron and another 3p electron makes a transition to the 3d levels resulting in $3s^2 3p^4 3d^6$ configurations.

However, going from Mn and Fe to Co and Ni compounds an unambiguous data decomposition in terms of exchange interactions is not possible and models based on exchange and intra-atomic screening are no more sufficient. In addition to exchange interactions, final state screening effects have also been considered in the deconvolution of 3s photoemission spectra of ionic 3d TM compounds. In particular Veal and Paulikas [6] have shown that the extra atomic relaxation is a general, and often dominant, source of satellite structures in 3d TM fluorides. As it will be shown later, this effect is even more remarkable in TM oxides.

Presently it is well known, from the large amount of XP data available on TM compounds, that the magnitude of the final state screening is determined by the relative magnitude of the energy of the CT fluctuation and the energy of the coulomb interaction between the core-hole and the 3d electrons [7-9,13]. In particular, referring to 2p core levels, two kinds of charge fluctuations have usually been studied: the Mott-Hubbard, U_{dd}, and the charge transfer, Δ, fluctuations. Only recently [11,12] exchange (EX) and spin-orbit (SO) interactions have been included in the impurity-cluster configuration-interaction model. However, due to the high number of configurations derived from the interaction between the 2p core-hole and the 3d valence electrons, the exchange energies can not be easily obtained from the experimental data. So far, they have been separately calculated through Hartree-Fock techniques and then, after a scaling to account for solid state effects, introduced in the model as fixed values.

The interpretation of 3s XP data presents two advantages with respect to 2p XP data. Namely, there is no spin-orbit splitting and the rules to calculate the intensities in the sudden approximation are simpler because the 3s hole creation does not change the orbital quantum number in the final state.

In the earlier studies on 3s XP spectra [3-6] the CT fluctuation has never been explicitly included. Only recent studies on the 3s XP spectra in 3dTM compounds [1,2] have shown that the magnitude of the EX energy with respect to the CT energy is a key issue in approaching the EX correlation problem in the late 3dTM oxides and halides. Kinsinger *et al.*[1] have proposed a decomposition of 3s XP spectra of several 3dTM halides in which the CT energy has been fully accounted for. In this case the CT interaction is assumed to be predominant, while the EX contribution is treated as a perturbation of CT-derived levels. On the contrary, recent *ab initio* calculation results [2] have shown that the EX and CT energies must be treated on an equal footing. This result will be one of the guidelines in formulating the model aimed to interpret the XP data.

THE MODEL

The XP core-level spectroscopy deals with the response of the outer shell electrons to the sudden creation of a core hole. It is well known that such a strong perturbation of the neutral N electrons state, switches on several electronic interactions in the N-1 electrons excited final state.

In the case of TM^{2+} compounds, the main final state effects are: (a) core hole screening by the valence electrons, (b) interaction between the core hole and the electrons in the 3d open shell. Each of the screening effects produces an electronic redistribution and it needs to be described by a configuration interaction approach. To choose the best configurations it is important to estimate the energies involved in both the electronic processes mentioned above.

In principle, several charge fluctuations, such as polar type inter atomic d-d charge fluctuations, ligand-metal charge transfer and metal 4s-metal 3d inter atomic charge fluctuations, may be invoked to account for screening effects. Starting from the formal $3d^n$ configuration of TM^{2+} all the three charge fluctuations are allowed. However, being the 4s-3d charge transfer energetically very expensive, it is usually neglected leaving the problem with the usual CT and Hubbard charge fluctuations. Once the dominant charge fluctuations is identified, the choice of the initial state configuration can be made. In the following we will refer to the Ni^{2+} case, being the extension to the other TM^{2+} ions straightforward. In our case, three wave functions could be used. They are denoted as $3d^8$, for the unscreened configuration, and $3d^9\underline{L}$ and $3d^{10}\underline{L}^2$ for the single-screened and double screened configurations respectively, with $\underline{L}$ indicating that a hole has been created on the ligand 2p level because of the electron transferred to the 3d orbitals. The N electrons initial state system can be treated as a three level system with a Hamiltonian $\hat{H}$ written as a 3x3 matrix [7]. The CT charge fluctuation is parametrised by the energy Δ, i.e. the amount of energy required to transfer an electron from the ligand bands to the Ni 3d levels, while the d-d charge fluctuation by the parameter U. Taking as a reference the energy of the 3d⁸ configuration, the Hamiltonian for the initial state of the photoemission process is

$$\hat{H} = \begin{pmatrix} 0 & \sqrt{2}T & 0 \\ \sqrt{2}T & \Delta & \sqrt{2}T \\ 0 & \sqrt{2}T & 2\Delta + U \end{pmatrix} \quad (5)$$

where $\Delta = \langle d^9\underline{L}|\hat{H}|d^9\underline{L}\rangle - \langle d^8|\hat{H}|d^8\rangle$ represents the energy difference between the two configurations, U the repulsion between the transferred electrons and $T = \langle d^8|\hat{H}|d^9\underline{L}\rangle = \langle d^9\underline{L}|\hat{H}|d^{10}\underline{L}^2\rangle$ is the Ni 3d - O 2p hybridisation term. In a first approximation T varies as d⁻⁴, where d is a measure of the Ni-O distance. The diagonalisation of the Hamiltonian $\hat{H}$ gives three eigenvalues. The one with the lowest energy, E_{GS}, corresponds to the ground state (GS). The GS eigenvalue has the general form:

$$\psi_{GS} = a|3d^8\rangle + b|3d^9\underline{L}\rangle + c|3d^{10}\underline{L}^2\rangle \quad (6)$$

with a²+b²+c²=1. The values a, b and c represent the weights of the configurations in the GS. In the case of Ni²⁺, the number of 3d electrons in the GS can be calculated according to the formula:

$$n_{GS} = 8 \cdot a^2 + 9 \cdot b^2 + 10 \cdot c^2 \quad (7)$$

The final state is described in a similar way once the core hole effect has been included in the model. To this purpose a new parameter $Q_{\underline{c}d}$, describing the coulomb interaction between the core hole and the electrons transferred from the O 2p orbitals to the Ni 3d orbitals, is introduced. The three configurations for the final state are: $\underline{c}3d^8$, $\underline{c}3d^9\underline{L}$ and $\underline{c}3d^{10}\underline{L}^2$, where $\underline{c}$ represents the core hole. The Hamiltonian, referred to the energy $\langle \underline{c}d^8|\hat{H}|\underline{c}d^8\rangle = 0$ for the $\underline{c}3d^8$ configuration, is:

$$\hat{H} = \begin{pmatrix} 0 & \sqrt{2}T & 0 \\ \sqrt{2}T & \Delta - Q_{\underline{c}d} & \sqrt{2}T \\ 0 & \sqrt{2}T & 2(\Delta - Q_{\underline{c}d}) + U \end{pmatrix} \quad (8)$$

The diagonal term $\Delta - Q_{\underline{c}d}$ represents the energy difference between the energy (Δ) required to transfer an electron from the ligand to the metal, and the energy ($Q_{\underline{c}d}$) gained by the system because of the coulomb attraction between the electron transferred from the ligand and the core hole on the metal cation. If Q_{cd} is greater than Δ, the coulomb attraction stabilises the state with a prevailing $\underline{c}3d^9\underline{L}$ character. This means that this state will have, in the photoemission spectrum, a smaller binding energy while, the one with a prevailing $\underline{c}3d^8$ character has a greater binding energy. Matrix diagonalisation gives the eigenvalues, i.e. the final state energies E_i, and the corresponding eigenvectors $\left|\psi_i^f\right\rangle$.

In the framework of the present model, considering the sudden approximation, the cross section $\rho(e_k)$ for photoionisation out of a core state $\left|\psi_c\right\rangle$ into a continuum state $\left|\psi_k\right\rangle$ is proportional to:

$$\rho(e_k) \propto \left|\langle\psi_c|\hat{r}|\psi_k\rangle\right|^2 \sum_{i=1}^{n}\left|\langle\psi_i^f|\underline{c}\psi_{GS}\rangle\right|^2 \delta(h\nu - e_k - E_i) \qquad (9)$$

where the first term represents the dipole transition matrix element for a core level ionisation, $h\nu$ the photon energy, e_k the photoelectron kinetic energy, E_i the eigenvalues for the system in the final state and n the number of configurations involved. The states $\left|\psi_i^f\right\rangle$ represent the Hamiltonian eigenvectors associated with the E_i eigenvalues and $\left|\underline{c}\psi_{GS}\right\rangle$ represents the state obtained by annihilating a core electron in the GS, keeping all the other electrons frozen.

This model represents the building block of the impurity cluster calculations. It can give satisfactory results even in this simple form as it is shown in Fig. 2 where the XP spectra calculated for a Ni^{2+} ion are reported. As it can be observed the model is very sensitive to the change of the parameters. In this case the value of Δ has been changed to study the effect of CT energies on the satellite weight. The results show that an increase of Δ reduces the weight of the satellites. This trend is in agreement with the experimental results on Ni halides reported by Zaanen et al.[7] where a decrease of the XP satellite structures is observed going from $NiBr_2$ to NiF_2 in parallel with the increase of Δ.

To account for the exchange energies, the model so far described has to be extended. The most suitable term to describe the exchange interactions is Q_{cd}. In the early works,[7] Q_{cd} was assumed to describe only the coulomb part of the electron-electron interaction. In the present case Q_{cd} can be written as $Q_{cd}=Q_{coul}+Q_{exch}$. In the case of a $\underline{pd}$ interaction, Q_{coul} can be written in terms of the Slater integral F_0 and Q_{exch} in terms of F_2, G_1 and G_3. For a $\underline{dd}$ interaction Q_{coul} represents again the contribution of F_0 while Q_{exch} that of F_2 and F_4. The coulomb part is the same for all the multiplet terms, while their energy separation is given by the exchange part of the electron-electron interaction.

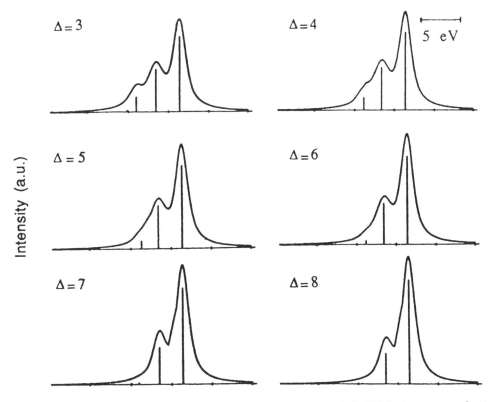

Figure 2. Calculated X-ray photoemission spectra relative to the core level of a TM divalent compound with U = 4.5 eV, Q_{cd}= 7.0 eV, T = 2.0 eV and Δ ranging from 3.0 to 8.0 eV.

In the case of the photoemission from the 3s core states, from the initial configuration, here denoted as $^{2S+1}\Gamma$, two final state configurations arise, $^{2S}\Gamma$ for the low spin state and $^{2S+2}\Gamma$ for the high spin state. A three-configuration Hamiltonian is associated to each of the two final state subsystems and the final spectra is calculated adding the contributions from the two Hamiltonians. The difference between the two matrixes is given by a diagonal term which can be related to the exchange energy [16].

The final state Hamiltonian accounts for the two spin configurations through the diagonal E_c term which is related to the exchange energy. Namely:

$$\hat{H}_{HS} = \begin{pmatrix} E_c\left(^{2S+2}\Gamma, d^n\right) & \sqrt{2}T & 0 \\ \sqrt{2}T & E_c\left(^{2S+2}\Gamma, d^{n+1}\underline{L}\right) + (\Delta - Q_{coul}) & \sqrt{2}T \\ 0 & \sqrt{2}T & 2(\Delta - Q_{coul}) + U \end{pmatrix}$$

(10)

for the HS ($S_f=2S+2$) final state, and

$$\hat{H}_{HS} = \begin{pmatrix} E_c(^{2S}\Gamma, d^n) & \sqrt{2}T & 0 \\ \sqrt{2}T & E_c(^{2S}\Gamma, d^{n+1}\underline{L}) + (\Delta - Q_{coul}) & \sqrt{2}T \\ 0 & \sqrt{2}T & 2(\Delta - Q_{coul}) + U \end{pmatrix} \quad (11)$$

for the LS ($S_f=2S$) final state. The coulomb part, Q_{coul}, of the interaction between the s-hole and the 3d-electrons is the same for both HS and LS states.

Choosing as a reference the HS energies ($E_c(^{2S+2}\Gamma, d^n) = E_c(^{2S+2}\Gamma, d^{n+1}\underline{L}) = 0$), the LS energy differs from the HS energy for the diagonal terms of the Hamiltonian matrix. One defines:

$$\langle 3d^n | \hat{H}_{HS} | 3d^n \rangle - \langle 3d^n | \hat{H}_{LS} | 3d^n \rangle = E_c(^{2S}\Gamma, d^n) \quad (12)$$

and

$$\langle 3d^{n+1}\underline{L} | \hat{H}_{HS} | 3d^{n+1}\underline{L} \rangle - \langle 3d^{n+1}\underline{L} | \hat{H}_{LS} | 3d^{n+1}\underline{L} \rangle = E_c(^{2S}\Gamma, d^{n+1}\underline{L}) \quad (13)$$

The energy E_c takes different values in dependence on the electronic configuration (e.g. ^2F for $3s^1 3d^8$ and ^1D for $3s^1 3d^9 \underline{L}$). In the case of Ni^{2+} 3s electrons, the ^4F-^2F energy splitting relative to the unscreened $3d^8$ configuration is $0.6F^2(3s,3d)$, while the ^3D-^1D energy splitting relative to the screened $3d^9\underline{L}$ configuration is $0.4F^2(3s,3d)$ and therefore results to be reduced with respect to the unscreened value [2]. This remark can be extended to the other compounds.

RESULTS OF DATA FITTING (I)

When the decomposition of the experimental data is performed with the present model a good fit of the spectral features is obtained (see, e.g., Ref.16 for NiO and CoO) and a series of parameter values can be consequently determined (Table 2).

For each compound, the parameter values $E_c(d^n)$ of the unscreened configuration are greater than the values $E_c(d^{n+1}\underline{L})$ of the CT-screened configuration, indicating that the charge transfer produces a screening of the exchange interaction. The trend for these values is in agreement with that predicted by the van Vleck's theorem (see, e.g., Ref. 15 and 17) for solid state systems and it is close to the trend estimated measuring the splitting between the two most intense spectral components of 3dTM fluorides 3s XP spectra [17]. However, as it was previously remarked, the measured energy separation is usually a factor 2 smaller than the predicted free-atom exchange integral calculations. Moreover, the intensity ratio I_{HS}/I_{LS}

Table 2. Parameters values obtained from the experimental data fitting (all energies in eV).

	Δ	U	Q_{sd}	E_c (d^n)	E_c (d^{n+1})	T	I_{HS}/I_{LS}	S+1/S
NiO	5.3	5.6	6.75	3.3	2.6	2.15	2.44	2.00
CoO	5.7	5.1	5.8	5.5	3.6	2.0	1.92	1.67
FeO	7.0	4.0	3.2	6.5	4.2	1.9	1.49	1.50
MnO	9.0	4.5	3.6	6.7	4.8	1.7	1.48	1.40

between the HS and LS contributions to the spectral weight deviates from the ratio S+1/S calculated through the multiplicity ratio rule (4). This effect has usually been attributed to intra-shell correlation effects [5] which have a relevant weight in Mn and Fe compounds. For Co and Ni compounds different mechanisms, i.e. CT effects, must be invoked to account for the deviation from the multiplicity ratio rule.

However the most remarkable result of the model calculations is the reduced value of E_c with respect to the van Vleck rule. To explain this result, one is forced to consider a sort of screening effect on exchange interactions produced by the solid-state like environment. The other possibility is that neglecting in the model the intra-shell correlation effects due to internal redistribution (IR) of M-shell electrons leads to a reduction of exchange-related E_c values since these values must, in some way, account for IR effects following the photoemission process. The second hypothesis seems to be the most appealing. Therefore CI-impurity cluster calculations were performed including also configurations derived from the internal redistribution of M-shell electrons.

RESULTS OF DATA FITTING (II)

The basic features of the model extension are summarised in the following. To make this approach simpler, only one charge transfer derived configuration ($3d^{n+1}\underline{L}$) was used instead of the two employed in the previous calculations ($3d^{n+1}\underline{L}$ and $3d^{n+2}\underline{L}^2$). However, for the LS subsystem, two new configurations were added. They are denoted as $3s^2 3p^4 3d^{n+1}$ and $3s^2 3p^4 3d^{n+1}\underline{L}$. The most general matrix for the low-spin subsystem is therefore:

$$\begin{pmatrix} E(3s^1 3p^6 d^n) & \tau & T & 0 \\ \tau & E(3s^2 3p^4 d^{n+1}) & 0 & 0 \\ T & 0 & \Delta - Q_{coul} + E(3s^1 3p^6 d^n \underline{L}) & \tau' \\ 0 & 0 & \tau' & \Delta - Q_{coul} + E(3s^2 3p^4 d^{n+1} \underline{L}) \end{pmatrix} \quad (14)$$

where τ and τ' represent the hybridisation integrals between the IR derived states, T is the hybridisation integral between the ionic and CT-derived configurations and the diagonal terms represent the energy of the four configurations. These values, reported in Table 3, were calculated through *ab initio* Hartree-Fock techniques [18].

Table 3. Calculated values for the hybridisation (τ, τ') and diagonal term in the LS subsystem in presence of IR ($3s^23p^4$) derived configurations (all energies in eV).

	τ	τ'	$3s^13p^6$		$3s^23p^4$	
			Atomic (d^n)	Atomic ($d^{n+1}\underline{L}$)	Atomic (d^n)	Atomic ($d^{n+1}\underline{L}$)
NiO	12.2	-9.7	8.1	5.1	36.8	31.4
CoO	-12.1	11.4	10.4	7.4	33.4	29.7
FeO	12.1	-11.2	12.5	9.4	32.8	29.4
MnO	15.2	-11.1	15.2	11.2	31.7	28.6

The results of the fit are presented in Fig. 3, where the calculated HS and LS contributions and the total calculated spectral weight are reported and compared to the expermental XP spectra. For NiO, the A component is well reproduced by the HS subsystem main line and the A' structure is fitted by the LS subsystem main line. The HS satellite contributes to peak B intensity. In the high-BE region, peak C is fitted by the LS satellite. The CoO fitting shows that the A component is well reproduced by the HS subsystem main line, while the LS subsystem main line and the HS satellite contribute to the central region of the spectrum. Finally, peak C is fitted by the LS satellite. In FeO the HS and LS configuration main lines mainly contribute to the spectral intensity, though the weight of the satellites is still effective in determining the intensity of peak B and the shoulder C at high BE.

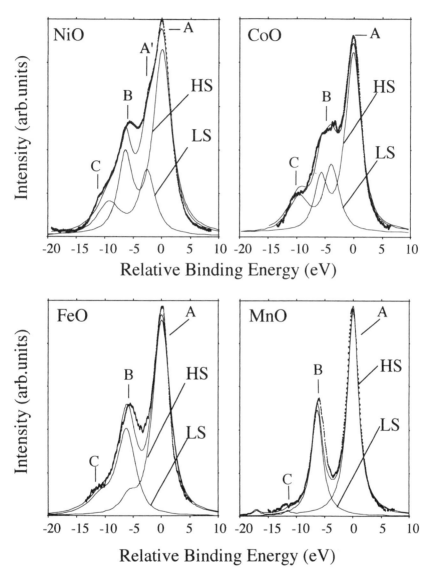

Figure 3. Experimental 3s XP spectra (dots) and calculated spectra (solid line) of NiO, CoO, FeO, and MnO. For the calculated 3s XP spectra, the HS and LS contributions as well as their sum are reported.

In MnO the XP spectrum is determined by the exchange-splitted A and B structures and only minor contributions from the satellites are detected at high BE (C). Moreover the energy separation, ΔE, and the intensities of the main features (A and B) are only slightly affected by the magnitude of Δ. On the contrary, the assignment of the C structure depends on Δ. In particular, when $\Delta \cong 10$ eV the C structure can be attributed to the satellite of the LS subsystem, while when $\Delta \cong 15$ eV the C structure is fitted by the satellite of the HS subsystem. On the basis of the present results it is not possible to determine whether the C structure has either a LS origin or a HS origin [18]. Nevertheless, recent *ab initio* cluster calculations [19] indicate that the C structure has a HS origin. Therefore, in the case of MnO, the fit results for $\Delta = 15$ are reported.

The studies on Ni^{2+} XP data [7,8] have shown that for low Δ compounds (e.g., $NiBr_2$, $NiCl_2$, NiO) the contribution of the CT-derived $3d^{10}\underline{L}^2$ configuration to the spectral weight is still relevant. The results shown in Fig. 3 are, in the case of NiO, slightly different from the results of Ref. 16 because in the present case the $3d^{10}\underline{L}^2$ configuration has not been used. Since the weight of this configuration becomes smaller as Δ increases, its effect on the spectral features decreases going from CoO to MnO where it is negligible. Table 4 reports the parameter values obtained from the fitting procedure. The intensity ratios between the HS and LS contributions to the total spectral weight and the energy separation ΔE between the HS and LS main lines are also reported.

Table 4. Parameter values, Δ, Q_{coul} and T, obtained through the fitting procedure, energy separation ΔE between the HS and LS main lines, intensity ratio I_{HS}/I_{LS} between the HS and LS contributions to the spectral weight, intensity ratio S+1/S calculated according to the multiplicity rule (all energies in eV).

	Δ	Q_{coul}	T	ΔE	I_{HS}/I_{LS} [1]	S+1/S
NiO	4.8	7.0	2.15	2.63	2.11	2.00
CoO	5.7	5.8	2.00	3.82	1.70	1.67
FeO	7.0	3.5	1.90	6.17	1.53	1.50
MnO	15.0	4.5	1.70	6.33	1.40	1.40

[1] This value has been calculated including also the contribution to the intensity from the IR-derived states lying 35-45 eV from the main line [18].

The main results can be summarised in the following:
- the hybridisation integral T as $1/d^4$, where d is the metal-oxygen distance. This scaling has been used as a parameter constraint;
- Δ increases from NiO to MnO, reflecting the bond electronegativity increase.

- the fitting-derived intensity ratio I_{HS}/I_{LS} is very close to the ratio calculated on the basis of the final state degeneracy (S+1/S). in particular, in case of NiO and CoO there is an improvement with respect to the values reported in Table 2.

With the aid of Fig. 3 a consistent picture of the interplay between exchange, coulomb and charge transfer energies and their effect on the spectral features can be drawn. Starting from MnO, one can observe two well separated peaks (6.33 eV). The satellites (indicated by dots), for both HS and LS configuration have a negligible intensity. Going from MnO to FeO a small increase in the satellites intensity can be detected, and the separation between the two main peaks (6.17 eV) is reduced with respect to MnO. The CoO case represents a turning point with respect to the previous cases. In fact, the energy separation between the HS and LS main lines is greatly reduced (3.82 eV) and the satellites intensities with respect to the main lines intensities are increased. In the case of NiO, the trend observed in the CoO case is fully established. Again, the energy separation between the HS and LS main lines is reduced (2.63 eV) and, because of the reduction of CT energy, the satellite intensities are increased.

CONCLUSIONS

In conclusion, an analysis of 3s XP data in 3dTM oxides, where exchange and CT energies are treated on the same footing, has been performed. The results show that the CT effects strongly characterise the NiO spectrum, while, on the opposite side, MnO spectrum is dominated by EX effects, the CT fluctuations being prevented from their high energy with respect to thr EX energy. The transition from the NiO behaviour to the MnO behaviour is gradual, FeO and CoO representing the intermediate steps. The degree of consistency of the model is strengthened by the values obtained for Δ, T and Q_{coul}. The Δ trend for each TM ion follows the bond electronegativity increase from Ni to Mn whereas, as expected, Q_{coul} decreases going from Ni to Mn oxides.

Moreover, it has been explained why the measured exchange integral (i.e. the HS-LS main line separation) is roughly half of the calculated atomic value. This reduction is due to final state effects related to the creation of the 3s core-hole which originates a redistribution of M-shell electrons. When IR-derived configurations are considered, the calculated atomic value of the exchange integral can be used. In the light of this result, the values of the exchange-related parameters E_c used in the first fitting could be considered as "dressed" values since they must account for the atomic correlations effects which in that case were neglected.

REFERENCES

1. V. Kinsinger, R. Zimmermann, S. Hüfner, and P.Steiner, Z.Phys.B - Condensed Matter 89: 21 (1992).
2. P.S. Bagus, G. Pacchioni, and F. Parmigiani, Chem.Phys.Lett. 207: 569 (1993).
3. C.S. Fadley and D.A. Shirley, Phys.Rev. A2: 1109 (1970).
4. P.S. Bagus, A.J. Freeman, and F. Sasaki, Phys.Rev.Lett. 30: 850 (1973).
5. E.-K.Viinikka and Y.Öhrn, Phys.Rev. B11: 4168 (1975).
6. B.W. Veal and A.P. Paulikas, Phys.Rev.Lett. 51: 1995 (1983).
7. J. Zaanen, C. Westra, and G.A. Sawatzky, Phys.Rev. B33: 8060 (1986).
8. G. Lee and S.-J. Oh, Phys.Rev. B43: 14674 (1991).
9. J. Park, S. Ryu, Moon-sup Han, and S.-J. Oh, Phys.Rev. B37: 10867 (1988).
10. J. Zaanen and G. A. Sawatzky, J. of Solid State Chem. 88: 8 (1990).
11. K. Okada and A. Kotani, J.of the Phys.Soc.of Japan 60: 772 (1991).
12. K. Okada and A. Kotani, J.of the Phys.Soc.of Jap. 61: 449 (1992).
13. F. Parmigiani and L. Sangaletti, J.Electron Spectrosc. Relat. Phenom. 66: 223 (1994).
14. M.A. van Veenendaal and G.A. Sawatzky, Phys. Rev. Lett. 70: 2459 (1993).
15. C.S. Fadley, Basic concepts of X-ray photoelectron spectroscopy, in "Electron Spectroscopy: theory, techniques and applications (Vol.2)", C.R. Brundle and A.D. Baker, eds., Academic Press, New York (1978).
16. F. Parmigiani and L. Sangaletti, Chem.Phys.Lett. 213: 613 (1993).
17. D.A. Shirley, Many-electron and final state effects: beyond the one-electron picture, in "Photoemission in solids I - General Principles", M. Cardona and L. Ley, eds., Springer Verlag, Berlin (1978).
18. P.S. Bagus, F. Parmigiani and L. Sangaletti, to be published.
19. P.S. Bagus, private communication.

CONTRIBUTORS

Arenholz, E.	131	Mills, D.L.	61
Baumgarten, L.	131	Navas, E.	131
Braicovich, L.	41	Panaccione, G.	181
Carra, P.	203	Parmigiani, F.	249
Colliex, C.	213	Rampe, A.	235
Dose, V.	173	Reese, M.	235
Fabrizio, M.	203	Rose, H.B.	85
Günterhodt, G.	235	Rossi, G.	181
Hartmann, D.	235	Roth, Ch.	85
Hillebrecht, F.U.	85	Sangaletti, L.	249
Hopster, H.	103	Siegmann, H.C.	1
Johnson, P.D.	21	Sirotti, F.	181
Kaindl, G.	131	Starke, K.	131
Kisker, E.	85	van der Laan, G.	153
Mathon, J.	113	Weber, W.	235

INDEX

Alkali atoms,
  D_2 spectrum, 184
Angle resolved,
  photoemission, 21,43
  inverse photoemission, 47
    in Cu(100), 48
    in Ni(100), 48
    in Co(100), 48
    in Fe(100) ,48

Band ferromagnetism, 173
Band mapping,
  of empty states, 43
Band structure, 21
  spin resolved,
    of Co/Cu(001), 124
  exchange split, 21
  in ferromagnetic systems, 21
Bloch waves, 205
Brillouin scattering, 78

Cascade polarization, 9
Charge-transfer, 53
Charge-transfer insulator, 52, 54
Circular Dichroism, 12
Circularly polarized radiation, 93
Clebsch-Gordon coefficients, 34
Cluster model, 53
Co
  3p spin resolved spectrum, 91
  spin polarization of low energy cascade electrons, 8
  transport spin polarization, 10
Co/W(100)
  surface magnetization, 2
Configuration Interaction calculation,
  of 3s transition metal spectra,253
Core level splitting, 190
  atomic model, 190

Core level splitting (continued),
  in Fe 3p,190
  in Ni 3p,190
Core levels photoemission spectroscopy,33
  Co 3p, 193
    fine structure,194
    spin selected spectra,194
  Fe 3s, 33
  Fe 3p, 95,191, 193
    fine structure, 194
    in bcc-Fe, 183
    in β-FeSi$_2$, 183
    spin selected spectra,194
  Ni 3p, 191
  initial state effects,34
  final state effects,34
  exchange interacion,35
  spin resolved, 85
  3s transition metal oxides, 249
    configuration interaction calculation,249
Correlation effects, 48
Correlation energy, 133
Convolution, d-d, 178
Coulomb effect, 182
Coulomb integral, intra-atomic,121
Cu,
  spin dependent attenuation length of 3d photoelectrons, 104
  BIS spectra, 45
  DOS curves, 45
  scattering total cross section, 5
Cu(100),
  angle resolved inverse photoemission, 46,47
Cu/Co(001),
  MCD, 31
  magnetic multilayers, 25
  photoemission spectra of Cu, 27
  quantum well states,29, 31
    effective mass, 33

267

Cu films, photoemission spectra, 29

Dichroism, 46
Dipole-selection rules, 135
Dirac equation, 67
Direct valence photoemission, 42
  weakly correlated limit, 42
Dyson equation, 115

ECS, 115
Elastic scattering,
  of electrons
    from Fe,
      on magnetic atoms, 9
    from magnetic materials, 61, 67
    spin dependent, 62, 107
      in Fe(110), 69
      in Fe/W(100), 69
    spin-orbit effect, 62
    exchange interaction, 62
Electron attenuation lenght, 104
  overlayer method, 104
  spin-dependent, 104
Electronic correlation, 249
Electron diffraction,
  spin polarized, 62
Electron-electron scattering, 5
Electron energy loss spectroscopy (EELS), 77, 213
  from 1s electrons, 220
  from 2p electrons, 220
  core edges, 220
    FeO, 221
    Fe_3O_4, 221
    α-Fe_2O_3, 221
    γ-Fe_2O_3, 221
    limits of detection, 221
    fine structure, 222
    transition matrix elements, 222
    molecular orbital cluster model, 222
    Mn oxides, 222
    band theory calculation, 223
    TiO_2, 224
    CeO_2, 224
    multiple scattering description, 223
    spatially resolved, 225
      carbon nanotubes, 227
      SiO_2-TiO_2 multilayers, 227
      Si nanowires, 227
      n-type nanoporous Si, 227
      Ni/Au multilayers, 227
  core level excitations,
    from Fe-M_{23}, 215
    from C-K, 215

Electron energy loss spectroscopy, (EELS), (continued),
  core level excitations, (continued),
    from O-K, 215
    from Fe-L_{23}, 215
  cross section,
    SPEELS contribution, 80
  loss function,
    in the low energy range, 66
    in core losses, 219
  transmission scattering, 213
  plasmon losses, 214
  spectrum,
    of $SiFe_2O_4$, 214
    of Si 2p, 217
    of C 1s, 218
  Spin polarized SPEELS, 62, 78
  theory, 217
Electron scattering, 65
  from Fe/Cu(100), 69
  from surfaces, 65
  spin-dependent mean free path, 69
Electric transport,
  in magnetic materials, 3
Electron subtraction spectroscopy, 49
Electron addition spectroscopy, 49
Electronic structure,
  of solids, 21
Empty Electron States,
  Core X-Ray Absorption Spectroscopy, 41
  Inverse Photoemission Spectroscopy, 41
  BIS in 3d elements, 46
  band mapping, 46
Exchange coupling, 61
  high spin coupling, 34
  low spin coupling, 34
  oscillatory, 31
  in core level spectroscopy, 35
  mediated by itinerant electrons, 104
Exchange field,
  transfer of, 2
Exchange hamiltonian, 115
Exchange interaction, 86, 96, 113, 131, 182
  in magnetic surfaces, 113, 118
  oscillatory, 114
    quantum well theory, 120, 235
    long period, 127
      in Co/Cu(001), 129
    short period, 127
      in Co/Cu(001), 129
  quantum well theory, 114, 121
    in Co/Cu(001), 114, 120, 123
  RKKY theory, 121, 235
  in Fe 2p core levels, 132

Exchange interaction (continued),
  in Gd 4f levels, 132
  in magnetic layer structures, 113
  local, 114
  interlayer, 122
Exchange splitting,
  in core level spectroscopy,35
Fano effect, 131,181
Faraday effect, 131
Fe,
  spin polarization of low energy cascade electrons, 8
  transport spin polarization, 10
  3s core level, spin polarized spectrum, 34
  3p spin resolved spectrum, 91
    from clean Fe(001), 36
  3p absorption specttroscopy,98
Fe-oxides,
  valence band spectra, 56
  final state effects, 58
    CI cluster model,58
Fe/W(100),
  surface magnetization, 2
Fermi surface spanning vectors,30
Ferromagnetism, 21
Ferromagnetic layers, 121,123
  Fe, 121
  Co, 121
  Ni, 121
Ferromagnetic materials, 173
Final state effects, 48
  in photoemission, 182
  in Fe-oxides, 48

Gd
  spin polarization of low energy cascade electrons, 8
Giant magnetic moment,
  in Cr(100), 58
Giant magnetoresistance effect,114
Green function,
  local spin waves, 115
  one electron, 122

Haas-van-Alphen oscillations,121
Hubbard hamiltonian, 30
  one band model, 80
Hybridization, 49
Hybridization angle, 51

Impurity model,
  in CoO,56
  in transition metal oxides,254
Inelastic scattering, 5

Inelastic scattering (continued),
  of electrons,213
    mean free path,214
    spin dependent, 61, 77,103
  of light (see also Brilluoin scattering),78
  of X-ray, resonant, 203
    Compton scattering, 203
    by valence electrons, 203
    by collective ion excitation, 203
    scattering rate, 204
    Fermi's golden rule, 204
Insertion devices, 23
Interfacial coupling,
  of Ag d-bands in Ag/Fe(001), 27
Interface states, 235
  of Co(0001), 237
  of paramagnetic overlayers in Co(0001), 235
  in Ag/Fe(001), 26
  in Au/Co(0001), 237
  in Rh/Co(0001), 236,241
  in Pd/Co(0001), 236, 239
  in Pt/Co(0001), 239
  in Ru/Co(0001),236,241
Inverse photoemission, 9,42
  band mapping,
    in Cu(100), 46
    in Ni(100), 46
    in Co(100), 46
    in Fe(100), 46
  bulk sensitivity, 46
  weakly correlated limit, 42
  matrix elements, 42
  joint density of states, 42
    energy distribution, 42
  resonant processes, 44
    in Ce-compounds, 44
  spin resolved, 46
    Ni d-states, 46
  strongly correlated limit, 48
  surface sensitivity, 46
Isochromat spectroscopy, 43

Kerr effect, 131,181
  in bcc-Fe(100)/Au(100), 236
  in Au/Co(0001)/Au(111), 236

LAPW, spin resolved calculation,
  on F2 2p, 177
  on Ni 2p, 177
LEED,
  from Au/Co(0001), 236
  from Pd/Co(0001), 236
  from Ru/Co(0001), 236
  from Pt/Co(0001), 236

LEED (continued),
  from Rh/Co(0001), 236
Light,
  circularly polarized, 131
  polarized, 131
  s-polarized, 97
  p-polarized, 97
Linearly polarized radiation, 99
Long range ferromagnetic order,
  Tb 4f core level, 133
Low energy electron spectroscopy, 3
  in Gd/Ni, 8
Low energy cascade electrons, 5
  spin polarization, 5, 7, 8
    in ferromagnetic metals, 8

Magnetic alignment,
  in ferromagnetic layers, 3
Magnetic Circular Dichroism (MCD),
  angular momentum coupling theory, 154
  geometrical dependance, 154, 160
  single particle theory, 153
  many body theory, 153
  on Ni, 192
  applications, 142
    X-ray polarimeter, 143
    surface magnetometer, 145
  in Cu/Co multilayers, 31
  perspectives, 148
  in X-ray absorption, 131
  in resonant photoemission, 140,153
    from Gd 4d->4f in Gd(0001)/W(110),140
  in photoemission, 131,183
    from Gd 4f levels, 134
    from Gd 4d levels, 139
    from Tb 4f levels, 134
    from Dy 4f levels, 134
    from Fe 2p core levels,131
    from Ni 2p in ferromagnetic Ni(110),154
    from valence band, 198
Magnetic Circular X-Ray Dichroism, 131,174,181
Magnetic coupling,
  in Cr/Fe(100), 92
  interlayer,
    in Fe/Ag/Fe, 107
    in Co/Cu/Co(111), 236
    in Co/Rh/Co, 243
    in Co/Ru/Co, 243
    in Co/Pd/Co, 243
Magnetic domains imaging, 16
Magnetic interfaces, 114
Magnetic layers,
  ferromagnetic alignment, 113
  anti-ferromagnetic alignment, 113

Magnetic Linear Dichroism (MLD), 93,181
  geometrical dependance, 154
  in Fe 3p core levels, 93,191
  in directional photoemission,181
    from core levels, 181
    from valence band, 181
Magnetic Linear Dichroism in the Angular
  Distribution of photoelectrons (MLDAD),
    93, 153, 162,195
  in Fe 3s core level, 184
  in Fe valence band, 184
  in Fe 3p core levels, 93, 184
    in polycristalline Fe, 186
    in amorphous rcp-Fe, 186
    temperature dependance, 189
  in Fe 2p, 98
  in Co 3p, 98
    in bcc-Co/Fe(100), 187
  from Cr 3p, in
    Cr/Fe(100), 188
  in Fe/Cr/Fe, 188
  in Ni 3p, 98
  in Gd 4f, 98
  state multipoles,193
  as surface magnetometer,188
  valence band photoemission,196
    from Co(100), 198
    from Fe(100), 198
    from Pd(111)/Co(0001),244
Magnetic materials, 61
  electric transport, 3
Magnetic moments,
  in Cr, 2
  in Fe(001), 23
  in Pd, 2
  recoupling, 156
  in Ru, 2
  in V, 2
  in clean surfaces, 23
Magnetic multilayers, 18, 25
  Cu/Co, 25
  exchange coupling, 28
  Fe/Ag, 25
Magneto optical recording, 147
Magnetic order, 92
Magnetic overlayers, 118
Magnetic susceptibiliy,
  giant in 2D ferromagnets, 2
Magnetic surfaces, 62,114
Magnetism,
  in non magnetic elements, 2
  in Tc, 2
  in Ru, 2
  in Rh, 2

Magnetism (continued),
  in Pd, 2
  in Os, 2
  in Ir, 2
  in Au(100) substrates, 2
  in Ag(100) substrates, 2
  very near surfaces
Magnetization,
  metastable in liquid Fe, 16
  local, 104
    temperature dependence, 104
Magnetometry,
  with cascade electrons, 4
  with low energy electrons, 15
Magneto-optical effect, 131
Magnetoresistance,
  giant, 3
    two current model, 3
Many body effects, 22
Mean free path,
  spin dependent, of electrons in ferromagnets, 103
  low energy range, 103
Microscopy,
  spin polarized electrons, 1
Molecular beam epitaxy, 89
  Fe films on Ag(100), Au(100), W(110), 89
  Co/Cu(100), 89
Mott- detector, 65
Mott-Hubbard energy, 249
Mott-Hubbard insulator, 54
  satellites in, 55
Mott-scattering,
  spin analysis, 86
Muffin-tin, 68
Multiple scattering theory, 61

Nanostructures, 235
Ni
  2p levels,
    LAPW spin resolved calculation, 177
    SXAPS, 177
  3p levels,
    MLDAD, 98
    spin polarization of photoelectrons from, 91
    splitting, 190
    photoemission spectrum, 191
    spin resolved photoemission spectrum, 91
  3d states,
    spin resolved spectrum, 46
  ferromagnetic layers, 121
  MCD, 192
  spin dependent density of states, 177
  spin polarization of low energy cascade
    electrons, 8

Ni (continued),
  transport spin polarization, 10
Ni(100),
  angle resolved inverse photoemission, 46, 47
Ni(110),
  MCD in 2p levels, 131
  spin polarized EELS, 110
    spectrum of secondary electrons, 108
Ni(111),
  photoelectrons spin polarization, 13
Ni/Au multilayers,
  spatially resolved EELS, 227
Ni/W(110), spin filter, 11
$Ni_{80}Fe_{20}$, ultrathin layer, 3
NiO,
  3s core level photoemission spectrum, 249
Non-magnetic spacer layers, 25

Oscillatory exchange coupling, 31

Photocurrent, 21
  matrix elements, 21
Photoelectron diffraction, 85
Photoemission, 21, 105
  transition probability, 155
  angle resolved, 21
    theory, 159, 162
  angle integrated spectra,
    theory, 156
  dichroic, 153
    from core levels, 153
  from core levels, 33
    of transition metal oxides, 249
    of CoO 3s, 249
    of FeO 3s, 249
    of MnO 3s, 249
    of NiO 3s, 249
  from valence band, 42
    weakly correlated limit, 42
  final-state multiplet components, 135
    atomic calculation in RE 4f levels, 136
    many body calculation in RE 4f levels, 136
  inverse, 41
  many body effects, 22
  magnetic circular photoemission, 131
  resonant, in Gd 4d->4f, 136
    atomic multiplet calculation, 141
    many body calculation, 141
  polarization, 104
  from rare earth materials, 131
  selection rules, 21
  self energy, 22
  spectral function, 22, 41
    weakly correlated limit, 41

Photoemission (continued),
  spectral function (continued),
    strongly correlated limit, 41
  spectra,
    in Fe(001), 23
    of valence band of Fe-oxides, 56
    of Gd 4f levels, 49
    of Cu in Cu/Co(001), 27
    from Tb 4f,134
    from Tb(0001) 4f surface states, 133
    from Eu/Gd(0001), 132
    from Yb/Mo(110), 132
    from Ag/Fe(001), 26
    from Co(0001), 238
    from Au/Co(0001), 238
  spin resolved spectra,
    from Fe(001), 23
    from Cu films, 29
    from Au/Co(0001),239
    from Pt/Co(0001), 239
    from Pd/Co(0001), 239
    from Rh/Co(0001), 241
    from Ru/Co(0001), 241
  spin polarized, B2, 107
    theory, 158
    on bcc-Fe, 107
    on Co/W(110), 107
    on Fe/Ag/Fe, 107
    intensity, 105
Photon helicity, 183
Polarimeter, figure of merit of, 22
Polarization of photoelectrons, 104

Quantum filter, 75
Quantum well states,235
  in Ag/Fe(001), 26
  in Co(0001), 237
  in Au/Co(0001), 237
  in Pd/Co(0001), 239
  in Pt/Co(0001), 239
  in Rh/Co(0001), 241
  in Ro/Co(0001), 241
  in Cu/Co(001), 29, 31
    effective mass, 33
  spin polarized, 235
  spin-orbit interaction, 244
  d-like, 235
  in metallic,
    overlayers, 235
    substrates, 235
  in semiconductor heterostructures, 235
  as mediators of the oscillatory interlayer
    exchange coupling, 235

Radiation,
  circularly polarized, 93
  linearly polarized, 93
Random phase approximation,
  of spin excitation,80

Sandwich structures, 113,119
Scattering,
  classical theory, 63
  cross section, 65
  current of, 65
  electron-electron, 5
  elastic of electrons on magnetic atoms, 9
  electron-surface,
    spin dependence, 62
  inelastic cross section, 5
    spin dependent, 6
    in transition metals, 6
    into the unoccupied surface states, 9
  mean free path, 3
  multiple, theory of, 61
  quasi-elastic, 9
  total cross section, 3, 5
    in Ag, 5
    in Au, 5
    in Cu, 5
  spin dependent, 6, 68
  spin independent, 6, 9
  t-matrix, 68
Scattering Transmission Electron Microscopy,
  in EELS, 216
Self-energy, 22, 68
Self interaction correction (SIC) in LDA,58
Sherman function, 22
SMOKE, 115
Spacer layers,114
  non magnetic, 25,121
Spectral density, 122
  of majority spin for Co/Cu(001),125
  of minority spin for Co/Cu(001), 128
  adlayers method of computation, 125
Spectral function, 22, 41
  weakly correlated limit, 42
  strongly correlated limit, 48
Spectroscopy,
  electron energy loss, 213
    unoccupied states, 220
  final state effects, 47
  inverse photoemission, 41
  Mössbauer,115,116
  low energy electrons, 3
  probing depth, 4
  spin polarized of core levels, 2
  spin polarized electrons, 1

Spectroscopy (continued),
  spin polarized inverse photoemission, 2
  X-ray absorption, 41
Spin dependent density of states, 30,174
  Fe, 177
  Ni,. 177
Spin dependent electron attenuation lenght,104
  of Cu 3d, 104
  from Fe/Cu(100), 104
Spin detection, 181
Spin exchange scattering, 14
Spin excitation,
  random phase approximation,80
Spin filter, 3, 7, 108,131
  in ferromagnetic metals, 8
  in Fe-wisker, 12
  in Fe/W(110), 11
  in Co/W(110), 11
  in Ni/W(110), 11
  with spin polarized electrons, 11
Spin flip, 13,80
Spin magnetic moment, 7
Spin-orbit interaction, 22, 61, 86,96,131,182
  in Fe 2p core levels, 132
  in rare earths 4f levels, 132
  on non-magnetic substrates, 63
Spin polarimeter, 86
Spin polarimetry, 86
  in Fe(100), 86
Spin polarization, 1,22, 34, 86, 87, 96, 181
  of photoelectrons, 85
    cross section, 85
    in Gd, 13
    in Fe, 13
      3p core levels, 91
    in Co, 13
      3p core levels,91
    in Ni, 13
      3p core levels, 91
    in Au/Fe(110), 236
    in Ag/Fe(100), 18
    in Cr/Fe(100), 18
    in Gd/Fe(100), 14
    in Ru/Fe(100), 18
    in V/Fe(100), 17
    from Co(0001), 13
    from Ni(111), 13
  of secondary electrons, 108,116
    spectrum from Ni(110), 108
    in ferromagnetic metallic glasses,108
    mean free path, 109
      spin dependent, 108
  of the excited electrons,131
Spin polarized Auger spectroscopy,174

Spin polarized core level spectroscopy,
      2, 13, 85,174
  of Cr 3s core level,92
  of Cr 3p core level,92
  dipole selection rules,174
  of Fe 3s,89
    atomic model, 98
    bandlike model,90
  of Fe 3p core level, 92, 95
  of 3p level in 3d ferromagnets, 91
  X-ray emission,174
  X-ray absorption,174
Spin polarized DF/LDA calculations,65
Spin polarized electrons,
  source, 175
Spin polarized electron diffraction SPLEED,
      62, 67, 107, 181,189
  multiple scattering theory, 67
Spin polarized EELS, 62, 109, 115, 114
  loss processes, 110
  from Fe(110), 110
  from Ni(110), 110
  from Cr(110), 110
Spin polarized electron microscopy, 1
Spin polarized electron scattering, 63
Spin polarized electron spectroscopy, 1, 86
  sources, 1
  magnetometry, 15
  from secondary electrons,115
Spin polarized photoemission, 22,173
  from bcc-Fe,107
  from Co/W(110),107
  from Fe/Ag/Fe, 107
  with synchrotron radiation, 23
Spin polarized inverse photoemission, 2
Spin polarized overlayer technique, 5
Spin polarimetry,
  to photoelectrons, 181
  to secondary electrons,181
Spin resolved appearance potential spectroscopy,
SXAPS, 174
  Lander's model, 175
  Fe $2p_{3/2}$ core levels,177
  Ni $2p_{3/2}$ core levels,177
Spin resolved AES,203
Spin resolved PES,203
Spin-spin interaction,205
Spin valves, 3
Spin waves, 61, 78,80
  density of states, 115
State multipoles, 193
Stoner excitation, 9, 61,78,80
Stoner factor, 30

Stronlgy correlated limit,
  in inverse photoemission, 48
  in X-ray absorption spectroscopy, 48
Superlattices, 113,119
  Fe/Cr,113
  Co/Cr, 113
  Co/Ru, 113
  Co/Cu, 113
Surface core level shift,133
Surface geometry,
  in Fe(110), 69
  in Fe/W(110), 69,75
  of Eu/Gd(0001), 147
Surface magnetic domains,
Surface magnetic moment, 74
  in Fe(110), 74
  in Cr(110), 75
  in Cr/Fe(100), 75
Surface magnetism, 1, 23, 62, 63,181
  dynamics, 181
  in Fe(110), 69
  in Fe/W(100), 69,75
  in ferromagnetic surfaces,117
  magnetization, 9,118
    classical spin wave theory,116
    RKKY theory, 120
    ferromagnetic transition metals multilayers,120
    in Gd(0001), 145
  magnetization profiles, 4
    in Fe/W(100), 2
    in Co/W(100), 2
Surface spin dependent states,23,25
Surface density of spin-waves states,117
Synchrotron radiation, 23,43

Tight-binding hamiltonian, 30,121
  spin-dependent calculation, 30
Thin films,
  adsorbed on surfaces, 63

Thomas precession, 63
Transport spin polarization, 7
  in Co, 10
  in Fe, 10
  in Ni, 10

Undulators, 23
Unoccupied electron states spectroscopy, 220

Van Hove formula, 80
VLEED,86,88

X-ray Absorption Spectroscopy, 41, 54
  bulk sensitivity, 46
  in Fe 3p core levels, 98
  strongly correlated limit, 48
  surface sensitivity, 46
  weakly correlated limit, 44
X-ray Photoemission Spectroscopy, 43
  photoemission cross section, 43
  in Fe-oxides, 56
X-ray resonant inelastic scattering,203
  correlation functions,208
  electric dipolar transitions,208
  Compton scattering,204
  by valence electrons,204
  by collective ion excitations,204
  scattering rete,204
  Fermi's golden rule,204

Wannier's functions,205
Weak correlation limit,
  in X-Ray absorption spectroscopy, 44
  in inverse photoemission, 44
Wigner-Eckart theorem,135

Zeeman, anomalous effect,182
  in alkali atoms,183
    D_2 spectrum,183